Biomass in Small-Scale Energy Applications
Theory and Practice

Energy Systems: From Design to Management

Series Editor

Vincenzo Bianco

Published Titles

Analysis of Energy Systems
Management, Planning and Policy
Vincenzo Bianco

Solar Cooling Technologies
*Sotirios Karellas, Tryfon C. Roumpedakis, Nikolaos Tzouganatos,
and Konstantinos Braimakis*

Biomass in Small-Scale Energy Applications
Theory and Practice
Mateusz Szubel and Mariusz Filipowicz

For more information about this series, please visit: www.crcpress.com/Energy-Systems/book-series/CRCENESYSDESMAN

Biomass in Small-Scale Energy Applications
Theory and Practice

Edited by
Mateusz Szubel
Mariusz Filipowicz

CRC Press
Taylor & Francis Group
Boca Raton London New York

CRC Press is an imprint of the
Taylor & Francis Group, an **informa** business

CRC Press
Taylor & Francis Group
6000 Broken Sound Parkway NW, Suite 300
Boca Raton, FL 33487-2742

© 2020 by Taylor & Francis Group, LLC
CRC Press is an imprint of Taylor & Francis Group, an Informa business

No claim to original U.S. Government works

Printed on acid-free paper

International Standard Book Number-13: 978-0-367-25105-5 (Hardback)

This book contains information obtained from authentic and highly regarded sources. Reasonable efforts have been made to publish reliable data and information, but the author and publisher cannot assume responsibility for the validity of all materials or the consequences of their use. The authors and publishers have attempted to trace the copyright holders of all material reproduced in this publication and apologize to copyright holders if permission to publish in this form has not been obtained. If any copyright material has not been acknowledged, please write and let us know so we may rectify in any future reprint.

Except as permitted under U.S. Copyright Law, no part of this book may be reprinted, reproduced, transmitted, or utilized in any form by any electronic, mechanical, or other means, now known or hereafter invented, including photocopying, microfilming, and recording, or in any information storage or retrieval system, without written permission from the publishers.

For permission to photocopy or use material electronically from this work, please access www.copyright.com (http://www.copyright.com/) or contact the Copyright Clearance Center, Inc. (CCC), 222 Rosewood Drive, Danvers, MA 01923, 978-750-8400. CCC is a not-for-profit organization that provides licenses and registration for a variety of users. For organizations that have been granted a photocopy license by the CCC, a separate system of payment has been arranged.

Trademark Notice: Product or corporate names may be trademarks or registered trademarks, and are used only for identification and explanation without intent to infringe.

Library of Congress Cataloging-in-Publication Data

Names: Szubel, Mateusz, editor. | Filipowicz, Mariusz, editor.
Title: Biomass in small-scale energy applications : theory and practice / Mateusz Szubel and Mariusz Filipowicz, editors.
Description: Boca Raton : Taylor & Francis, CRC Press, 2019. | Series: Energy systems : from design to management | Includes bibliographical references.
Identifiers: LCCN 2019016191 | ISBN 9780367251055 (hardback : alk. paper) | ISBN 9780429286063 (ebook)
Subjects: LCSH: Small power production facilities. | Biomass energy. | Electric power plants--Fuel.
Classification: LCC TK1071 .B56 2019 | DDC 662/.88--dc23
LC record available at https://lccn.loc.gov/2019016191

Visit the Taylor & Francis Web site at
http://www.taylorandfrancis.com

and the CRC Press Web site at
http://www.crcpress.com

Contents

Preface vii
Editors ix
Contributors xi

PART I Perspectives of Using Raw and Processed Biomass in Small-Scale Energy Systems

Chapter 1   Thermochemical Processing of Solid Biomass 3

Tomasz Mirowski and Eugeniusz Mokrzycki

PART II Economic Issues of Biomass-Based Small-Scale Energy Systems

Chapter 2   Economic Issues of Biomass-Based Small-Scale Energy Systems 21

Artur Wyrwa, Wojciech Suwała and Marcin Pluta

PART III Environmental Issues of Biomass-Based Small-Scale Energy Systems

Chapter 3   Environmental Impacts of Biofuel-Fired Small Boilers and Gasifiers 41

Jozef Viglasky, Juraj Klukan and Nadezda Langova

Chapter 4   Emissions of Pollutants from Biomass Combustion: Relevant Regulatory Measures and Abatement Techniques 99

Robert Kubica

Chapter 5   Monitoring and Modeling of Pollutant Emissions from Small-Scale Biomass Combustion 127

Janusz Zyśk

Chapter 6   Calculation Approach to Fulfill Emission and Energy Efficiency Limits of Individually Built Tiled Stoves (Kachelofen) 145

Thomas Schiffert

v

PART IV Heat and Power Generation Systems Based on Biomass Thermochemical Conversion

Chapter 7 Novel and Hybrid Biomass-Based Polygeneration Systems............ 157

Rafał Figaj, Maria Di Palma and Laura Vanoli

Chapter 8 Application of Thermoelectric Power Generators in Small-Scale Heating Devices... 185

Mariusz Filipowicz, Krzysztof Sornek, Mateusz Szubel and Maciej Żołądek

Chapter 9 Straw-Fired Boilers as a Heat Source for Micro-Cogeneration Systems 215

Krzysztof Sornek, Mariusz Filipowicz and Karolina Papis

Chapter 10 Straw Drying: The Way to More Energy Production in Straw—Fired Batch Boilers.. 231

Wojciech Goryl

Chapter 11 Assessing the Feasibility of Renewable Energy Sources for Treatment of Biomass from Wastewater Treatment 247

Simona Di Fraia, Adriano Macaluso, Nicola Massarotti and Laura Vanoli

Chapter 12 Biomass-Based Low-Capacity Gas Generator as a Fuel Source...... 271

Tomasz Chmielniak, Aleksander Sobolewski, Joanna Bigda and Tomasz Iluk

Chapter 13 Production of Generator Gas from Biomass and Fuels from Waste on a Small Scale 283

Danuta Król and Sławomir Poskrobko

PART V Computational Fluid Dynamics as a Modern Tool in Studies of Biomass-Based Small-Scale Energy Devices

Chapter 14 Computational Fluid Dynamics as a Modern Tool in Studies of Biomass-Based Small-Scale Energy Devices 317

Mateusz Szubel, Maciej Kryś and Karolina Papis

Index.. 341

Preface

Indisputably, a growing interest in renewables has been recently observed. Biomass is especially popular, having great potential for development owing to several advantages, such as possibilities for storage and application in many different forms, for heat and electricity generation, and for fuel production.

For the last few years, advances in biomass-based techniques have been significant and have resulted in various systems, as well as conceptual and laboratory research.

Principles that rule the thermochemical conversion of biomass are already determined and well known; however, it is still difficult to design devices that provide the possibility to utilize flexibly any solid biofuel. Many factors, such as the origin and type of biomass, harvesting and storing methods, or further processing, determine the necessity of considering each biomass-based energy system individually. This assumption has resulted in the development of various advanced technologies dedicated to combustion, gasification, and even preprocessing (i.e., drying) in small-scale applications. A significant proportion of the problems related to biomass-based micro-gasification or micro-cogeneration systems has been completely or partially solved; however, the issue of biomass thermochemical treatment still faces many challenges. Thus, different researchers' points of view on these problems, based on energy, economic, and environmental approaches, are presented here.

This book is intended to be a text for engineering master's courses dealing with small/medium biomass-based energy systems. It is based on the authors' experiences in running classes over several years. The emphasis here is toward applications of biomass in terms of energy generation and usage. The approach is to demonstrate how the fundamental topics related to biomass, its composition, devolatilization, combustion, gasification, and others (biomass properties and phenomena taking place during the process of the biomass thermochemical treatment), can be applied to develop a purpose-designed energy system. Political, economic, and environmental aspects are included to give a full picture of the related problems.

The processes of thermochemical conversion of biomass are considered in Chapters 1, 12, 13, and 14. The cogeneration systems based on biomass are presented in Chapters 7, 8, 9, and 12. Environmental issues are discussed in Chapters 3–5. Biomass pretreatments, such as drying and other processes, are presented in Chapters 10 and 11. Moreover, advanced computer-aided simulation methods, supporting currently experimental and analytical research of the biomass-based energy technologies, are presented in Chapters 4 and 12. Other issues, such as economics and calculations for stoves, are presented in Chapters 2 and 6. Of course, single chapters can be used independently. The scope of the volume covers a significant part of problems related to biomass applications. It can be considered as an attempt to complete the description of the problems connected with biomass thermochemical conversion in small-scale energy applications.

Editors

Mateusz Szubel (MSc), is assistant professor at AGH University of Science and Technology, Faculty of Energy and Fuels, Krakow, Poland. Professor Szubel is a member of a research group focused on the issues of renewable energy technologies in the Department of Sustainable Energy Development. He is a research specialist on the conditions for the development of sustainable energy. He is also involved in the investigation of the possibilities of using computational fluid dynamics (CFD) in optimization of renewable energy technologies and increasing the energy efficiency of energy devices. He is focused on numerical modeling of biomass thermochemical treatments. Professor Szubel is the academic teacher for parts of the courses that are related to the practical aspects of renewable energy technologies' applications and application of commercial CFD codes to analyze the elements of these systems. He is author and coauthor of more than 100 scientific and popular science papers.

Mariusz Filipowicz (DSc, Eng.), is associate professor at AGH University of Science and Technology, Faculty of Energy and Fuels, and head of the Department of Sustainable Energy Development, Krakow, Poland. Professor Filipowicz graduated with an MSc (technical physics) in 1991, a PhD in 1998, and habilitation in 2010 in the field of nuclear physics–nuclear fusion catalyzed by negative muons. From 1994–1995, he completed a postgraduate course, Energy and Environment, under the Tempus-Joint European Project (JEP) Postgraduate Course on Energy and Environment. Professor Filipowicz's main areas of research are related to renewable energy technologies (mainly biomass, solar, and wind), energy efficiency, and nuclear physics (nuclear fusion reactions in the range of ultra-low energy). He is author and coauthor of more than 250 scientific papers and leader of scientific projects at AGH, such as, "BioEcoMatic: Construction of small-to-medium capacity boilers for clean and efficient combustion of biomass for heating" and "BioORC: Construction of cogeneration system with small to medium size biomass boilers" with the support of KIC InnoEnergy.

Contributors

Joanna Bigda
Institute for Chemical Processing
of Coal
Zabrze, Poland

Tomasz Chmielniak
AGH University of Science
and Technology
Krakow, Poland

Simona Di Fraia
University of Naples Parthenope
Naples, Italy

Maria Di Palma
University of Naples Parthenope
Naples, Italy

Rafał Figaj
AGH University of Science
and Technology
Krakow, Poland

Wojciech Goryl
AGH University of Science
and Technology
Krakow, Poland

Tomasz Iluk
Institute for Chemical Processing
of Coal
Zabrze, Poland

Juraj Klukan
Technical University in Zvolen
Zvolen, Slovakia

Danuta Król
Silesian University of Technology
Gliwice, Poland

Maciej Kryś
MESco sp. z o.o.
Tarnowskie Góry, Poland

Robert Kubica
Silesian University of Technology
Gliwice, Poland

Nadezda Langova
Technical University in Zvolen
Zvolen, Slovakia

Adriano Macaluso
University of Naples Parthenope
Naples, Italy

Nicola Massarotti
University of Naples Parthenope
Naples, Italy

Tomasz Mirowski
Mineral and Energy Economy Research
Institute of the Polish Academy
of Sciences
Krakow, Poland

Eugeniusz Mokrzycki
Mineral and Energy Economy Research
Institute of the Polish Academy
of Sciences
Krakow, Poland

Karolina Papis
AGH University of Science
and Technology
Krakow, Poland

Marcin Pluta
AGH University of Science
and Technology
Krakow, Poland

Sławomir Poskrobko
Bialystok University of Technology
Białystok, Poland

Thomas Schiffert
Austrian Kachelofen
(Tile Stove) Association
Research Institute
of Kachelofen Builders
Vienna, Austria

Aleksander Sobolewski
Institute for Chemical Processing
of Coal
Zabrze, Poland

Krzysztof Sornek
AGH University of Science
and Technology
Krakow, Poland

Wojciech Suwała
AGH University of Science
and Technology
Krakow, Poland

Laura Vanoli
University of Naples Parthenope
Naples, Italy

Jozef Viglasky
Technical University in Zvolen
Zvolen, Slovakia

Artur Wyrwa
AGH University of Science
and Technology
Krakow, Poland

Maciej Żołądek
AGH University of Science
and Technology
Krakow, Poland

Janusz Zyśk
AGH University of Science
and Technology
Krakow, Poland

Part I

Perspectives of Using Raw and Processed Biomass in Small-Scale Energy Systems

1 Thermochemical Processing of Solid Biomass

Tomasz Mirowski and Eugeniusz Mokrzycki

CONTENTS

1.1 Introduction 3
1.2 Energy Resources from Biomass..4
1.3 Biomass Conversion Technologies 9
   1.3.1 Combustion..10
   1.3.2 Gasification 11
   1.3.3 Pyrolysis...13
1.4 The Hydrothermal Carbonation Process 13
1.5 Conclusions...14
References..16

1.1 INTRODUCTION

There is no clear definition of biomass. In the scientific and technical literature there are various definitions of biomass, depending on the needs of the issue being discussed. Under EU legislation, "biomass" means the biodegradable fraction of products, waste, and residues from agriculture (including vegetal and animal substances), forestry, and related industries, as well as the biodegradable fraction of industrial and municipal waste (Directive, 2009). The law in individual EU member states clarifies the definition of biomass, taking into account the specificity of the resources possessed and the possibilities for their management.

In Poland, biomass is understood as a biodegradable fraction of products, waste, or residues of biological origin from agriculture (including vegetal and animal substances), forestry, and related industries, including fisheries and aquaculture, processed biomass, in particular in the form of briquettes and pellets, of biocarbon, and the biodegradable fraction of industrial or municipal waste of plant or animal origin, including waste from waste treatment installations and waste from water and wastewater treatment, in particular sewage sludge, in accordance with the provisions on waste concerning classification of the energy recovered from thermal treatment of waste (Act, 2015). The introduction of new concepts to the definition of biomass in Poland, such as biocarbon, has enabled wider use of humid biomass with a low calorific value (6–13 MJ/kg).

The energy value of plant biomass comes from solar energy used in the process of photosynthesis. It should be emphasized that biomass has a significant share in the energy consumed in developing countries. In addition, developed countries have shown considerable interest in biomass in relation to waste management and air protection.

The aim of this chapter is to discuss biomass conversion technology. Owing to space constraints of the publication, attention is focused on thermochemical processes of biomass processing only.

1.2　ENERGY RESOURCES FROM BIOMASS

There is an increased interest in biomass as a potential source of energy for the following reasons (Dreszer et al., 2003):

- Economic—acquiring biomass (not energy from biomass) is relatively cheap.
- Striving for energy self-sufficiency owing to the allocation of other energy sources (especially fossil fuels).
- Environmental protection, in particular the reduction of the greenhouse effect.
- Reclamation of fallow and waste lands.

Biomass resources for energy purposes are diverse and can be divided into the following:

- Primary energy sources, mainly used to produce biomass: wood, straw, and energy crops.
- Secondary energy sources: liquid manure, organic waste, or sewage sludge.
- Processed energy sources: biogas, bioethanol, biomethanol, biodiesel, and bio-oil.

The form of solid biomass (wood, straw) that is fed to the boiler is of decisive importance for the technical solutions used for its combustion, and thus significantly affects the economic performance of the investment.

The energy efficiency of cereal straw depends primarily on its moisture content, the composition associated with the type of straw (cereals, rapeseed, maize), and vegetation conditions of the plants. The average net calorific value of straw with a moisture content of 15% is 14–15 GJ/Mg. The coefficient of energy concentration (MWh) in 1 m^3 of straw depends on its form: loose straw—0.07 to 0.16; chaff—0.13 to 0.19; perpendicular bales—0.23 to 0.43; round bales 0.19 to 0.29; conventional bales—0.16 to 0.36; and briquettes—0.99 to 1.48.

Wood is a chemically heterogeneous substance (cellulose, hemicellulose, and lignin), characterized by a high variability of thermophysical features that make it impossible to determine the actual energy value. It includes wood waste from forest industry and chips of fast-growing perennial tree species. Wood is used in various forms, including logs, billets, chips, cuttings, shavings, sawdust, and wood dust. The net calorific value of wood with natural humidity (50%–60%) is 6–8 GJ/Mg, while after drying to air-dry condition (10%–20%) it is 14–16 GJ/Mg.

Wood can also be in the form of briquettes and pellets. Briquettes are formed as a result of particle aggregation. The durability of an agglomerate depends on a number of factors that deserve special attention, including internal forces between individual particles, adhesion joints, cohesion, and mechanical forces. Properly crushed raw material of vegetable origin subjected to external and internal forces (pressure agglomeration) is compacted, and the product obtained has a constant geometric form (agglomerate). It should be emphasized that the complexity and diversity of problems occurring in the agglomeration process, carried out in systems of different consistency, causes a number of technical, technological, and operational difficulties (Hejft, 2012). Dry wood parts can also be ground in suitable mills (particle size 0.8–1.0 mm) for combustion in fluidized bed boilers. Biomass from tree and bush cultivation is also used as energy sources (shrub willow, poplar, etc.). Willow is considered to be a future source of biomass for conversion to gas for the production of electricity and bioethanol, which can be converted into hydrogen fuel for fuel cells. Energy crops include annual plants with a high sugar and starch content (cereals, potatoes, beets, corn for grain) and oilseeds (rapeseed, sunflower, giant flax, and some grass species) may be of great importance.

Secondary energy sources include biomass of plant and/or of animal origin (liquid manure, manure), organic products (pomace, sludge, fats, etc.), or sewage sludge. Liquid manure and manure are used to produce biogas.

All by-products have only local significance as energy sources and in most cases require individual technologies and techniques of use. Sewage sludge is incinerated in municipal wastewater treatment plants.

Processed energy sources include a large group of products from biomass. They are, among others, biogas, ethanol, methanol, bio-oil, and biogas.

Biomass is not the preferred fuel for the power industry due to its physicochemical properties affecting its storage, transportation, and combustion (especially in coal-fired boilers). The physicochemical properties of biomass can be improved as a result of the torrefaction process.

Torrefaction is the thermal method of biomass treatment—a thermolysis process at 250–300°C, at about atmospheric pressure, without air access. As a result of such a process, a solid product, biochar, is obtained. It is characterized by (Zuwała et al., 2015) the following:

- More favorable physicochemical properties than raw biomass.
- Homogeneity.
- Hydrophobic properties.
- Resistance to biological agents in comparison with raw biomass.
- Lower transport costs than for unprocessed biomass; according to Koppejan et al. (2012), the total transport costs (maritime and road transport, and storage costs) for raw biomass are around 4.11 USD/GJ, and about 2.40 USD/GJ in the case of pellets from torrefied biomass.

The main components of biomass are carbohydrates (disaccharides, simple sugars), starch (amylose, amylopectin), and lignin (biopolymer, which consists of unsaturated alcohols and phenols). Carbohydrates and starch are the products of agricultural crops and serve as food for humans and animals; in addition, they are used in the production of ethanol. The remaining part of the biomass—cellulose, hemicellulose, and lignin—are very good energy sources; their approximate composition is as follows: cellulose 40%–60%, hemicellulose 20%–40%, and lignin 10%–25%.

The composition of biomass depends on many factors, including the type of biomass (plant species, part of the plant), growth process, plant age, fertilizers and pesticides used, distance from sources of pollution, time and techniques of harvesting, contamination during collection, transport, and mixing of different types of biomass.

The chemical structure of solid biomass can be presented in a simplified way using the following formula: $C_1H_{1.45}O_{0.7}$ (Kubica, 2003).

Figure 1.1 shows a comparison of the average composition of different types of biomass and coal. When comparing the properties of fossil fuels and biomass, it should be stated that the elemental composition is qualitatively the same, while the differences occur in the proportions of individual elements and chemical compounds. Biomass contains, on average, four times more oxygen and double less carbon, as well as less sulfur and nitrogen. As a result of this composition, biomass is characterized by a high content of volatile matter and high reactivity.

Table 1.1 presents the technical and elementary analysis of solid biomass (straw, chips, pellets, and willow). The essential properties of biomass-based solid fuels include moisture content, net calorific value, density, particle size, and ash content. The importance of other properties of solid biofuels, including the content of N, S, Cl, alkaline elements (Na, K), and heavy metals (Cd, Zn, and Pb), depends on the type of fuel, the combustion conditions in the combustors, and the reduction of pollution.

The most important feature of energy resources from biomass, in relation to combustion and other thermochemical processes, are the moisture content, the type of biomass (its composition), and the fuel production method. The net calorific value is between 5 and 19 MJ/kg, and is usually almost double lower than the net calorific

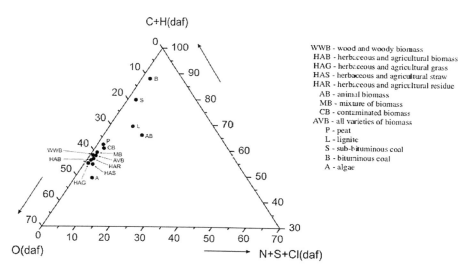

FIGURE 1.1 A comparison of the average composition of the different types of biomass and coal. (From Vassilev, S. V. et al. 2010. *Fuel*, 89, pp. 913–933.)

TABLE 1.1
Technical and Elemental Analysis of Biomass

Element	Unit	Yellow Straw	Gray Straw	Chips	Pellets/ Briquettes	Willow
Moisture content	%	10–20	10–25	20–50	7–12	50–60
Volatile matter	%	70–80	70–80	76–86	>70	>70
Ash content	%	5	3	0.8–1.4	0.4–1.5	1.1–4.0
C[a]	%	45–48	43–48	47–52	48–52	47–51
H	%	5–6	5–6	6.1–6.3	6–6.4	5.8–6.7
O	%	36–48	36–48	38–45	40	40–46
Cl	%	0.97	0.14	0.02	0.02–0.04	0.02–0.05
N	%	0.3–0.6	0.3–0.6	<0.3	0.3–0.9	0.2–0.8
S	%	0.05–0.2	0.05–0.2	<0.05	0.04–0.08	0.02–0.1
K	%	1.3	0.7	0.02	N.M.	0.2–0.5
Ca	%	0.6	0.1	0.04	N.M.	0.2–0.7
Gross calorific value	MJ/kg	17.4	17.4	19.2–19.4	16–19	18.4–19.2
Density	kg/m^3	100–170	100–170	250–350	500–780	120
Melting point	°C	800–1,000	800–1,000	1,000–1,400	>1,120	n.m.[b]

Source: Data from Rybak, W. 2006. *Spalanie i współspalanie biopaliw stałych (Combustion and co-combustion of solid biofuels)* (in Polish). Publishing House of the Wrocław University of Technology, Wrocław.

[a] Elemental analysis (dry basis).
[b] n.m.—not marked.

---

value of coal (25 MJ/kg) (Ściążko and Zieliński, 2003). Figure 1.2 presents changes in the net calorific value of wood depending on the moisture content.

Such a low net calorific value is connected with incurring significant financial costs for its displacement from the place of production to the place of use and providing adequate storage facilities. Figure 1.3 compares chemical energy concentration in a unit of volume and mass for various fuels.

The volatile matter content in solid biofuels is high (Figure 1.4). The majority of thermal energy is emitted during the combustion of volatile matter. Therefore, effective biomass combustion requires specially designed combustion chambers. Figure 1.4 shows a comparison of the average content of ash, volatile matter, and bound carbon in various types of biomass and coal.

There are two types of densities of solid biofuels, that is, grain density and bulk density. The grain density describes the density of the material itself and is important for the combustion process (evaporation temperature, energy density), biomass feeding (pneumatic/mechanical equipment), and storage. Bulk density plays a role in transport and storage (trade and supplies).

The shape and size of particles and their quantitative share in solid biofuels are parameters that are important when distributing loads between generation units.

Ash is a secondary product, obtained by the action of high temperature on the mineral matter. The chemical and mineral composition of biomass ash is significantly

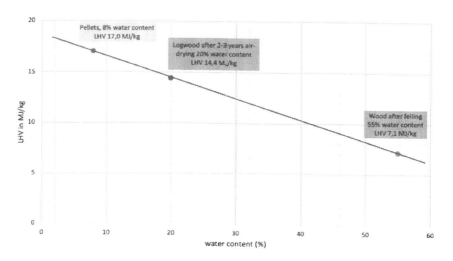

FIGURE 1.2 Changes in the calorific value of wood depending on the moisture content. (Based on Tiser, A. and Edöcs, O. 2006. *Przewodnik—Aspekty techniczne finansowania inwestycji w odnawialne źródła energii i efektywność energetyczną* (*Guide—Technical Aspects of Financing Investments in Renewable Energy Sources and Energy Efficiency*) (in Polish). Project: Financial Institutions Personnel Training in the Concepts of Renewable Energy and Energy Efficiency Technologies for the Evaluation of Relevant Projects—FiP-TREET, Contract No. HU/04/B/F/PP-170031; topic 5, pp. 111–139.)

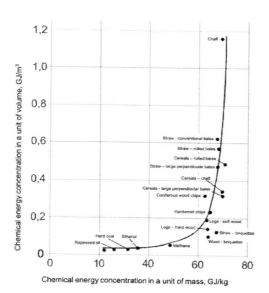

FIGURE 1.3 Comparison of chemical energy concentrations in a unit of volume and mass for various fuels. (From Dreszer K. A. et al. 2003. *Energia odnawialna—możliwości jej pozyskiwania i wykorzystania w rolnictwie* (*Renewable energy—the possibilities of its acquisition and use in agriculture*) (in Polish). Polskie Towarzystwo Inżynierii Rolniczej, Kraków–Lublin–Warszawa.)

Thermochemical Processing of Solid Biomass

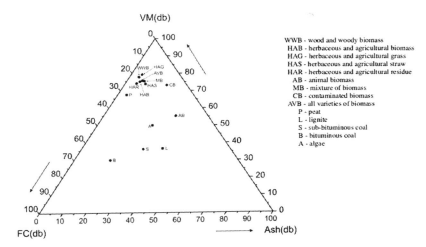

FIGURE 1.4 Comparison of the average content of ash, volatile matter (VM), and fixed carbon (FC) in various types of biomass and coal. (From Vassilev S. V. et al. 2010. *Fuel*, 89, pp. 913–933.)

different compared to ash from hard coal combustion. Ashes from the combustion of biomass from green plants and grains for energy purposes are characterized by different softening temperatures (from 750 to 1000°C), which may increase the deposition of chemical compounds on heated surfaces, contributing to failures and high costs of repairs. In some technologies, ash can play a positive role (as it accumulates part of the heat) creating a heating surface and providing heat for the final stage of combustion—burning of coke, thus protecting the grate from excessive flame radiation.

1.3 BIOMASS CONVERSION TECHNOLOGIES

Recently, there has been an increased interest in the use of energy from renewable sources, owing to the growing demand for energy and European Union (EU) requirements related to the reduction of atmospheric emissions of harmful compounds (Galvez-Martos and Schoenberger, 2014).

Biomass is the third world's largest renewable energy source(after solar and wind energy) (IEA, 2018). The amount of carbon dioxide emitted into the atmosphere during the combustion of biomass is equal to the amount of carbon dioxide absorbed by plants producing biomass during photosynthesis.

According to Vassilev et al. (2013), the global production of biomass, which potentially can be used in the power sector, is estimated at about 7 billion tons per year, including 3 billion tons of forest waste, between 1.1 and 3.1 billion tons of agricultural waste, and about 1.1 billion tons of municipal waste plus sewage sludge and other biomass resources. Biomass has therefore become a viable alternative to solid fossil fuels. From 8 to 15% of the world's total production of electricity, heat, and fuels for transport is produced from biomass (Heinimö and Junginger, 2009; Abbasi and Abbasi, 2010).

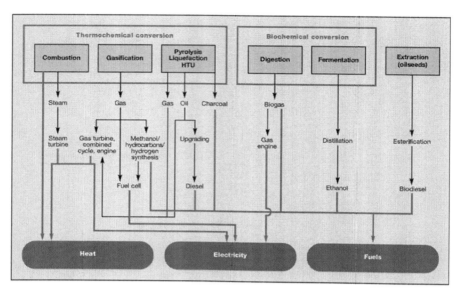

FIGURE 1.5 Simplified biomass conversion scheme. (From Turkenburg, W. C. et al. 2000. *World Energy Assessment—Energy and the Challenge of Sustainability.* J. Goldemberg (Ed.), UNDP and UN-DESA, New York, NY, USA, WEC, London, UK, pp. 219–272.)

Biomass conversion technologies can be divided into the following groups (Figure 1.5):

- Thermochemical processes: combustion, gasification, and pyrolysis.
- Biochemical processes: anaerobic digestion, ethanol fermentation.
- Physicochemical processes: esterification.

Owing to the complexity of biochemical and physicochemical processes and the limited publication space, attention is focused on thermochemical processing only. As already mentioned (Figure 1.5), thermochemical conversion includes all processes based on thermal energy, that is, direct combustion, gasification, and pyrolysis.

1.3.1 Combustion

The use of biomass as a fuel is a very complex issue, which has been studied by many research units around the world.

The energy use of biomass is associated with a number of problems (Nowak and Wesołowska, 2013) including difficulties in maintaining stable boiler operation resulting from the diversity of physicochemical properties of biomass, agglomeration and defluidization (especially in fluidized bed boilers), surface pollution as a result of deposition (soot, pitch), and high temperature corrosion.

The selection of the most appropriate technology for biomass utilization depends mainly on the biomass type to be used in a given process, the availability of technologies, and economic considerations (Nowak and Wesołowska, 2013). In addition, the density of these fuels, that is, how much energy can be obtained from fuel collected from a specific area, should be taken into account, as it may turn out that the costs of such fuels will be significantly higher than traditional fuels (e.g., coal). This also applies to the specific volume of these fuels (e.g., energy from 1 m^3 of coal is equivalent to about 13 m^3 of straw in bales—see Figure 1.3).

Direct combustion is a well-known thermochemical conversion process. The solid biomass can be burned without processing, and the resulting gases can be used as heat and steam sources. Combustion takes place in four stages: drying, pyrolysis, combustion of volatile matter, and reduction. The process of proper combustion requires a high temperature and a sufficient amount of air and time for full combustion.

Cogeneration, that is, simultaneous generation of electricity and heat from biomass as an energy carrier, is of great importance. Cogeneration can be carried out in combined heat and power plants or can be set up in the vicinity of customers (distributed cogeneration). Cogeneration biomass-fired units (Nowak and Wesołowska, 2013) may use a reciprocating engine, steam turbine, Integrated Gasification Combined Cycle (IGCC) gas turbine, Stirling engine, screw engine, microturbine, fuel cell, biomass valorization (biocarbon), or liquid fuels (methanol, ethanol, or synthetic fuels).

The most common method of obtaining heat from biomass is its direct combustion, differentiated depending on the method of fuel preparation (wood chips, pellets, briquettes, etc.). The combustion of biofuel is similar to coal combustion technology. Burning wood for heating purposes in open fireplaces with low efficiency causes significant emissions of CO_2 and smoke. Nowadays, heating devices with a closed combustion chamber complying with emission standards and a high energy efficiency are used increasingly often for domestic heating. It should be emphasized that the wood pellet market for domestic heating purposes is developing dynamically.

Figure 1.6 presents an overview of the biomass processing technologies and their current state of development (IFC, 2017; Mirowski et al., 2018).

The combustion of biomass (and other fuels) in the power industry results in the emission of pollutants; however, it is much lower than the emissions from plants burning hard coal and lignite. Also, emissions of heavy metals, such as cadmium, mercury, and lead, are lower than those from lignite and hard coal combustion. Nevertheless, the combustion of biomass in the power industry resulted in an increase in the domestic emission of PCDD-F in 2011 and 2013 (National Emission Inventory, 2014). In addition, the sum emissions of organic compounds from combustion of biomass may be higher than in the case of hard coal (Wielgosiński, 2009).

1.3.2 Gasification

Gasification is the conversion, by partial oxidation, of a carbonaceous feedstock (e.g., biomass) into a gas energy carrier. The gasification process takes place in two stages. The first stage involves partial combustion of biomass and the production of gas and charcoal. In the second stage, water and carbon dioxide produced in the first stage are reduced (chemically) by charcoal. As a result of this reaction, carbon monoxide and hydrogen are obtained (Figure 1.7). The produced gas (syngas) should not contain

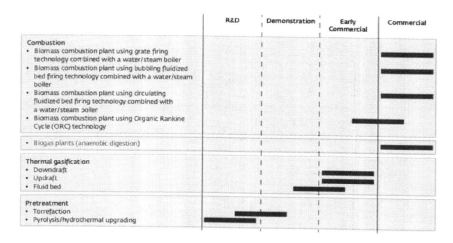

FIGURE 1.6 Overview of biomass processing technologies and their current state of development. (Adapted from Mirowski T. et al. 2018. *The Bulletin of the Mineral and Energy Economy Research Institute of the Polish Academy of Sciences* (105), pp. 63–74.)

residual tars and higher hydrocarbons, therefore the gasification process requires temperatures of around 800°C or higher. Syngas contains hydrogen (18%–20%), carbon monoxide (18%–20%), carbon dioxide (8%–10%), CH_4 methane (2%–3%), trace amounts of higher hydrocarbons (ethane, ethene), and various impurities (small particles of charcoal, ash, tar, and oils).

Partial oxidation is carried out using air, oxygen, steam, or a mixture of these three components. As a result of oxidation of biomass with air, a producer gas is

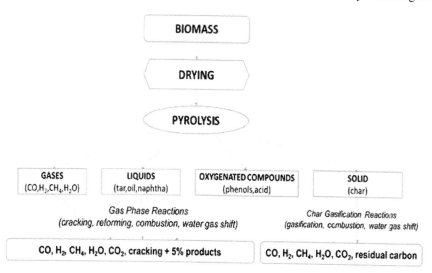

FIGURE 1.7 The biomass gasification process. (From Sikarwar, V. S. et al. 2016. *Energy & Environmental Science* 9(10), pp. 2939–77.)

Thermochemical Processing of Solid Biomass

obtained, which can be used in boilers or turbines. Oxygen gasification allows for the obtainment of synthesis gas, which may be suitable—due to the higher energy density than the producer gas—for pipeline transport.

The oxidation of biomass (lignin and cellulose) with a mixture of air, oxygen, and water vapor makes it possible to obtain a pyrolysis gas.

1.3.3 Pyrolysis

The process of thermal conversion of biomass without oxygen is called pyrolysis. As a result of the pyrolysis process, solid products, liquid products, and gases are obtained. The process takes place in the temperature range of 400–800°C, where cellulose, hemicellulose, and partly lignin are broken down into smaller, lighter particles, some of which, after cooling, form pyrolysis oil (the main product of the process). The pyrolysis process can be carried out in the presence of a small amount of oxygen (gasification), water (steam gasification), and hydrogen (hydrogenation).

1.4 THE HYDROTHERMAL CARBONATION PROCESS

The hydrothermal carbonization (HTC) process involves the conversion of coal into liquid and solid products. Depending on the pressure and temperature, the following processes take place: carbonization, liquefaction, and gasification (Figure 1.8). Figure 1.9 presents a simplified scheme of the hydrothermal carbonization process.

The conditions for hydrothermal carbonation for the production of biocarbon are as follows (Grönberg and Wikberg, 2015):

- The typical temperature range: 180–250°C.
- Required pressure: 20–50 bar.

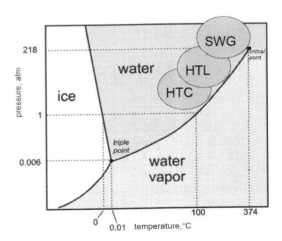

FIGURE 1.8 Hydrothermal processes. HTC, hydrothermal carbonization; HTL, hydrothermal liquefaction; SWG, supercritical water gasification. (From Grönberg, V. and Wikberg, H. 2015. Hydrothermal carbonization for biochar production. VTT Technical Research Centre of Finland [available at http://www.fibrafp7.net].)

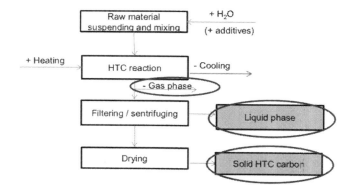

FIGURE 1.9 The simplified diagram of the hydrothermal carbonization process (HTC). (From Grönberg, V. and Wikberg, H. 2015. Hydrothermal carbonization for biochar production. VTT Technical Research Centre of Finland [available at http://www.fibrafp7.net].)

- Reaction in water (without the pretreatment drying phase).
- All biomass and organic waste can be used as raw materials.
- Required time: several hours.
- pH below 7.
- Catalyst: citric acid.
- Reaction: exothermic.
- Water: recycling in the process.

The raw materials that can be used for the HTC process are shown in Table 1.2.

1.5 CONCLUSIONS

1. In recent years, a growing interest in biomass as a source of energy has been observed. This applies to both developing and developed countries and is related to air pollution on a global scale.
2. The biomass resources that can be used for energy production are varied and include primary residues, secondary residues, and tertiary residues.
3. In most cases, energy sources from biomass are not suitable for direct use and should be processed into another type of solid, liquid, or gas fuel. Biomass conversion technologies can be divided into thermochemical, biochemical, and physicochemical processes.
4. In the case of thermochemical conversion, combustion is the most widely used technology, while gasification is a promising solution for the needs of electricity production. Pyrolysis, which enables the production of liquid fuel for generating units, is the solution for the future. Advanced conversion processes (gasification, pyrolysis) are electricity generation technologies that are more efficient than methods based on combustion and steam turbines.
5. The devices used for direct combustion of solid biomass are of various sizes, starting from small boilers (5–50 kW), through to small and medium ones

TABLE 1.2
Biomass Raw Materials for the Hydrothermal Carbonization (HTC) Process

No.	Type	Advantages	Disadvantages
1.	Wood polymers Lignin	• Inexpensive • Attractive to customers	• Hard to carbonize in HTC • Contains impurities (e.g., S, Na) which are unwanted in some applications
	Cellulose	• Very useful in producing high value added solid and liquid products	• Different prices
	Hemicellulose	• Very useful in producing high value added solid and liquid products	• Different prices
2.	Wood residues (e.g., wood chips, bark, leaves, etc.)	• General interest in renewable bioresources	• Raw material variations might need pretreatment prior to HTC (e.g., grinding)
3.	Agricultural residues (e.g., straw, grasses, brewer's spent grain (BSG), bagasse, animal manure, etc.)	• Inexpensive large global markets	• Collection infrastructure is missing/patchy
4.	Municipal sewage sludge	• Inexpensive waste (gate fees) includes nutrients	• Large (seasonal) variations of the raw material for low value products mainly (e.g., energy, landfill)
5.	Algae	• Inexpensive	• Low concentrations in the feed
6.	Forest industry side streams (e.g., black liquor, dissolving pulp side stream, drinking water, biosludge, etc.) • Black liquor • Prehydrolyzate from kraft dissolving pulp process	• Inexpensive • Availability • Very useful in producing high value added solid and liquid products	• Contains impurities which are unwanted in some applications, • availability

(50–500 kW and 500 kW–5 MW), and ending with large boilers used in power plants and combined heat and power plants (>5 MW boilers).

6. The main factors affecting the efficiency of installations using biomass are raw material costs, installation size and technology used, the use of generated heat, the price of electricity and/or heat, and the number of operating hours per year.

A renewable energy source such as biomass can be a good solution to problems related not only to environmental protection, but also to social, economic, and energy security issues. An ever-growing stream of waste from separate collection contains biomass in various forms. Efficient use of this energy carrier is currently a major challenge for countries (including Poland) where waste collection systems are developed and legal regulations require the energy use of such waste. The diversification of energy supply, taking into account biomass fuels, increases energy security at the level of local communities. This is also the perception of biomass in EU policy until 2030, where the emphasis is on supporting the distributed generation and more environmentally sustainable use of biomass, as well as increasing the promotion of heating and cooling, including system heating, using renewable energy.

REFERENCES

Abbasi, T. and Abbasi, S.A. 2010. Biomass energy and the environmental impacts associated with its production and utilization. *Renewable and Sustainable Energy Reviews*, 14(3), pp. 919–937.

Act. 2015. Ustawa z dnia 20 lutego 2015 r. o odnawialnych źródłach energii (Dz. U. 2015, poz.478) [tekst ujednolicony 21.09.2018]. (The Act of 20 February 2015 on renewable energy sources. Journal of Laws of 2015, item 478 [consolidated text 21/09/2018]) (in Polish).

Dreszer, K. A., Michałek, R., and Roszkowski, A. 2003. *Energia odnawialna—możliwości jej pozyskiwania i wykorzystania w rolnictwie (Renewable energy—The possibilities of its acquisition and use in agriculture)* (in Polish). Polskie Towarzystwo Inżynierii Rolniczej, Kraków–Lublin–Warszawa.

Directive. 2009. Directive 2009/28/EC of the European Parliament and of the Council of 23 April 2009 on the promotion of the use of energy from renewable sources and amending and subsequently repealing Directives 2001/77/EC and 2003/30/EC.

Galvez-Martos, J-L. and Schoenberger, H. 2014. An analysis of the use of life cycle assessment for waste co-incineration in cement kilns. *Resources, Conservation and Recycling*, 86, pp. 118–131.

Grönberg, V. and Wikberg, H. 2015. *Hydrothermal carbonization for biochar production*. VTT Technical Research Centre of Finland (available at http://www.fibrafp7.net).

Heinimö, J. and Junginger, M. 2009. Production and trading of biomass for energy—An overview of the global status. *Biomass & Bioenergy*, 33, pp. 1310–1320.

Hejft, R. 2012. Innowacyjność w granulowaniu biomasy (Innovation in biomass granulation) (in Polish). *Czysta Energia (Clean Energy)*, (6), pp. 32–34.

International Energy Agency (IEA). 2018. Renewables 2018—Market analysis and forecast from 2018 to 2023. (available at https://www.iea.org/renewables2018/).

International Finance Corporation (IFC). 2017. Converting biomass to energy. A guide for developers and investors. 2121 Pennsylvania Avenue, N.W. Washington, D.C. (available at https://openknowledge.worldbank.org/handle/10986/28305).

Koppejan, J., Sokhansani, S., Melin, S., and Madrali, S. 2012. *Status overview of torrefaction technologies*. IEA Bioenergy Task 32 Report, Enschede 2012, pp. 1–54.

Kubica, K. 2003. Przemiany energochemiczne węgla i biomasy (Energetic and chemical transformation of coal and biomass). In: *Termochemiczne przetwórstwo węgla i biomasy (Thermochemical Processing of Coal and Biomass)* (in Polish). Ściążko, M. and Zielinski, H. (Eds). Wydawnictwo Instytutu Chemicznej Przeróbki Węgla, Kraków–Zabrze, pp. 145–197.

Mirowski, T., Mokrzycki, E., Filipowicz, M., and Sornek, K. 2018. *Characteristic of selected biomass technologies in distributed energy sector. The Bulletin of the Mineral* and *Energy Economy Research Institute of the Polish Academy of Sciences* (105), pp. 63–74 (available at https://min-pan.krakow.pl/wydawnictwo/wp-content/uploads/sites/4/2018/10/mirowski-i-inni.pdf).

National Emission Inventory, 2014. *Krajowy bilans emisji SO₂, NOₓ, CO, NH₃, NMLZO, pyłów, metali ciężkich i TZO za lata 2011–2012 w układzie klasyfikacji SNAP.* 2014. Raport syntetyczny. (National emission inventory: SO_2, NOx, CO, NH_3, NMVOC, dust, heavy metals and POPs in 2011–2012 in the SNAP classification system. Synthetic report) (in Polish). The National Centre for Emissions Management (KOBiZE), version 2, March 12, 2014.

Nowak, W. and Wesołowska, M. 2013. Uwarunkowania techniczne spalania biomasy w kotłach energetycznych (Technical conditions of biomass combustion in power boilers). [W:] *Biomasa na cele energetyczne* ([In:] *Biomass for Energy Purposes*) (in Polish). Gołos, P. and Kaliszewski, A. (Eds). Prace Instytutu Badawczego Rolnictwa (Works of the Forest Research Institute) Sękocin Stary, pp. 216–224.

Rybak, W. 2006. *Spalanie i współspalanie biopaliw stałych (Combustion and co-combustion of solid biofuels)* (in Polish). Publishing House of the Wrocław University of Technology, Wrocław.

Ściążko, M. and Zieliński, H. (Eds.) 2003. *Termochemiczne przetwórstwo węgla i biomasy (Thermochemical Conversion of Coal and Biomass)* (in Polish). The Publishing House of the Institute for Chemical Processing of Coal, Kraków–Zabrze.

Sikarwar, V. S., Zhao, M., Clough, P., Yao, J., Zhong, X., Memon, M. Z., Shah, N., Anthony, E. J., and Fennell, P. S. 2016. An overview of advances in biomass gasification. *Energy & Environmental Science*. 9(10), pp. 2939–77. The Royal Society of Chemistry. doi: 10.1039/C6EE00935B.

Tiser, A. and Edöcs, O. 2006. Zastosowanie bioenergii (The use of bioenergy). [W:] *Przewodnik—Aspekty techniczne finansowania inwestycji w odnawialne źródła energii i efektywność energetyczną* ([In:] *Guide – Technical aspects of financing investments in Renewable Energy Sources and Energy Efficiency*) (in Polish). Project: Financial Institutions Personnel Training in the Concepts of Renewable Energy and Energy Efficiency Technologies for the Evaluation of Relevant Projects—FiP-TREET, Contract No. HU/04/B/F/PP-170031; topic 5, pp. 111–139.

Turkenburg, W. C., Beurskens, J., Faaij, A., Fraenkel, P., Fridleifsson, I., Lysen, E., Mills, D. et al. 2000. Renewable energy technologies. In: *World Energy Assessment—Energy and the Challenge of Sustainability*. J. Goldemberg (Ed.) UNDP and UN-DESA, New York, NY, USA, WEC, London, UK, pp. 219–272.

Vassilev, S. V., Baxter, D., Andersen, L. K., and Vassileva, C. G. 2010. An overview of the chemical composition of biomass. *Fuel*, 89, pp. 913–933.

Vassilev, S. V., Baxter, D., Andersen, L. K., and Vassileva, C. G. 2013. An overview of the composition and application of biomass ash. Part 1. Phase–mineral and chemical composition and classification. *Fuel* 105, pp. 40–76.

Wielgosiński, G. 2009. Czy biomasa jest paliwem ekologicznym? (Is a biomass an ecological fuel?) (in Polish). [W:] *Materiały III Ogólnopolskiego Kongresu Inżynierii Środowiska*. Monografia Komitetu Inżynierii Środowiska Polskiej Akademii Nauk red. naczelny prof. dr hab. Lucjan Pawłowski, vol. 61/1, s. 347-356. Łódź [In:] *Materials of the 3rd National Congress of Environmental Engineering*. Monography of the Environmental Engineering Committee of the Polish Academy of Sciences. Editor-in-chief prof. dr hab. Lucjan Pawłowski, vol. 61/1, pp. 347-356, Lodz.

Zuwała, J., Kopczyński, M., and Kazalski, K. 2015. Koncepcja systemu uwierzytelniania biomasy toryfikowanej w perspektywie wykorzystania paliwa na cele energetyczne (The concept of torrefied biomass certification system with a view to use as fuel for energy purposes) (in Polish). *Polityka Energetyczna—Energy Policy Journal*, 18(4), pp. 89–100.

Part II

Economic Issues of Biomass-Based Small-Scale Energy Systems

2 Economic Issues of Biomass-Based Small-Scale Energy Systems

Artur Wyrwa, Wojciech Suwała and Marcin Pluta

CONTENTS

2.1 Economic Analysis of Small-Scale Heating Installations 21
    2.1.1 Cost of a Small Biomass Boiler 22
    2.1.2 Capital Expenditures (C1) 22
    2.1.3 Operation Costs (C2) 23
    2.1.4 Maintenance Costs (C3) 23
    2.1.5 Other Costs (C4) 23
    2.1.6 Salvage Value (C5) 23
    2.1.7 Evaluation of the Investment Project's Profitability 24
        2.1.7.1 Physical and Economic Lifetimes 24
        2.1.7.2 Simple Payback Time 25
        2.1.7.3 Dynamic Metrics for Profitability Calculations 25
        2.1.7.4 Discount Rate 25
        2.1.7.5 Real and Nominal Discount Rates 27
        2.1.7.6 Net Present Value 27
        2.1.7.7 Net Present Value Quotient 28
        2.1.7.8 Internal Rate of Return 28
        2.1.7.9 Levelized Cost of Heat 28
        2.1.7.10 Life Cycle Cost 29
2.2 Experience Curves 30
2.3 Economic Evaluation of Space Heating Technologies: Case Study for Poland 31
2.4 Summary 36
References 36

2.1 ECONOMIC ANALYSIS OF SMALL-SCALE HEATING INSTALLATIONS

The main aim of the economic evaluation of the investment project is to indicate if it brings a positive economic return from an investment, that is, if it is profitable. In the case of investments in space heating technology, the problem can be formulated as choosing the technology giving the lowest overall space heating costs. In such a case, a relative profitability analysis can be performed to give a clear guidance to select the best option over existing alternatives. The profitability metrics described in this chapter are widely used in real biomass-based investment projects. For instance, in ref. [1] the net present value (NPV) and life cycle cost (LCC) indicators were used in a feasibility study for Grafton County to examine economic issues involved in replacing the current fossil fuel heating systems with a central wood-fired heating. Among others, NPV and internal rate of return (IRR) were used by Wang et al. [2] in their economic analysis of residential wood pellet boilers. In Vávrová et al.[3], a geographic information system (GIS)-based model employing NPV was developed for the evaluation of locally available biomass competitiveness for decentralized space heating in villages and small towns. In all investment projects, costs remained one of the most important issues. We now examine a sample cost breakdown of a small biomass-based energy system.

2.1.1 Cost of a Small Biomass Boiler

The total cost related to the lifetime of space heating equipment can be broken down into a number of components. It was proposed by Wang et al. [2], to distinguish capital expenditures, operating cost, maintenance cost, and cost of decommissioning a depleted installation. Sometimes the depleted installation or its parts have a value and can be sold, thus bringing a revenue at the end of the system's lifespan.

2.1.2 Capital Expenditures (C1)

Capital expenditures include the equipment cost, installation cost, and fuel storage cost. For small biomass boilers, such as the pellet boiler, equipment costs typically include the cost of boiler and fuel feeding system enabling automatic stocking, cost of control equipment such as the room regulator, temperature sensors, as well as the cost of heat storage and the expansion tank. The installation cost includes the labor cost, and the cost of materials such as pipes, fittings, water pumps, valves, and deaerators. Finally, a storage place also may be needed for the fuel. In the case of the existing buildings that have used solid fuels for space heating, the existing storage area could be used as biomass storage. In order to recalculate the capital cost for different capacities of the system, having given the base one, the following equation, based on the scaling factors, can be used [4]:

$$C_a = C_b \cdot \left(\frac{Bs_a}{Bs_b}\right)^\alpha \quad (2.1)$$

where:
C_a—capital cost of the boiler "a"
C_b—capital cost of the boiler "b"
Bs_a—thermal capacity of the boiler "a"
Bs_b—thermal capacity of the boiler "b"
α—scaling factor—exponent value, ranging from 0.4 to 0.8 for processing equipment

Economic Issues of Biomass-Based Small-Scale Energy Systems

2.1.3 Operation Costs (C2)

Operation costs are directly linked with the operation of the boiler. Their main component is the cost of fuel, which is a factor of fuel consumption and fuel price. Also, the labor cost of fuel unloading and the fuel storage room should be considered. The cost of electricity is related to the operation of all auxiliary equipment such as water pumps, the fuel feeder, primary air fan, and boiler igniter.

2.1.4 Maintenance Costs (C3)

The maintenance costs include ash disposal, boiler cleaning to maintain high heat conversion efficiency, replacement of wearing parts such as flame deflectors, and cleaning of the boiler room.

2.1.5 Other Costs (C4)

Other costs may be related to the unforeseen boiler failures, for example, as a result of power outages, failures of mechanical elements of the fuel feeding system caused by the use of inappropriate fuel, and so on.

2.1.6 Salvage Value (C5)

At the end of the economic lifetime of the boiler, elements of the heating system may still have some value and the owner can get revenue from their sale. For instance, the boiler may be offered for collection of scrap metal; however, labor and other costs may be incurred during system liquidation. Nevertheless, the allocation of the infrastructure and facilities—the "brown field" site—has a positive value for a further installation.

Table 2.1 shows the cost breakdown for a 25 kW biomass-based space heating system in selected countries.

The first dataset presented in Table 2.1 is given for a house located in New York State [2]. The data for Poland were provided by one of the authors of this chapter, who recently installed a 25 kW pellet boiler. In addition, a review of techno-economic parameters of pellet boilers offered on the Polish market has been done. The data for Denmark were based on the newest technical data provided by the Danish Energy Agency [5] for a pellet boiler with automatic stoking in an existing one-family house. They include the total costs of establishing the technology for the consumer, including equipment, as well as engineering and civil works. They were provided for a boiler with a 12 kW thermal capacity and equal to €7000. The costs for a 25 kW boiler were calculated using Equation 2.1 and split into equipment and labor, at a ratio of 80% and 20%, respectively. The exponent value α was calibrated based on the data in ref. [6] and equaled to 0.55.

$$C_{25kW} = 7000 \cdot \left(\frac{25}{12}\right)^{0.55} \text{ [EUR]} \tag{2.2}$$

TABLE 2.1
Example of a Pellet Boiler Cost Breakdown for Selected Countries [€]

Cost Types	Items	Cost USA[a]	Cost Poland[a]	Cost Denmark
Capital cost (C1)	25 kW Boiler	12048–13358	2092–4183	8385
	Pellet feeder			
	Temperature sensors			
	Thermal energy storage tank		465–1162	
	Expansion tank		50–116	
	System piping and fittings	2183–3056	465–930	
	System components	1746–2619	465–697	
	Pellet storage	873–1746		
	Total equipment	16850–20779	3537–7088	8385
	Labor	3929–4365	465–930	2096
Operation costs (C2)	Fuel costs	210 (per ton)	163–279 (per ton)	290 (per ton)
	Electricity	35 (per month)	14–20 (per month)	16 (per month)
Maintenance costs (C3)	Cleaning, ash disposal, etc.	1% of C1	1% of C1	500
Other costs (C4)		1% of C1	1% of C1	
Salvage value (C5)		602–668	74–112	
Total cost		C1+C2+C3+C4+C5		

[a] The conversion from US$ and Polish Zloty (PLN) to € has been done using InforEuro currency converter.

One should bear in mind that the data provided in Table 2.1 should be considered only as example of how cost breakdown can look like. In fact as indicated in [7] investment costs for heating systems can vary not only between different countries, but also within countries. This is because of the different characteristics of individual buildings, price differences between producers/retailers/installation companies and economies of scale.

2.1.7 Evaluation of the Investment Project's Profitability

2.1.7.1 Physical and Economic Lifetimes

Economic calculations are performed using the information on the economic lifetime of the equipment. The economic lifetime, as opposed to the physical lifetime, is the period over which the equipment remains fully useful to the owner. Taking a laptop as an example, its physical lifetime can reach 20 years; however, owing to the development of new software with higher hardware requirements, its economic lifetime will be limited to just a few years before becoming obsolete. In the case of biomass boilers, the economic lifetime may depend on environmental regulations that change over time. Although, in any given moment that a boiler is technically efficient, it may need to be replaced owing to noncompliance with new emissions standards.

2.1.7.2 Simple Payback Time

The payback period method is used to calculate the time (usually the number of years) it takes to recover the capital invested in the project. This static method does not take into account the change in the value of money over time. Simple payback time (SPBT) is calculated according to the following formula:

$$SPBT = \frac{I_0}{\overline{CF_a}} \qquad (2.3)$$

where:
I_0—initial capital expenditures
$\overline{CF_a}$—the annual average net cash flow of the project

2.1.7.3 Dynamic Metrics for Profitability Calculations

Dynamic metrics take into account the change of the value of money over time. The present value of money today is viewed as higher than the expected value of money in the future. This phenomenon of change in the value of money over time results from the psychological inclination of people toward current consumption, which means that even assuming zero inflation, people consider receiving the same amount of money today more valuable than getting it later. Additionally, there is the impact of risk, as money received today is certain and therefore has a greater value than the money we will receive in the future. The concept of valuing the value of goods and money over time is introduced by the so-called discounting process employing a discount rate. Therefore, this section starts with a description of the discount rate.

2.1.7.4 Discount Rate

Discount rate r is the real (not current, i.e., without inflation) rate of return on investment in the economy. It is also a factor for calculating the cost of capital employed. Typically, it is assumed to hold a value between 5% and 10%, depending on the risk of the investment. The discount rate is sometimes also treated as an alternative to cost of capital. This reasoning assumes that revenues should be discounted at a rate of return that can be achieved with comparable investments. A higher rate is assumed for higher risk investments [8]. As described above, the construction of a biomass boiler requires initial capital expenditure, it is hoped, in exchange for a stream of costs savings in relation to other heating technologies in the future. Therefore, in order to calculate the present value of future savings, the discounting operation is used. The lower the discount rate the higher the return value of the project's future costs and benefits. However, the higher the discount, the lower is the future return value. The present value (PV) determines the actual value of the amount received in the future. Its calculation is based on the following formula:

$$PV = FV_t \cdot (1+r_t)^{-t} \qquad (2.4)$$

where:
FV—future value of money
PV—current value of money
r—discount rate
t—a year in which the money is obtained in the future

For example, if the present value of €1 received after 10 years is equal to €0.61 at a discount rate of 5%, then with a discount rate of 10% it is only c. €0.39 (Figure 2.1). The present value is often referred to as the discounted value. The process to obtain a *PV* value is called discounting and is the reverse process to capitalization.

It should be noted that the discount rate *r* is usually taken individually by each investor, who can set its value at different levels according to their own preferences. However, too high a discount rate may lead to all investments being considered unprofitable, and too low to accept investments that multiply the capital employed at a much slower rate than the existing market opportunities in a given market. If the investment project is financed from many sources (e.g., equity and debt capital) the calculations shall assume a discount rate equal to the weighted average cost of capital (WACC). WACC is the product of the shares of particular types of capital in the financing of the project and the corresponding cost for each of its types.

$$WACC = \frac{D}{D+E} \times C_d + \frac{E}{D+E} \times C_e \tag{2.5}$$

where:
WACC—weighted average cost of capital
D—share of credit in the structure of investment financing
E—share of own funds in the structure of investment financing
C_d—cost of credit
C_e—cost of equity capital

The example below shows how the *WACC* is calculated with the assumption that the interest rate on borrowed capital is 6.5% per year and the cost of equity (own capital) is one and a half times higher, whereas its share in the financing structure of the project is 30%.

$$WACC = \frac{0.7}{0.7+0.3} \times 6.5 + \frac{0.3}{0.7+0.3} \times (1.5 \times 6.5) = 7.5\% \tag{2.6}$$

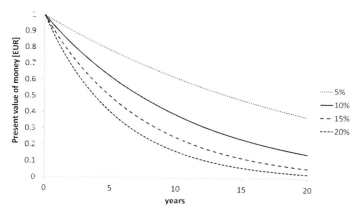

FIGURE 2.1 Present value of €1 obtained in different years in the future.

Economic Issues of Biomass-Based Small-Scale Energy Systems

2.1.7.5 Real and Nominal Discount Rates

Let us assume that purchasing a panel of goods and services today (year "0") costs €100. If purchasing the same panel of goods and services after 1 year equals €105, then the purchase power of the money has decreased. It means that the inflation rate is 5%. Thus, the costs of goods and services in year "1" can be expressed as €105 current or €100 constant for the year "0." Typically, economic calculations are done in constant € of year "0" and the nominal discount rate is corrected to a real discount rate. Therefore, the results are expressed in constant € for the year "0." The year "0" assumed for discounting is typically a year when the investment is completed and production starts.

The most well-known dynamic metric to evaluate project profitability is net present value (NPV). Apart from NPV, in the text below we also describe other relevant metrics such as net present value quotient (NPVQ), internal rate of return (IRR), levelized cost of heat (LCOH), and life cycle cost (LCC).

2.1.7.6 Net Present Value

NPV is one of the most popular metrics to evaluate project profitability. NPV presents the algebraic difference between the present value of a series of discounted benefits (positive cash flows) and the present value of a series of discounted costs (negative cash flows) as they occur over the economic lifetime of the project. For instance, if we consider an investment generating a revenue with a short construction period, the following formula can be used:

$$NPV = \sum_{t=1}^{T_f}(CF_t)*(1+r_t)^{-t} - I_0 \qquad (2.7)$$

where:
t—time index, years
CF_t—net cash flows occurring in year t [€/years]
T_f—year of the end of the project
I_0—capital expenditure incurred in the year "0"

In the case of energy efficiency investments, the annual net saving resulting from this investment can be expressed as:

$$B_t = S_t \cdot E_t - OM_t \qquad (2.8)$$

where:
B_t—annual net saving [€/year]
S_t—energy saved [kWh/year]
E_t—energy cost [€/kWh]
OM_t—additional operation and maintenance cost [€/year]

For such investments, we can calculate NPV using the following formula:

$$NPV = \sum_{t=1}^{T_f}(B_t)*(1+r_t)^{-t} - I_0 \qquad (2.9)$$

With the assumption that the investment brings equal annual savings during its economic lifetime (i.e., $B = B_1 = B_2 = B_t$) and liquidation costs as well as the salvage value equal zero, Equation 2.9 can be simplified to the following form (for more explanation see Equation 2.25):

$$NPV = B \cdot \frac{1-(1+r)^{-T_f}}{r} \quad (2.10)$$

where:
B—annual net saving [€/year]

NPV is the principal criterion for making decisions about the implementation of the project. In the case where NPV > 0, the project should be implemented. If NPV < 0, the project should be rejected because it is unprofitable. When NPV = 0, the project is neither profitable nor unprofitable and simply enables the investor to cover only the costs of capital (expressed by the discount rate) without any additional income.

2.1.7.7 Net Present Value Quotient

NPVQ is the ratio of NPV to the initial capital expenditures. It is often used to rank the alternative investment options with limited investment resources to maximize their productivity.

$$NPVQ = \frac{NPV}{I_0} \quad (2.11)$$

2.1.7.8 Internal Rate of Return

Another metric for assessing the profitability of potential investments is the internal rate of return (IRR). IRR is the discount rate that makes the NPV of all cash flows equal to 0.

$$\sum_{t=1}^{T_f} (CF_t) * (1+r_t)^{-t} - I_0 = 0 \quad (2.12)$$

2.1.7.9 Levelized Cost of Heat

The necessity of comparing the costs of various technologies forced the creation of universal measures, which, among others, include the levelized cost of heat (LCOH) and life cycle costs (LCC). These two metrics are helpful in economic evaluation of investment projects that do not generate revenue and which rather aim at minimization of the costs. The concept of LCOH is similar to a very popular metric used for comparison of power generation technologies, that is, levelized cost of electricity (LCOE). Both LCOE and LCOH take into account all costs associated with a given technology from the commencement of construction to its decommissioning after operation. LCOE is more meaningful, as typically power technologies sell electricity and LCOE corresponds to the price of electricity, which allows for obtaining revenues that guarantee investment efficiency at the level of 0 extraordinary profits, but with profits covering the cost of capital. To express this in another way, such a price set over the entire period of operation from installation guarantees an income making the NPV of the project equal to 0. In the case of small-scale space heating technologies, the heat is not sold to anybody but it is used to maintain the thermal comfort of the residents of the house. The use of LCOH makes it possible to compare easily the heating costs expressed per unit of heat generated, which facilitates economic comparison of technologies using different fuels with different proportions of fixed and variable costs, and finally different installation lifetimes. In general, the formula for LCOH presented below is in line with cost breakdown presented in Section 2.1.

$$LCOH = \frac{\sum_{t=T_s}^{T_0}(I_t)*(1+r_t)^t + \sum_{t=1}^{T_f}(OM_t + F_t + L_t)*(1+r_t)^{-t} - SV_{T_f}*(1+r_{T_f})^{-T_f}}{\sum_{t=0}^{T_f} H_t*(1+r_t)^{-t}}$$

(2.13)

where:
T_s—year of the start of the investment process
T_0—year of completion of construction and start of production
F_t—total fuel costs incurred in a year t
L_t—total liquidation costs incurred in a year t
SV—the salvage value of the installation after its operation
H_t—useful heat production in year t

The choice of the "0" year deserves some discussion. In the case of some heating technologies, the construction of the heating system may take more than 1 year (e.g., shallow geothermal). We opt for the solution that treats the period of construction and operation separately. The period of construction lasts from $t = T_s$, that is, the starting time of the construction up to $t = T_0$, that is, finalization of the construction. Note that this part of the formula would make the total costs at the time the technology enters into service higher than expenditures incurred during construction (as the cost of capital given by r will be added). The operation period includes all years until decommissioning and operation costs are discounted at the end of each modeling year.

2.1.7.10 Life Cycle Cost

The life cycle cost method, expresses the total discounted (present worth) cash flow for an investment and future costs during operation and dismantling (i.e., the salvage value that may be positive or negative—if liquidation costs are higher than the value of technology at the end of its life).

LCC is calculated according to the following equation:

$$LCC = I_t + \sum_{t=1}^{t=T_f} \frac{C_t}{(1+r)^t} - \frac{SV}{(1+r)^{T_f}}$$

(2.14)

As one can see, the *LCC* metric is very similar to the nominator of LCOH.

2.2 EXPERIENCE CURVES

Future prices of biomass-based small-scale energy systems will also change as the results of the technology learning and discussing this phenomenon deserves some explanation. There is empirical evidence that the unit costs of technologies exponentially decline with the cumulative level of their production as manufacturers accumulate experience [9]. These learning mechanisms include, among others, R&D, learning-by-doing, effects of upscaling, and experiences in an installation's operation [10]. It is clear that for the first installations of a given type, some parts are unique and therefore have high costs owing to the fixed R&D and production costs in general. Regulatory costs, such as safety or environmental impact assessments, are also higher owing to the need for exceptional research. Overcoming the problems related to the lack of experience with such installations allows for improvements and allocation of fixed costs to a larger number of installations. If X_0 is a cumulative number of units produced for the price C_0, then the progress ratio parameter pr can be defined as a ratio of final to initial costs associated with a doubling of cumulative output:

$$pr = \frac{C_f}{C_0} = \left(\frac{2X_0}{X_0}\right)^b = 2^b \tag{2.15}$$

where:
C_f—unit investment cost of the technology after doubling its production
b—elasticity of the progress ratio representing the speed of learning

The learning rate lr, which represents the proportional cost savings made for a doubling of cumulative output, can be linked with pr through the following equation:

$$lr = 1 - pr \tag{2.16}$$

At a learning rate lr of 10% (pr of 90%), the cost of a technology decreases by 10% for every doubling of its cumulative production. Unit investment cost for a given technology in time t after a cumulative number of units has been produced can be calculated using the formula:

$$C_t(x_t) = C_0 \left(\frac{X_t}{X_0}\right)^b \tag{2.17}$$

where
C_t—unit investment cost of the technology subjected to process improvement in time t
C_0—the initial unit investment cost of the technology
X_t—cumulative number of units produced until time t
b—elasticity of the progress ratio representing the speed of learning

Economic Issues of Biomass-Based Small-Scale Energy Systems

Expressing Equation 2.3 in a logarithmic form will result in a linear equation:

$$\log C_t(x_t) = \log C_0 + b \cdot \left(\frac{X_t}{X_0}\right) \qquad (2.18)$$

One Equation 2.18 has been plotted on a double-logarithmic scale, b represents the slope of the line indicating the rate at which the unit costs of technology decrease with increase of its production. The learning curves are constructed by fitting the learning curve with the use of the historical data, which contain information on the technology cumulative production level and respective costs. However, often, owing to lack of costs data, they are replaced with price data, which is more readily available but can also vary owing to market influences. Once the learning curve has been fitted, then it can be extraplolated to predict future costs of development. The learning rates for biomass energy technologies vary between 7% and 10% [11]. Currently, about 28 million tons of pellets are produced globally, mainly for residential heating in pellet stoves and pellet boilers [12]. Suppose that a biomass boiler of 20 kW heat capacity consumes 5 tons of pellets annually, which gives 120 GW of installed capacity worldwide. Let us assume that the unit investment cost of a small biomass heating installation is €158/kW$_{th}$. How will the unit cost change if the capacity is tripled, assuming a learning rate of 10% ($pr = 90\%$)? To answer this question, let us substitute these values into Equations 2.1 and 2.3:

$$b = \frac{\log(0.9)}{\log(2)} \cong -0.152 \qquad (2.19)$$

$$C_t(360) = 158 \cdot \left(\frac{360}{120}\right)^{-0.152} = 133.7 \left[\frac{EUR}{KW_{th}}\right] \qquad (2.20)$$

As one can see, tripling the current biomass capacity will bring the technology costs down by c. 15%.

2.3 ECONOMIC EVALUATION OF SPACE HEATING TECHNOLOGIES: CASE STUDY FOR POLAND

In this case study, we demonstrate how to use selected dynamic metrics, that is, LCOH and LCC, to evaluate the profitability of space heating technologies for one family house located in Poland. The annual useful heat demand of the house is equal to 20 MWh. To meet the peak heat demand of the house, 15 kW thermal capacity technology was chosen. Five heating technologies are considered by the house owner (Table 2.2). The owner evaluates the investments based on a discount rate of 10% per year.

The heating technologies include a condensing natural gas boiler, hard coal boiler with automatic stocking meeting the emission norms set by the ecodesign

TABLE 2.2
Technical and Economic Parameters of Heating Technologies in the Case Study

Fuel	Efficiency [%] (A)	Efficiencies of Transmission, Regulation and Accumulation Process [%] (B)	Investment Costs (€kW^{-1}) (C)	Operation and Maintenance (% of Investment Costs) (D)	Fuel Price (€MWh^{-1}) (E)	Lifetime (years) (F)
Natural gas (condensing)	105	84	127	2	59	20
Hard coal (ecodesign, automatic)	91	84	127	2	33	15
Electricity (heat pump, horizontal)	350	84	635	2	131	25
Fuel oil (condensing)	105	84	158	2	85	20
Biomass (ecodesign, automatic)	91	84	158	2	42	15

Directive [13], ground source horizontal heat pump, fuel oil condensing boiler, and a pellet boiler. Each technology has been characterized by the set of techno-economic parameters based on ref. [14] (Table 2.2). The differences in efficiencies between technologies mean that the final energy consumption in each case is different leading to different capacity factors, as presented in Table 2.3.

As in our case, we assume that establishing of the technology for the owner does not take much time and can be done in year "0"; also, liquidation costs and salvage value are not considered, so that Equation 2.13 can be expressed in the following form:

$$LCOH = \frac{I_0 + \sum_{t=1}^{T_f}(OM_t + F_t)*(1+r_t)^{-t}}{\sum_{t=0}^{T_f} H_t *(1+r_t)^{-t}} \quad (2.21)$$

One can note that with these assumptions the nominator in Equation 2.21 is equal to LCC (see Equation 2.14). We also make an assumption that fuel prices remain constant during the economic lifetime of the technology. Then, the annual fuel cost for each technology F_{tech} can be calculated using the formula:

$$F_{tech} = \left(H_{tech}^F\right)\cdot(E_{tech})\cdot 1000^{-1}\left[\frac{EUR}{yr}\right] \quad (2.22)$$

Economic Issues of Biomass-Based Small-Scale Energy Systems

TABLE 2.3
Final Energy Demand and Capacity Factor for Heating Technologies

	Overall Efficiency [%]	Final Energy Demand [kWh]	Boiler Size [kW]	Hours of Operation with Nominal Capacity [hr/yr]	Capacity Factor
	(G)	(H^F)	(I)	(J)	(K)
Type	[A × B]	[20,000/(A × B)]		[H × I^{-1}]	[J × 8760^{-1}]
Natural gas	88.20	22675.74	15	1511.72	0.173
Hard coal	76.44	26164.31	15	1744.29	0.199
Electricity	294.00	6802.72	15	453.51	0.052
Fuel oil	88.20	22675.74	15	1511.72	0.173
Biomass	76.44	26164.31	15	1744.29	0.199

Taking an example of a natural gas-based technology, the annual fuel cost is equal to:

$$F_{gas} = 22,675.74 \left[\frac{kWh}{yr}\right] \cdot 59 \left[\frac{EUR}{MWh}\right] \cdot 1000^{-1} = 1337.87 \left[\frac{EUR}{yr}\right] \quad (2.23)$$

Similarly, we can calculate the annual costs of fuel for other technologies:

$F_{hard\ coal} = €863.42$
$F_{electrcicty} = €891.15$
$F_{fuel\ oil} = €1927.44$
$F_{biomass} = €1098.90$

Noting that if OM_t and F_t hold fixed values over years, we can rewrite part of Equation 2.21 as follows:

$$\sum_{t=1}^{T_f}(OM_t + F_t)\cdot(1+r_t)^{-t} = (OM_t + F_t)\cdot \sum_{t=1}^{T_f}(1+r_t)^{-t} \quad (2.24)$$

With a fixed r_t, the sum involving discounting factors can be replaced by the sum of a geometric series:

$$\frac{1}{(1+r)^1} + \frac{1}{(1+r)^2} + \cdots \frac{1}{(1+r)^{T_f}} = \frac{1-(1+r)^{-T_f}}{r} \quad (2.25)$$

The reversal of the obtained term is called the capital recovery factor (CRF):

$$CRF = \frac{r}{1-(1+r)^{-T_f}} \quad (2.26)$$

The same procedure can be applied to the denominator of Equation 2.21. As a result, Equation 2.21 takes the following form:

$$LCOH = \frac{I_0 + (OM_t + F_t) \cdot CRF^{-1}}{H_t \cdot CRF^{-1}} \quad (2.27)$$

Let us calculate the CRF for a natural gas boiler employing an economic lifetime of 20 years and discount rate of 10%:

$$CRF = \frac{0.1}{1-(1+0.1)^{-20}} = 0.11746 \quad (2.28)$$

Finally, putting all calculated values into Equation 2.27, we obtain:

$$LCOH = \frac{127\left[\frac{EUR}{kW}\right] \cdot 15[kW] + \left(127\left[\frac{EUR}{kW}\right] \cdot 15[kW] \cdot 0.02 + 1337.87\left[\frac{EUR}{yr}\right]\right) \cdot CRF^{-1}}{20{,}000[kWh] \cdot CRF^{-1}} \quad (2.29)$$

$$LCOH = 0.080\left[\frac{EUR}{kWh_{th}}\right] \quad (2.30)$$

As mentioned before, the LCC is equal to the nominator in Equation 2.29:

$$LCC = 127\left[\frac{EUR}{kW}\right] \cdot 15[kW]$$
$$+ \left(127\left[\frac{EUR}{kW}\right] \cdot 15[kW] \cdot 0.02 + 1337.87\left[\frac{EUR}{yr}\right]\right) \cdot CRF^{-1} \quad (2.31)$$

$$LCC = 13{,}619.4[EUR] \quad (2.32)$$

The results of the LCOH and LCC metrics for other space heating technologies are presented in Table 2.4.

TABLE 2.4
Results of LCOH and LCC Metrics for Heating Technologies from the Case Study

Metric	Natural Gas (Condensing)	Hard Coal (Ecodesign, Automatic)	Electricity (Heat Pump, Horizontal)	Fuel Oil (Condensing)	Biomass (Ecodesign, Automatic)
LCOH [€/kWh$_{th}$]	0.0800	0.0576	0.1065	0.1126	0.0694
LCC [€]	13619.4	8762.0	19343.2	19182.9	10553.1

Economic Issues of Biomass-Based Small-Scale Energy Systems

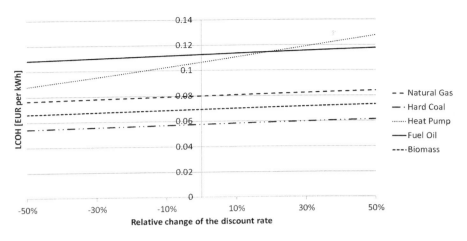

FIGURE 2.2 Change in LCOH as a result of change in the discount rate.

As one can see a hard coal-fired boiler is the cheapest option from all the technologies considered in the analysis. It is worth checking how the change in selected parameters can change the value of LCOH. Figure 2.2 shows the impact of the owner's discount rate on the value of LCOH.

As expected, the discount rate has the biggest impact on the most capital intensive technology, that is, the heat pump. When the discount rate decreases by 50% (i.e., is equal to 5%), the LCOH of the heat pump drops to c. €0.088 per kWh$_{th}$ (by 17.5%). The impact of discount rate on the LCOH of other technologies is not significant. Figure 2.3 shows how a change in fuel prices impacts the results. One can see that LCOH for all technologies considered is affected by the change in fuel price.

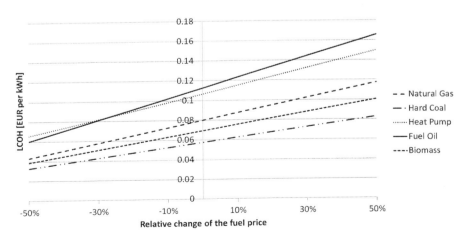

FIGURE 2.3 Change in LCOH as the result of change in fuel prices.

2.4 SUMMARY

This chapter provides guidance on methodologies for evaluation of an investment project's profitability. The most commonly used metrics, such as SPBT, NPV, and IRR were presented and discussed. In the context of biomass-based small-scale energy systems, two additional metrics have been demonstrated that are helpful in the comparison of space heating costs. These are LCOH and LCC. They can both be used to support an investment decision. The use of these metrics is demonstrated in a case study carried out for an existing one family house in Poland. The owner of the house considered installation of a new space heating system and the alternatives included natural gas-, hard coal-, fuel oil-, and biomass-fired boilers, as well as a ground source heat pump. The authors guide the readers through the calculation process and provide relevant examples. With the underlying assumptions on techno-economic parameters, the technology with the lowest LCOH is a hard coal-fired boiler followed by the biomass boiler. The reader should treat the results as demonstrative. As noted before, costs for heating systems can vary not only between different countries, but also within one country. This is because of the different characteristics of individual buildings, price differences between producers/retailers/installation companies, and economies of scale. This notwithstanding, the case study demonstrated the usefulness of both metrics in economic comparison of space heating technologies. To make the results more meaningful, the sensitivity analysis was performed showing the change in the value of results owing to the change in given input parameters, notably the discount rate and fuel prices. Sensitivity analysis revealed that the change in fuel prices had more significant impact on the results than a change in discount rate. This chapter can be regarded as a concise compendium of knowledge in the field of profitability analysis of biomass-based small-scale energy systems.

REFERENCES

1. Biomass Heating Feasibility Study for the Grafton County Complex, Biomass Energy Resource Center, 2009, https://www.biomasscenter.org/images/stories/Biomass_Heating_Grafton_County_Complex_Feasibility_Study.pdf [Accessed:15.01.2019].
2. Wang, K., Zhang, Y., Sekelj, G., Hopke, P., Economic analysis of a field monitored residential wood pellet boiler heating system in New York State, *Renewable Energy* 133, 2019, 500–511.
3. Vávrová, K., Knápek, J., Weger, J., Králík, T., Beranovský, J., Model for evaluation of locally available biomass competitiveness for decentralized space heating in villages and small towns, *Renewable Energy* 129, 2018, 853–865.
4. Swanson, R.M., Platon, A., Satrio, J.A., Brown, R.C., Techno-economic analysis of biomass-to-liquids production based on gasification, *Fuel* 89, 2010, 11–19.
5. The Danish Energy Agency, Technology Data for Individual Heating Installations, August 2016.
6. Connolly, D. et al. Heat Roadmap Europe 2: Second Pre-Study for the EU27, *Department of Development and Planning, Aalborg University*, May 27, 2013.
7. Knobloch, F., Mercure, J-F., Pollitt, H., Chewpreecha, U., Lewney, R., A technical analysis of FTT: Heat—A simulation model for technological change in the residential heating sector, *European Commission, Directorate-General for Energy*, December 2017.
8. Khatib, H., Review of OECD study into "Projected costs of generating electricity—2010 Edition," *Energy Policy* 38, 2010, 5403–5408.

9. Grübler, A., Nakićenović, N., Victor, D.G., Dynamics of energy technologies and global change, *Energy Policy* 27, 1999, 247–280.
10. OECD/IEA, Experience Curves for Energy Technology Policy, IEA Publications, 2000. http://www.wenergy.se/pdf/curve2000.pdf [Accessed: January 16, 2019].
11. Henkel, J. Modelling the diffusion of innovative heating systems in Germany—decision criteria, influence of policy instruments and vintage path dependencies. PhD diss., TU Berlin, 2012.
12. World Energy Council, World Energy Resources – Bioenergy, World Energy Council, 2016. https://www.worldenergy.org/wp-content/uploads/2016/10/World-Energy-Resources-Full-report-2016.10.03.pdf [Accessed: January 16, 2019].
13. Commission Regulation (EU) 2015/1189 of April 28, 2015 implementing Directive 2009/125/EC of the European Parliament and of the Council with regard to ecodesign requirements for solid fuel boilers.
14. Wyrwa, A., City-level energy planning aimed at emission reduction in residential sector with the use of decision support model and geodata, *IOP Conference Series: Earth Environmental Science* 214, 2019, 1–12.

Part III

Environmental Issues of Biomass-Based Small-Scale Energy Systems

3 Environmental Impacts of Biofuel-Fired Small Boilers and Gasifiers

Jozef Viglasky, Juraj Klukan and Nadezda Langova

CONTENTS

3.1	Introduction	42
3.2	Bioenergy Systems	45
	3.2.1 Biomass Heating: A System of Systems	45
	3.2.1.1 Handover Phase	45
	3.2.1.2 Operation Phase	46
	3.2.2 Biomass Combustion Systems	46
	3.2.2.1 Choosing the Right Fuel	48
	3.2.2.2 Fuel Feeding and Handling Systems	50
3.3	Biomass Combustion and Emissions	51
	3.3.1 Wood Fuel Composition	51
	3.3.2 Fuel Characteristics	52
	3.3.2.1 Woodchip Fuel Characteristics	52
	3.3.3 Fuel Standards and Testing	54
	3.3.3.1 Fuel Standards	54
	3.3.3.2 CEN/TC 335	54
	3.3.3.3 BS EN 14961-1 (2010)	54
	3.3.3.4 ÖNORM	55
	3.3.4 Biomass Combustion	55
	3.3.5 Combustion Air	57
	3.3.6 Ash Characteristics and Slag Formation	57
	3.3.6.1 Ash Characteristics	57
	3.3.6.2 Slag Formation	58
	3.3.7 Pollution Emissions	59
	3.3.7.1 NO_x	59
	3.3.7.2 Particulate Matter	61
	3.3.7.3 Regulated Levels of PM and NO_x	62
	3.3.7.4 Carbon Monoxide (CO)	65
	3.3.7.5 Benzene and PAH	65
	3.3.8 The Effects of Burning Out-of-Specification Fuels	65
	3.3.8.1 Too Wet	65
	3.3.8.2 Too Dry	65
	3.3.8.3 Safety Considerations	66

3.4　Environmental Aspects of Heat Production Based on Biofuels 66
　　3.4.1　Solid Pollution Matter 66
　　　　3.4.1.1　Why Burning Wood Threatens the Health
　　　　　　　　of Europeans 67
　　3.4.2　Other Pollutant Emissions and Working Conditions 68
　　3.4.3　Conclusions 69
3.5　Case Study: A Potential Feedstock Resource 69
　　3.5.1　Amaranth (*Amarantus* L.) as Raw Material Resource for
　　　　　Biofuels Production 69
　　　　3.5.1.1　Introduction 69
　　　　3.5.1.2　Material and Methods 70
　　　　3.5.1.3　Results 70
　　　　3.5.1.4　Discussions 72
　　　　3.5.1.5　Conclusions 74
3.6　Environmental Aspects of Biomass Gasification 75
　　3.6.1　Introduction 75
　　　　3.6.1.1　Biomass Power System's Environmental Characterization ... 76
　　3.6.2　Methods of Systematic Environmental Assessment, Applied to
　　　　　Biomass Gasification Technologies 77
　　　　3.6.2.1　Environmental Aspects of Gasification Plant Operations ... 78
　　　　3.6.2.2　Hazards of Gasifier Plant Operation 81
　　3.6.3　Examples of Environmentally Advanced Biomass Gasification-
　　　　　to-Electricity Systems 83
　　　　3.6.3.1　Technology Overview 83
　　　　3.6.3.2　Environmental Discharges and Controls 85
3.7　Discussion 88
　　3.7.1　Potential Applications for Gasification 89
　　　　3.7.1.1　Small-Scale Downdraft Gasification Systems 89
　　　　3.7.1.2　Woody Generator Gas as an Alternative: Renewable
　　　　　　　　Energy Carrier 89
　　　　3.7.1.3　Direct-Fired Heating System of Wood Drying Kilns 90
　　3.7.2　Future Trends 94
3.8　Conclusions 95
References 95

3.1　INTRODUCTION

Energy is vital for people and industry. Although the consistent global increase in energy consumption during the coming decades will be supplied mainly by fossil fuels, an increasing part will originate from biomass. The increased use of solid biomass for energy production can only be achieved by broadening the scope of biomass used in the energy sector. Presently, forest and plantation wood, short rotation coppice such as willow and poplar, together with a narrow selection of agricultural products such as wheat and corn straw stalks constitute the basis for the energy sector. In the future, this scope must necessarily expand to include industrial biowaste, mixtures of different agricultural and wood waste, and even aquatic biomass such as microalgae and seaweed. The utilization of new types of biomass necessitates the development of new or improved pretreatment methods so that they can be used with existing fuel conversion technology. Mixtures of feedstocks in pellets and briquettes are another possibility for increasing the use of challenging biomass that have an undesirable chemical composition.

The characterization of physical and mechanical parameters demands common standards. The development of standards will help to expand the market for solid biofuels, which is necessary to meet the increasing demand. There are vast amounts of biomass or biomass feedstocks for solid fuel production available within several regions worldwide (Viglasky, 2002, 2009a; Viglasky and Langova, 2005).

With the increase in biomass used for energy production and the competition for biomass for a variety of products and uses, all available resources, including mixtures, will become important for energy conversion. Although the global potential biomass feedstock is far from being fully exploited and even heavily exploited areas such as, for example, Scandinavia, where biomass has been utilized on an industrial scale since the early 1970s, the potential of local biomass is great. Biomass can be categorized in several ways, for example, woody biomass, herbaceous biomass, industrial biomass waste, technology residues, aquatic biomass, and manure.

The fundamental physical and chemical characteristics that influence the utilization of the feedstock are ash content and composition (which depend on the type of biomass), moisture content, particle size and distribution, bulk density, and mechanical durability (for pellets and briquettes, which depends on the pretreatment). The characteristics of biomass feedstock vary widely and depend both on the kind and species of biomass, and on the applied pretreatment of the raw material (Table 3.1).

The physical and chemical parameters may vary considerably within a feedstock lot (which is a defined quantity of fuel for which the quality is to be determined). For example, the moisture content (MC) can vary within a single truckload of woodchips from approximately 20% to 40% (Møller and Esbensen, 2005). In contrast, the physical and chemical parameters for upgraded fuels such as pellets and briquettes have comparatively less variation, that is, the moisture and ash content in a shipload varies less compared to raw biomass. The requirements for fuel quality and characteristics vary among energy applications and combustion technologies. Therefore, relatively cheap, inhomogeneous fuels of low quality should be used as feedstocks in large combustion systems with high-quality management and control systems, which can compensate for the variations in the feedstock. High-quality fuels such as briquettes and pellets should be utilized in small-scale applications or as a high-density energy carrier for cocombustion in coal-fired reactors.

The size and size distribution of the particles in a feedstock significantly influence the combustion behavior and counter pressure. Several external and internal characteristics, such as bridging and the angle of repose, are closely related to particle size and shape (Viglasky and Langova, 2005; Viglasky et al., 2011). The particle size is thus important for securing the optimal conditions for in-feeding and the combustion of solid biofuels and must be adapted to the selected combustion technology. Moisture content is a key characteristic of biomass feedstock and is essential for the optimization of boiler settings. Feedstock with a high moisture

TABLE 3.1
Physical and Chemical Characteristics of Typical Biomass Feedstock for Energy Applications

Fuel	Typical Particle Size (mm)	Common Preparation Method
Briquettes	Ø > 25	Mechanical compression
Pellets	Ø < 25	Mechanical compression
Fuel powder	< 1	Milling
Sawdust	1–5	Cutting with sharp tools
Woodchips	5–100	Cutting with sharp tools
Hog fuel		Crushing with blunt tools

Wood composition (DM): C (47%–52% by weight), O_2 (38%–45%) and H_2 (6.1%–6.3%).

Structurally, wood is composed of cellulose (40%–45%), hemicellulose (20%–35%), and lignin (15%–30%). These proportions are based on the weight of the dry matter. The nitrogen content (N) is low (less than 0.5%) and the sulfur content is less than 0.05%. Generally, the mineral content is less than 1%. The key minerals are potassium (K), magnesium (Mg), manganese (Mn), sulfur (S), calcium (Ca), chlorine (Cl), phosphorous (P), iron (Fe), aluminium (Al), and zinc (Zn).

content has a higher flue-gas content and longer residence time in the boiler, which all influence the design and control of the boiler (Viglasky et al., 2015).

Mechanical durability describes the hardness of the fuel pellets and briquettes. Uniform particle distribution is a basic requirement for reliable conveying and combustion in heating systems based on pellets and high mechanical stability means that the numbers of fine particles and broken pellets will be low. Besides influencing the combustion systems, increased amounts of fine particles (PM2.5) can impose a variety of problems from the simple inconvenience of dust in the pellet storage to eye irritation and potentially leading to respiratory diseases from long-term exposure to dust (Dias et al., 2004; Fiedler, 2004; Madsen et al., 2004).

The ash content and composition varies considerably between feedstocks, ranging from below 0.5 weight % (dry basis—db.) in wood pellets produced from debarked stem wood to 5–10 weight % (db.) in agricultural residues, straw and *Amaranthus* or *Miscanthus* (Blaho and Viglasky, 1997; Viglasky, 2002; Viglasky et al., 2009b). The concentrations of the major ash-forming elements in biomass—silicon, calcium, magnesium, potassium, sodium, and potassium—are of great importance for combustion characteristics. Generally, magnesium and calcium increase the melting temperature of the ash whereas potassium and sodium tend to decrease the melting point. The ash melting point is defined as the point where the ash starts to flow and eventually melt leading to slag on the grate and in the bed. The ash-forming elements and the ash melting point vary considerably between biomasses, which should be considered when allocating biomass to different applications.

3.2 BIOENERGY SYSTEMS

Bioenergy systems from a few kilowatts to 5 MW have the potential for domestic heating or district heating plants and to provide heat, or heat and power cogeneration (CHP) at the village or community scale, or for a small industrial use.

3.2.1 BIOMASS HEATING: A SYSTEM OF SYSTEMS

With a biomass heating system, there are a choice of boilers and fuels, and although these can often be integrated with an existing distribution system, a number of additional components are required for the system to perform optimally. These will include fuel storage and handling, additional boilers, options for thermal storage, and a facility for ash extraction. For this reason we rarely speak of a biomass boiler in isolation and prefer instead to refer to a biomass system. This approach treats the installation as a "system of systems" (Figure 3.1) and is common in large or complex projects. In order to specify a system it is not necessary to have a detailed understanding of each subsystem; however, it is important to understand the relationships between each of the components and to ensure that the provider responsible for each subsystem has relevant information on dependent systems.

3.2.1.1 Handover Phase

Once the system is commissioned, the system is handed over to the user or the operator. Modern biomass systems are clean and efficient; however, the learning curve for new owners can be steep where there has been no previous experience of biomass or solid fuels. An important part of the handover process is to ensure that operators receive training in all aspects of operation and basic maintenance such as emptying ash bins, cleaning, and simple fault finding.

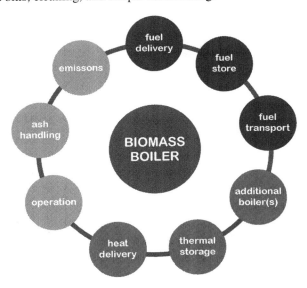

FIGURE 3.1 Biomass heating as a system of systems.

3.2.1.2 Operation Phase

Once the system is left to the client, there is inevitably a period of familiarization with the system. Operatives will gain insight into the way the system responds under different loads or to slight variations in fuel quality. A period of monitoring to ensure that the system is behaving as expected is highly recommended. Ensuring that the new system integrates as expected with existing building management systems and secondary boilers is important, and fuel quality and quantity, heat output, and ash volumes can all be easily monitored. Monitoring these performance and cost indicators will also be a good early indicator of any problems within the system.

3.2.2 BIOMASS COMBUSTION SYSTEMS

Biomass combustion is the main technology route for bioenergy and is responsible for over 90% of the global contribution to bioenergy. The selection and design of a biomass combustion system is determined by a number of factors including:

- The characteristics of the fuel to be used.
- The energy capacity and pattern of demand to be met.
- The costs and performance of the equipment.
- Local legislation relating to buildings and the environment.

A biomass system usually consists of the following components:

- Biomass boiler.
- Fuel storage.
- Chimney.
- Hydronic distribution system for the hot water produced by the boiler.
- Hydronic discharge systems (floor heating or radiators).
- A central control device with an outdoor temperature sensor.

In biomass heating systems, the fuel is transported from the storage facility to the combustion chamber, where it is combusted. A fan is installed to improve the heat transfer and supply sufficient air for an optimal combustion. The flue gas from the combustion process passes through a heat exchanger and transfers its energy to water. A circulation pump transports the heated water through the distribution system. To reduce heat losses to the boiler room, the boiler and all pipes should be highly insulated. Some of the features of a modern biomass heating include:

- High fuel efficiency (80%–90%).
- Ultra-low emissions.
- Fully automatic operation (automatic ignition and shutdown, fuel supply, ash removal, heat exchanger cleaning).
- Very high operation and fire safety standards.
- Low fuel costs.

Environmental Impacts of Biofuel-Fired Small Boilers and Gasifiers

Modern systems utilize a two-stage combustion process in order to combust the fuel as completely as possible. In the primary combustion zone, which is located on the grate, the drying and solid combustion takes place. Volatile gases are released and burned with air in the secondary combustion zone. The two-stage combustion process results in complete combustion and very low emissions of particulates because of the absence of unburned hydrocarbons in the flue gas. The particulate matter from the system is primarily inorganic, while emissions from lower technology stoves are mostly unburned organics. There are mainly three types of burner that vary according to the orientation of their fuel feeds:

- Horizontal feed burners: The combustion chamber is either fitted with a grate or a burner plate. The fuel is introduced horizontally into the combustion chamber. During the combustion, the fuel is pushed or moved horizontally from the feeding zone to the burner plate or grate. Both woodchips and pellets can be used.
- Underfeed burners (underfeed stoker or underfeed retort burners): The fuel is fed into the bottom of the combustion chamber or retort. These burners are most suited for fuels with low ash content such as woodchips with low moisture content or wood pellets.
- Stepped or moving grate (Figures 3.2 and 3.3): Designed for larger boilers and low quality, high moisture fuel. The wet fuel arrives at the top of an inclined, reciprocating grate and is shuffled toward the combustion zone, getting drier as it makes its way down.

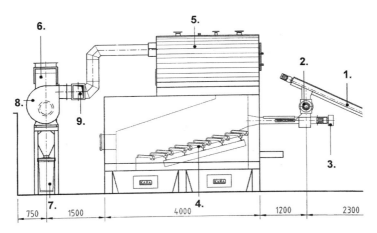

FIGURE 3.2 0.6 MW district heating plant. Schema of a hot water boiler with capacity 605 kWth; 1—screw conveyor of woodchips, 2—bio-fuel hopper - stern-valve (turnstile), 3—screw conveyor of bio-fuel into burning chamber, 4—hydraulic inclined moving grate system, 5—three pass hot water boiler, 6—multi cyclone cleaner, 7—ash bin, 8—flue gas blower, 9—flue gas outlet to a stack. (From Viglasky J. 2001. In: *Proceedings of Inter. Conference "Renewable Energy Sources on the Verge of the XXIst Century"*. PL - Warsaw, 10–11th December 2001, pp. 281–285.)

FIGURE 3.3 Boiler burner type and inclined moving grate system. (From Viglasky J. 2001. In: *Proceedings of Inter. Conference "Renewable Energy Sources on the Verge of the XXIst Century"*. PL - Warsaw, 10–11th December 2001, pp. 281–285.)

Top feed burners may also be found in small-scale units developed for pellet combustion. The fuel falls through a shaft onto a bed of fire consisting of either a grate or a retort. The feeding system and the fire bed are separated which makes it an effective protection against burn-back into the storage. The ash is removed mechanically by a dumping grate or manually.

3.2.2.1 Choosing the Right Fuel

The key fuel characteristics of biomass fuels are energy density, moisture content, particle size, and ash properties. Fuels are chosen in order to fulfill the technological and ecological requirements of the combustion technology and the most suitable combination of fuels and technologies can vary from case to case. In general, low quality fuels demonstrate inhomogeneous characteristics including high moisture content, variable particle size, and poor ash-melting behavior. Such fuels tend to be used in large-scale systems, while higher quality fuels are necessary for small-scale systems. This is due to the complexity and robustness of the fuel feeding systems, the combustion technology, and the management of emissions, all of which require economies of scale to be economically viable.

The majority of biomass systems burn woodchips or wood pellets, although agricultural residues and energy crops such as *Miscanthus* and short rotation coppice based on willow or poplar are also used. While some boilers are designed to take a range of fuels, others are more particular and regular switching between fuels is not practical as combustion settings have to be adjusted to cope with the different burning characteristics. Low-grade woodchips can have a moisture content of 50% or more and will have a relatively low energy density (630–860 kWh/m^3 depending on species), while high-grade chips will be expected to be around 30% moisture content (690–930 kWh/m^3). High quality wood pellets will be expected to be less than 10% moisture content and will have an energy density of around 3100 kWh/m³.

Key considerations will therefore be how much fuel storage is needed on site and how the fuel will be delivered. A range of vehicles, including lorries, tankers, and tractors, may make fuel deliveries and fuel suppliers should be consulted early on to provide assurance that deliveries to the site will be possible. Sites with lower or intermittent heat loads and limited space may opt for pellets, while nondomestic sites and those with ample space and larger heat loads are likely to find woodchips more economical. Boiler manufacturers will specify which fuels are suitable for each appliance according to predefined standards (see Table 3.2). European or national standards may be quoted and will govern all aspects of the fuel characteristics including energy density, moisture content, particle size, and distribution. Using incorrect or out of specification fuel can cause the system to operate poorly and will eventually lead to system failure.

3.2.2.1.1 Fuel Specifications

Fuel characteristics stated by the boiler manufacturer include the following (Viglasky, 2001):

- Fuel will consist of clean wood material in the form of woodchips or mixture of woodchips and sawdust.
- Required minimum of lower heating value (LHV) of woody fuel is 8.5 MJ·kg⁻¹, which is equivalent to maximum MC of the biofuel of approximately 50% on a wet basis (wb).
- Maximum ash content of the fuel is 2.0% on a dry basis (db).
- Maximum size of woodchips is 35 mm.

TABLE 3.2
Dendromass-Based Biofuels

Fuel	MC [%]	HHV [kJ·kg⁻¹]	Ash [%]	Burning Efficiency [%]	Bulk Density [kg·m⁻³]
Dendromass/wood	0	18,560	0.3–1.5	70–92	Spruce–beech
1. Wood fuel	8–40	19,000–10,100	0.3–1.0	75–92	SW HW
a. Woody chips	10	16,400			160–180
(Length l: 5–50 mm;	20	14,280			180–200
Width š: 5–30 mm;	30	12,180			200–220
Thickness h: 5–15 mm)	40	10,100			215–235
b. Sawdust/dust wood	8–10	17,600–15,000			120–180
(0.5–5.0 mm/0.004–0.5)					
c. Bark	40–48	12,100–10,500	1.5		300

Source: Viglasky J. 2001. In: *Proceedings of Inter. Conference "Renewable Energy Sources on the Verge of the XXIst Century"*. PL, Warsaw, 10–11th December 2001, pp. 281–285.

- Fuels that are not allowed (neither for cocombustion purposes) are:
  - contaminated wood with heavy metals, paint residues, plastics, and other waste materials; and
  - fossil fuels: coal (brown and hard), natural gas, etc.
- It is allowed to replace part of the woody fuel with bark and/or leaves as the total fuel mixture meets the fuel quality requirements (minimum LHV, maximum ash content, maximum size or MC), as specified above.

Theoretical higher heating value (HHV) of completely dry dendromass or phytomass is 20 MJ·kg^{-1} on average. Actual LHV of phytomass depends mainly on its MC, that is, water content and in smaller ranges on ash content and extraction matters content. MC in the range of 15%–25% is usually considered as optimal and technically obtainable. A dendromass of MC above 50% is already not suitable for burning due to several reasons. Basic parameters for chosen fuels based on biomass–dendromass are listed in Table 3.2.

The advantages of biofuels are: as produced from sufficiently dried dendromass or phytomass, fitted shape (form), and size for specific energy system, it is able fully to substitute brown coal regarding HHV. Ash content (0.3%–1.5%, incl. 0.01% sulfur) is negligible in comparison with brown coal (15%–30%, including 4% sulfur).

3.2.2.2 Fuel Feeding and Handling Systems

Information for fuel quality is available from the manufacturer of the boiler (see Figures 3.2 through 3.4). Biofuel mix—woodchips up to 35 mm and sawdust—are collected into a storage facility with a total volume of 1000 m^3. A tractor will be used to transport the biofuel to a walking floor system that acts as a day silo and ensures

FIGURE 3.4 Detail of fuel inlet to the burning chamber by a screw conveyor into the burning process. (From Viglasky J. 2001. In: *Proceedings of Inter. Conference "Renewable Energy Sources on the Verge of the XXIst Century"*. PL, Warsaw, 10–11th December 2001, pp. 281–285.)

proper feeding. During normal operation, the walking floor needs to be reloaded every one or two days. Biofuel—woodchips—are then conveyed to the fuel hopper of the combustion system. The combustor chamber is an inclined moving grate system, existing of horizontally moving cast iron grates between fixed grates. The stepwise movement ensures an optimal fuel distribution, dividing the combustion process into different phases. Together with appropriate primary and secondary air supply at each section, an optimal burnout is achieved. The robust technology allows application of inhomogeneous fuels: both sawdust and chips can be handled, with a moisture content up to 50%. The optimal fuel—woodchips—have a moisture content of 34% (Viglasky, 2001). The hot flue gases of about 900°C are directed into a three pass boiler, heating up the incoming water from 80 to 90°C. The boiler system is closed: the hot boiler water passes a heat exchanger, transferring the heat to the existing heating water pipeline system. After the boiler, a multicyclone gas cleaning system is installed to remove dust particles from the flue gases. The system will comply with the strict environmental legislation of the European Union (EU). An outlet dust concentration <150 mg·Nm^{-3} is guaranteed.

3.3 BIOMASS COMBUSTION AND EMISSIONS

Biomass for burning as a fuel includes wood, energy crops, agricultural crop residues, wood manufacturing byproducts, clean recycled wood, and farm animal litter. The selection of the fuel to be used can be complex as it depends on many factors. While cost is a key driver for fuel selection, the space available for fuel storage, access for fuel deliveries, and the method of delivery are all key considerations. While this chapter is concerned primarily with woodchips or wood pellets, the potential for using other solid biomass fuels should be borne in mind when considering fuel types. Burning of the following fuel types is not covered in this chapter: straw, grains (barley, oats, oilseed rape, wheat, rice, etc.), grain husks, oilseed rape, other agricultural residues, recycled wood containing glues and resins, farm animal litter, and distiller's draff.

3.3.1 Wood Fuel Composition

Wood comprises three main constituents:—cellulose, hemicellulose, and lignin—together with a number of trace elements. The three main constituents are complex structures of carbon, hydrogen, and oxygen. Cellulose—40%–50% of the wood—is essentially a polymerized sugar formed into long chains and gives wood its strength. Hemicellulose—20%–35% of the wood—is similarly formed of sugars; it is generally a smaller molecule than cellulose, but has many more branches in its chemical structure. Lignin—15%–35% of the wood—has a very complex chemical form incorporating numerous five and six carbon aromatic ring structures. It is the unwanted by-products formed during the combustion of lignin that gives rise to the characteristic smell of wood fires. In addition to this, there are numerous trace elements taken up during growth. These include calcium, chlorine, nitrogen, phosphorus, potassium, silicon, sodium, and sulfur. These minerals tend to be most prevalent in the bark, are concentrated in the ash during combustion, and will vary depending upon the elements' abundance (either naturally or as a contaminant) in the environment the tree was grown in. The ratios of the three main constituents differ between tree species, as do the ratios of subtypes within each constituent; however, as all are comprised of carbon, oxygen, and hydrogen, there tends to be little variation in the overall quantities of the individual elements. A representative approximation of the composition of wood, ignoring trace elements is:

- Carbon: 49% by weight;
- Oxygen: 45% by weight; and
- Hydrogen: 6% by weight;

giving rise to the generic chemical formula for wood of $C_1H_{1.4}O_{0.7}$.

3.3.2 Fuel Characteristics

Moisture content and calorific value are the two most significant characteristics of solid biomass fuels (i.e., all woody fuels), but there are a range of other important characteristics that have an impact on the design and operation of biomass systems which need to be taken into consideration.

3.3.2.1 Woodchip Fuel Characteristics

3.3.2.1.1 Moisture Content and Calorific Value

Woodchip is available with moisture contents between 20%–55%, the moisture reducing the calorific value of the fuel in two ways. It forms a noncombustible mass within the fuel and requires energy from the fuel to boil it and convert it into water vapor so that it can escape up the flue. The greater the moisture content, the less energy is available from the fuel, as the energy used to vaporize water is not available to provide heat from the boiler. For this reason, net calorific values are used for all wood fuels and reflect European continental practice. Furthermore, as the majority of available boiler plants are manufactured in Europe, this allows the performance of boilers to be assessed on a common basis.

Numerous woodchip calorific value calculators are available, all of which calculate the LHV of woody biomass fuels. The LHV of fully dried woodchips is usually between 18 and 18.5 MJ/kg (5.0 and 5.14 kW·h/kg), while there is no calorific value remaining at 88% moisture content. There is little practicable difference between softwoods and hardwoods. Calorific value is a linear function of the moisture content in the fuel (Kasimir and Nemestothy 2008):

$$\text{LHV} = \text{HHV}(1 - \text{MC}) - 2.447(\text{MC} - 9.01\text{H}(1 - \text{MC})) \tag{3.1}$$

where LHV is the lower heating value in MJ/kg, HHV is the higher heating value, or UHV is the upper heating value (usually taken as 20 MJ/kg for softwoods and 19.5 MJ/kg for hardwoods), MC is the moisture content in percent and H is the hydrogen content in percent (tree species dependent, but usually taken as 6%).

In practice, woodchip characterized by moisture content can be divided into two bands: 15%–35% moisture content; and 40%–65% moisture content. A wide range of boilers, with low levels of ceramic lining and rated at up to 500 kW, is available

Environmental Impacts of Biofuel-Fired Small Boilers and Gasifiers

to burn woodchips up to a maximum of 35% moisture content, while more thermally massive boilers are required to burn woodchips with a moisture content of 40% or above. Beyond 65%, moisture content combustion is very difficult to maintain.

3.3.2.1.2 Bulk Density

When storing woodchips, a significant proportion of the occupied volume is empty space between the woodchips. The bulk density is a measure of the mass of a quantity of woodchips divided by the occupied volume; the higher the bulk density, the more mass of fuel exists in a given volume. The greater the moisture content of the fuel, the greater the bulk density, which has a hyperbolic relationship to volume (Kasimir and Nemestothy 2008):

$$\text{Bulk density of softwood chips} = \frac{150}{(1-\text{MC})} \quad (3.2)$$

where MC is the moisture content in percent.

Bulk density, unlike density, is not intrinsic to a material as the same piece of wood would have different bulk densities if processed into woodchips, logs, or pellets. Moisture content also affects bulk density, as each particle has a greater mass but does not occupy more space; this is an important consideration, because fuels with higher moisture contents will have greater masses and, therefore, higher bulk densities.

3.3.2.1.3 Energy Density

Energy density is a measure of the energy contained within a unit volume of fuel. Energy density is expressed in MJ/m^3 and can be derived by multiplying the calorific value in MJ/kg by the bulk density in kg/m^3 or, alternatively, the calorific value in kW·h/kg by the bulk density in kg/m^3:

$$\text{energy density} = \text{calorific value} \times \text{bulk density} \quad (3.3)$$

In practice, the energy density does not vary greatly with moisture content over the most common range of woodchip moisture contents (20%–35%) because the calorific value curve runs in the opposite direction to the bulk density curve. Energy density is an important variable as it can help designers to assess volumetric fuel consumption rates, the size of fuel storage and the frequency of deliveries required, and the annual fuel requirement.

3.3.2.1.4 Particle Size and Dimensions

Most boilers will be able to accept a maximum particle size which is, typically, defined by a combination of cross-sectional area and maximum length for woodchips. In general, the smaller the boiler, the smaller the fuel feed system (usually based on augers) and the smaller the maximum chip size that can be accommodated. However, fuel feed systems incorporating a ram stoker can accept significantly oversized chips, as these systems have a cutting knife at the fuel inlet to the boiler. Furthermore, outsize particles, or the presence of an excessive proportion of "fines" (very small particles such as sawdust), are common causes of system blockages. For these reasons, fuel standards exist to assist in the specification of fuel quality.

3.3.3 FUEL STANDARDS AND TESTING

3.3.3.1 Fuel Standards

From autumn 2014, biomass fuel used by Renewable Heat Incentive (RHI) participants must meet a lifecycle greenhouse gas (GHG) emissions target of 34.8 g CO_2 equivalent per MJ of heat or 60% GHG savings against the EU fossil fuel average. Ofgem intends that biomass installations of <1 MW capacity will be able to use the default GHG emissions values outlined in the EU's *Report on Biomass Sustainability* (EC, 2010). Those installations above the 1 MW threshold will need to use actual values and are recommended to use the *UK Solid and Gaseous Biomass and Biogas Carbon Calculator* (Ofgem, 2012). From spring 2015, it was also planned that biomass fuel must meet land criteria, which currently differ for different types of biomass. For wood fuel, the criteria are outlined in the *UK Timber Standard for Heat and Electricity* (DECC, 2014).

3.3.3.2 CEN/TC 335

CEN/TC 335 allows all relevant properties of all forms of solid biofuels within Europe—including woodchips, wood pellets and briquettes, logs, sawdust, and straw bales the fuel—to be described. It includes both normative information that must be provided about the fuel, and informative information that can be included but is not required. The European Committee for Standardisation (CEN, under committee TC 335) published 27 technical specifications for solid biofuels during 2003–2006. As well as the physical and chemical characteristics of the fuel as it is, CEN/TC 335 also provides information on the source of the material. These technical specifications have now been updated to full European Standards (EN). The two primary technical specifications deal with classification and specification (BS EN 14961) and quality assurance for biofuels (BS EN 15234).

3.3.3.3 BS EN 14961-1 (2010)

EN 14961-1 covers the full range of solid biomass fuels including woodchips and wood pellets, which are the primary focus in this chapter. The classification of solid biofuels is based on their origin, and the fuel production chain should be clearly traceable from source to the point of use. Normative specifications for woodchips include:

- Origin
- Particle size (P16/P31.5/P45/P63/P100)
- Moisture content (M20/M25/M30/M40/M55/M65)
- Ash content (A0.7/A1.5/A3.0/A6.0/A10.0).

Normative specifications for chemically treated wood or used wood include:

- Nitrogen (N0.5/N1.0/N3.0/N3.0+).

Informative specifications for woodchips include:

- Net energy content (lower heating value (live)) as MJ/kg or kW·h/m³ loose
- Bulk density in kg/m³ loose
- Chlorine content (Cl0.03/Cl0.07/Cl0.10/Cl0.10+)
- Nitrogen (N0.5/N1.0/N3.0/N3.0+).

3.3.3.4 ÖNORM

While the EN 14961-1 standards have superseded all other European standards for solid biofuels across Europe, the Austrian ÖNORM standard is frequently referred to. ÖNORM is the Austrian Standards Institute and, while they are now adopting their own implementations of the EN 14961-1 standards, many Austrian boilers have been installed in the UK and specify fuel according to ÖNORM M 7133 for woodchips (*Woodchips for energy generation: quality and testing requirements* [PAS 111, 2012]) and ÖNORM M 7135 for pellets.

3.3.4 BIOMASS COMBUSTION

Solid organic fuels are not flammable under ambient conditions and a chain of thermochemical conversion processes need to take place in order for solid biofuels to burn (DGS, Ecofys, 2005):

- *Stage 1*: Warming of the fuel (less than 100°C). Biomass solid fuels are usually stored at between 0 and 25°C, so before reactions can begin the fuel needs to be warmed. Stage 1 requires the combustion chamber to be hot above and around where the fuel enters the grate and most biomass boilers contain some refractory material for this reason. Boilers designed to burn wet woodchips have substantial refractory linings in large combustion chambers. The greater the quantity of refractory lining, the less responsive the boiler to changes in heat demand, the longer the time taken to reach ignition temperature and the greater the residual heat that will need to be dissipated when the boiler is switched off.
- *Stage 2*: Drying of the fuel (100–150°C). Vaporization of water trapped inside the fuel occurs above 100°C.
- *Stage 3*: Pyrolytic decomposition of the wood components (150–230°C). Pyrolytic decomposition begins at approximately 150°C where the long-chain components of the solid fuels are broken down into short-chain compounds. The products of pyrolysis are liquid tars, gases such as carbon monoxide (CO), and gaseous hydrocarbons (CmHn). The pyrolytic decomposition of wood does not require any oxygen.

  Stages 1–3 are endothermic (heat-absorbing) reactions. They take place in any fire and prepare the fuel for oxidation. Once the flash point has been reached, approximately 230°C, exothermic (heat-producing) reactions commence with the input of oxygen. Wood can be externally ignited at approximately 300°C, and undergoes spontaneous combustion from 400°C.
- *Stage 4*: Gasification of the water-free fuel (230–500°C). The thermal decomposition of dry fuel under the influence of oxygen commences at the flashpoint of approximately 230°C, where gasification takes place mainly in the fire bed of a solid fuel fire. The oxygen supplied in the primary air produces sufficient heat in reaction with the gaseous pyrolysis products to affect the solid and liquid pyrolysis products such as carbon and tar.
- *Stage 5*: Gasification of the solid carbon (500–700°C). Combustible carbon monoxide is generated under the influence of carbon dioxide (CO_2), water vapor and oxygen (O_2). The gasification of solid carbon is an exothermic reaction producing both heat and light which can be seen as a visible flame.

- *Stage 6*: Oxidation of the combustible gases (700–1400°C). The oxidation of all combustible gases produced by the preceding process stages represents the end of the combustion reaction for the solid fuel. Most of the energy is produced at this stage when the combustible gases, mainly a mixture of carbon monoxide (CO) and hydrogen (H_2), are burned some distance away from the grate at a high temperature, while maintaining the temperature range on the grate required at Stages 1, 2, and 3 to convert the solid material to energy. The clean and complete combustion of the gas mixture is accomplished under the influence of secondary air.

Full conversion of solid fuels requires:

- Good mixing of the wood gas, produced at Stages 4 and 5, with the combustion air.
- The oxidation air at Stage 6 to be in excess of stoichiometric requirements (i.e., $\lambda > 1$).
- A sufficiently long dwell time for the wood gas/air mixture in the reaction zone.
- A sufficiently high combustion temperature (Figure 3.5) (Viglasky, 2001).

The necessary conditions for complete and even combustion of the fuel, and resulting low emissions are achieved by a spatial separation of the air supply to the fire bed (primary air) and that to the gas combustion zone (secondary air). The requirement for combustion air is in addition to that for general boiler house ventilation, the requirement for which should be separately assessed.

If a typical grated combustion system using a stepped grate automatic stoker is considered, then the area near the fuel feed will be predominantly drying, followed by the pyrolysis section with the carbon burnout occurring at the end of the grate. The transition between sections is gradual, and the depth of fire bed material will gradually reduce as mass is removed by pyrolysis and combustion of charcoal.

The surface area to volume ratio will affect the combustion rate. Fine fuels will have a faster rate of combustion and a more intense heat. Pellets have a greater mass within the surface area and will take longer to burn compared to the same sized chip. With all these variables, the feed rate of the grate must be set up correctly to avoid early burn off which will concentrate heat at one end of the grate.

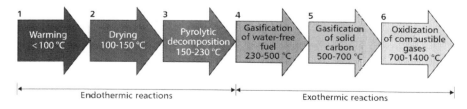

FIGURE 3.5 Stages of biomass combustion.

3.3.5 COMBUSTION AIR

The complete combustion of the fuel and the combustible gases requires a sufficient supply of oxygen in the combustion air. If just sufficient oxygen is supplied for the complete combustion of fuel and combustible gases, the fuel air mixture is deemed to be stoichiometric, and such a mixture is described as having an excess air ratio of 1. The excess air ratio is called λ, where $\lambda = 1$ for a stoichiometric mixture. Combustion air for wood fuelled boilers requires:

- *Primary air* which is introduced under the grate of a grated stoker or into the retort of an underfed stoker. This air rises up through the fire bed reacting with the fuel in a limited oxygen environment. λ is always less than 1 in a primary combustion zone. The reduced oxygen content allows control of the gasification process in the boiler, and reduces the flame and firebox temperatures.
- *Secondary air* which is introduced above the fire bed to react with the gases evolving from the fire bed.

Most of the energy in wood is converted to flammable gases by the pyrolysis process resulting in a larger requirement for secondary air in comparison with other solid or liquid fuels. Secondary air is likely to comprise some 80% of the total combustion air requirement, and an excess air ratio greater than 1 is always required with typical values of $\lambda = 1.5-2$. However, the provision of too much combustion air will result in lower flame temperatures and a cooler boiler. Heating the excess air to flue gas temperature also represents an energy loss, but less than that resulting from incomplete combustion. There must be sufficient time and turbulence in the combustion space above the fire bed for the combustion of the evolved gases to take place. The complete burnout of fuel and combustible gases requires temperature, time and turbulence. A minimum temperature of 850°C must be maintained for a minimum of 0.5 seconds in a turbulent environment with a minimum Reynolds number of 2300.

Separate control of primary air (from beneath the grate) and secondary air (into the gas oxidation zone) is required to maintain the lower grate temperature for Stages 1–3 while ensuring that a sufficiently high temperature and turbulence exist to oxidize the wood gases completely at Stage 6. A lambda sensor measuring flue gas O_2 content is used to ensure the correct supply of primary and secondary combustion air, with the boiler exhaust fan regulating combustion chamber pressure to ensure it is negative, that is, the primary and secondary air fans modulate on flue gas O_2 content with the boiler exhaust fan modulating to maintain negative pressure.

3.3.6 ASH CHARACTERISTICS AND SLAG FORMATION

3.3.6.1 Ash Characteristics

The ash content of solid biomass fuels is much lower than that of coal and is typically between 0.5% and 5% by weight of the dry fuel. Ash is a by-product of solid biomass combustion and derives principally from the minerals that predominate in the bark or, in the case of small diameter fuel such as straw, throughout the biomass material, together with some unburned carbon. Fuels with a high bark content, and those containing a high percentage of silica, will produce more ash. The exact proportion of ash produced is dependent on the chemical composition of the fuel. Wood pellets usually produce the lowest proportion of ash together with dry woodchips (~0.5%), while wetter woodchips and woodchips produced from whole trees, will produce up to 1.5% ash. Fuel derived from softwood grown in sandy areas will also produce higher ash outputs, while high silica fuels derived from agricultural residues and non-debarked trees could produce up to 5% ash. Approximately 98% of the ash produced is bottom ash from the grate with the remaining 2% emitted as fly ash. Any heavy metals present in the fuel will predominately precipitate out in the fly ash; this fact has been used as a method of decontaminating land containing heavy metal residues. Fly ash is usually captured by a flue gas clean-up system or by a fly ash drop-out chamber within the boiler. Under some circumstances, wood ash from combustion appliances can be spread on the soil under the relevant environmental regulators guidelines. This does not relate to domestic boilers for which the restrictions do not apply.

The quality of ash produced by a biomass system can provide a good indication of the health of a system. In particular, a change in the quantity of ash produced, its color, density, particulate size, and moisture content, can provide an early indication of a change in combustion conditions or a problem within the boiler. It is recommended that a sample of ash be taken when a boiler is first commissioned and is retained as a reference sample. Any subsequent change in boiler performance or the characteristics of the ash produced can then be evaluated against the reference sample.

3.3.6.2 Slag Formation

Depending upon the chemical constituents of the ash, it may soften, forming glasses from silica within it. The temperature at which this occurs is the ash fusion temperature, and varies between fuels as differing trace elements affect the melting point. Slag formation occurs when naturally occurring silica (sand), or unwanted silica in the fuel, converts to glass. For example, while pure silica melts at 1700°C, chlorides present in the fuel can reduce this to as low as 773°C, making it important to keep the temperature of the grate below 750°C in this instance. Once ash has melted it will cool and solidify to form a solid clinker. The fire bed temperature must be controlled to below the ash fusion temperature to prevent clinkering, and some types of boiler require water-cooled grates and/or flue gas recirculation to manage the fire bed temperature. High silica fuels, such as straw, present a particular problem as they often also contain high concentrations of chlorides, and specially designed boilers are usually required to allow such fuels to be burned. Biomass containing fresh, green matter such as leaves, needles, and woody residues will contribute to an increase in the concentration of chlorine and will affect the ash melting temperature in addition to producing much higher NO_x and pm emissions. Burning such material will increase the maintenance burden on the boilers and heat exchanger surfaces.

3.3.7 POLLUTION EMISSIONS

The principal emissions from biomass boilers include:

- The gas phase emissions of carbon monoxide (CO), carbon dioxide (CO_2), water vapor and nitric oxide (NO), and nitrogen dioxide (NO_2), collectively known as NO_x.
- Particulate matter including salts, soot, condensable organic compounds, and volatile organic compounds. Collectively these are known as PM.

Careful control of the air supplies to the primary and secondary combustion zones are required to minimize the formation of all NO_x and particulates. However, as complete oxidation of the wood gases requires a slightly higher combustion temperature than in fossil fuelled boilers, the quantity of NO_x produced by biomass boilers per unit of heat generated is usually greater than that from gas fuelled boilers.

Other airborne toxic fumes are created by the incomplete combustion of biomass fuel. Incomplete combustion produces benzene-related polyaromatic hydrocarbons (PAH), which give wood smoke its distinctive aroma; however, they are carcinogens and chronic exposure to them can ultimately be fatal.

3.3.7.1 NO_x

The reduction of fuel nitrogen to molecular nitrogen is favored in the fuel rich primary combustion zone. The minimum total fixed nitrogen (TFN = HCN + NH_3 + NO + NO_2 + $2N_2O$) emission from the primary combustion zone is reached for a stoichiometric ratio (λ) of 0.7–0.8 at a temperature of 700–750°C with a mean residence time of 0.5 s (Keller, 1994). After the reduction zone the combustion is completed in the secondary combustion (burnout) zone by injection of excess air.

The study of the effects of moisture content on flame temperature dealt with emissions, heat transfer, and combustion efficiency of biomass stoves and boilers. Effects of the moisture content (MC) of wood on the temperature of the flame (ToF) are presented in Table 3.3 (Alakangas et al., 2008).

The MC of the biofuel, firewood, has a direct effect on pyrolysis and combustion. The higher the MC, the slower the pyrolysis and combustion processes. The MC also has an effect on the amount of pyrolysis byproducts and residual charcoal. The evaporation of the MC requires heat, just like pyrolysis.

The affinity to water and water vapor is a negative feature of firewood. The relative MC of a freshly felled tree in the dormant season ranged from W = 35%–65%, depending on the wood species. Firewood in branch form stacked under cover

TABLE 3.3
Effects of the MC of wood on the ToF

MC, [w-%]	0	10	15	20	25	30	35	40	45	50	60
ToF, [°C]	1200	1160	1120	1100	1090	1070	1025	1000	975	915	805

or against a sheltering wall dries naturally to an air-dried state, that is, a MC of W = 18%–25%. Unprocessed wood waste with a MC of W = 10%, or biofuel in the form of briquettes and pellets made from wood with a specific size and MC, plays an important role in terms of the energy efficiency of boilers.

NO_x production increases with an increase in flame temperature. From this, it can be seen that the drier the biomass fuel, the hotter the flame and the greater the potential to produce NOx. For this reason, and to protect boiler internals from damage (particularly ceramic linings), a key feature of boiler design is to ensure that combustion temperatures are maintained at, typically, no more than 1200°C. Flue gas recirculation will permit the combustion of drier fuels without causing excessive NO_x formation in gases or clinkering of ash, and is essential if dry fuels are to be burned in a boiler designed to burn wetter fuels.

The efficiency of producing heat from firewood depends on the construction of the heat generator as well as on the energy properties of the firewood and the energy and environmental benefits delivered by the boiler. The energy properties of the firewood especially depend on its MC. The basic energy properties include the high heating value (HHV) and low heating value (LHV), but also the burning process in the furnace or boiler, including the flame temperature; amount of flue gases created; dew point temperature of flue gases; and emission production, which are all negatively affected by the firewood MC. The design and construction of the boiler heat exchanger affect the use of the calorific value of flue gases, namely the cooling rate of the flue gases before they are delivered into the atmosphere. Currently, the energy efficiency of mid-efficient energy firewood boilers is $\eta_B = 80\%–85\%$. On the other hand, the energy efficiency of progressive current biofuel boilers with guaranteed energy properties is $\eta_B = 92\%–96\%$.

The temperature of flue gases (t_{fg}) emitted to the atmosphere by boilers is dependent on the design and construction of the boiler heat exchanger and the temperature of the heated water, thermal oil, or water vapor produced. The temperature of the flue gases leaving the boiler ranges from $t_{fg} = 120–200°C$, according to prestigious manufacturers of firewood boilers such as KARA Energy Systems B.V., Viessmann, Justsen Energiteknik A/S, KWB The Biomass Heating System GmbH, and RIKA Innovative Ofentechnik GmbH.

For example, the atmospheric thermal load created by the heat of emitted flue gases from a natural gas boiler and a boiler fed with firewood at moisture contents of W = 10% and W = 60% into the atmosphere at $t_{fg} = 120–200°C$.

The volume of flue gases emitted to the atmosphere as a result of the production of the heat of 1 GJ from natural gas (51.6 MJ/GJ) is 2.4 times lower than the volume of flue gases produced and delivered to the atmosphere when 1 GJ of heat is produced from dried firewood with a moisture content of W = 10% (96.2 MJ/GJ). The volume value is four times lower than the value from the production of heat from wet wood with a MC of W = 60% (179.8 MJ/GJ). The values of the atmospheric thermal load created by the flue gases emitted in the production of 1 GJ of heat from natural gas with a temperature of $t_{fg} = 110°C$, and those of dried and wet firewood with a temperature of $t_{fg} = 120°C$, as mentioned above for this example.

The atmospheric thermal load created by flue gases resulting from the combustion process of dried wood, with a temperature of flue gases of $t_{fg} = 120°C$, is 1.8 times higher, and the atmospheric thermal load created by flue gases resulting from the combustion process of wood with a MC of W = 60% and temperature of flue gases of t_{fg} = 200°C is seven times higher than in the combustion of natural gases.

The above arguments on the influence of wood moisture on the warming of the atmosphere by flue gases emitted from a boiler illustrate the fact that the combustion of moist wood reduces the thermal efficiency of the boiler and increases the production of emissions. This is the reason for introducing the economically efficient pre-drying and seasoning of fuel wood. Such technologies currently used within Scandinavia include the technology of transpiration drying of the branches and tops of trees before the production of wood chips, as well as the natural drying of stored firewood in covered storehouses.

The effect of the MC of firewood on the atmospheric thermal load created by the heat of flue gases emitted from a firewood boiler can be found as the following:

- In the production of 1 GJ of heat from dried wood with a MC of W = 10%, a volume of flue gases of $V_{fg\text{-}1GJ}$ = 96.2–177.8 m_n^3 with a temperature of t_{fg} = 120–200°C is delivered to the atmosphere. In the combustion process of wet wood with a MC of W = 60%, a volume of flue gases of $V_{fg\text{-}1GJ}$ = 179.8–361.9 m_n^3 is emitted to the atmosphere.
- An increase of 1% in the MC of firewood results in an increase in the atmospheric thermal load created by the heat of flue gases with a temperature of t_{fg} = 120°C of ΔQ_{fg} = 1.7 MJ, on average. When the temperature of the flue gases is t_{fg} = 200°C, it increases by ΔQ_{fg} = 3.68 MJ.
- Comparing the values of the atmospheric thermal load created by the flue gases resulting from the combustion process of firewood to the thermal load caused by the combustion process of natural gas, we can state that the atmospheric load caused by the combustion of firewood ranges from 1.8–7 times higher.
- The heat of the water vapor from the evaporated water of the combusted wood, as well as the heat of the heated nitrogen and unoxidized oxygen in the combustion air delivered to the furnace of a boiler to dry firewood, are the reasons for the increasing volume of flue gases. This, therefore, causes the increasing atmospheric thermal load created by the heat of the flue gases resulting from the combustion of wood with a higher MC.
- Due to the increased moisture level of the used firewood, there is a thermal load on the atmosphere. This is due to the burning of moist biofuel and the discharge of these flue gases from the stack at an elevated temperature. Of course, due to the increased fuel moisture level, the boiler's energy efficiency is low and the heat production efficiency decreases, while emissions are increased. This justifies the use of economically efficient forms of pre-drying and seasoning of firewood, such as the technology of transpiration used to dry branches and tree tops before the production of wood chips, or the natural pre-drying of stored firewood logs in a sheltered position.

3.3.7.2 Particulate Matter

Particulate matter (PM) from biomass combustion are tiny particles which include fly ash, black carbon, tars, and benzopyrenes. Since the 1970s, PM from all sources has been known to have adverse health effects. PM smaller than 10 μm aerodynamic

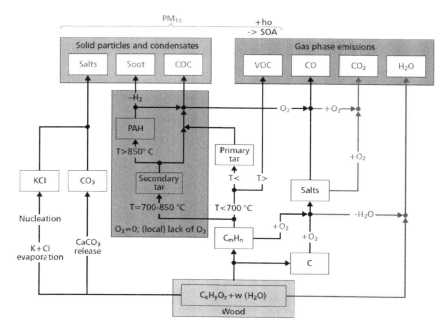

FIGURE 3.6 Pathways to the formation of PM.

diameter, PM_{10}, can settle in the bronchi and alveoli (and so is referred to as the thoracic fraction) while PM smaller than 2.5 μm ($PM_{2.5}$, the respirable fraction) can pass through the lungs and into the bloodstream. Many of the tars and benzopyrenes are carcinogens, and are now known to contribute to cardiovascular disease.

PM_{10}s are formed as a result of incomplete combustion (Nussbaumer, 2011). Salts, soot, condensable organic compounds (COC), and volatile organic compounds (VOC) are deemed primary aerosols, while secondary aerosols can also be formed as secondary organic aerosols (SOA) and secondary inorganic aerosols (SIA). Soot is formed from tars via PAH as an intermediate product. VOC resulting from incomplete combustion act as precursors for SOA, while NO_x and SO_x from nitrogen and sulfur in the fuel act as precursors for nitrates and sulphates. Figure 3.6 shows the mechanisms for PM production.

3.3.7.3 Regulated Levels of PM and NO_x

PM is formed in three ways:

- Incomplete combustion at low temperature can lead to COC as pyrolysis products and also called tar, which form primary organic aerosols and contribute to brown carbon in the ambient air.
- Incomplete combustion at high flame temperature but with a local lack of oxygen or flame quenching can lead to soot formation resulting in elemental carbon (EC) finally contributing to black carbon (BC) at ambient temperature.

- Mineral matter, including alkali metals and chlorine in the fuel, leads to the formation of inorganic fly ash particles mainly consisting of salts such as chlorides and oxides. In addition, other metals may also be available in certain concentrations.

The two graphs on the Figures 3.6 and 3.7 show the influence of moisture content on flame temperature, and how NOx production increases with increasing flame temperature. From these, it can be seen that the drier the biomass fuel, the hotter the flame and the greater the potential to produce NO_x.

Figure 3.6 also shows that herbaceous, which include straw and grasses or Amaranth (*Amarantus* L.), produce more NO_x than wood and that recycled materials containing adhesives, such as chipboard, produce even higher levels of NO_x.

Wood pellets are usually manufactured from heartwood and will produce the lowest NO_x emissions. Woodchips, which are manufactured from whole trees including the bark, and logs, will have higher emissions than wood pellets but lower emissions than herbaceous biofuels. NO_2 is known to cause sensitization of the respiratory system, hence exposure to combustion gases must be avoided at all times. NO_2 is a pollutant deriving from complete combustion of biomass fuel and is always present in flue gases (Nussbaumer, 1997).

In 2013, the UK introduced legislation which requires the maximum concentrations of 30 g/GJ for PM and 150 g/GJ for NO_x for boilers to be accredited for the RHI. Boiler manufacturers have to provide laboratory testing certificates to demonstrate boilers meet these standards, utilizing post-combustion cleaning if required, for a boiler to be accredited under the RHI. Boiler testing results are in normalized mg/m³

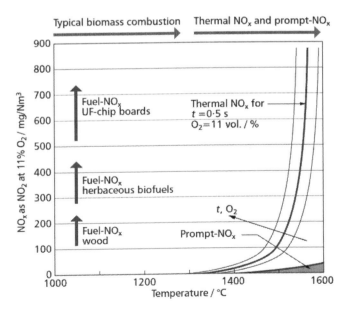

FIGURE 3.7 Influence of temperature on NO_x emissions for biomass combustion.

of flue gas (i.e., corrected to a flue gas density at 0°C and 1013.25 mbar at normal temperature and pressure). These data have to be converted to g/GJ of net heat input for RHI purposes, but the normalized densities are a function of excess O_2 content in the flue. The conversion can be carried out following the advice provided in the Defra report (Defra, 2011–2012) and by Ofgem (2013a).

The conversions from concentration (CN) in mg/m³ to emissions factor (EF) in g/GJ and vice versa can be carried out using the following formulae. A dilution factor (DF) is defined as:

$$DF = \frac{21\% - O_2}{21\%} \quad (3.4)$$

where O_2 is the oxygen concentration in the flue gas in %, giving:

$$EF = \frac{CN \times 253}{DF \times 1000} \quad (3.5)$$

and:

$$CN = \frac{EF \times 1000 \times DF}{253} \quad (3.6)$$

Table 3.4 shows concentrations of PM and NO_x corresponding to the emissions factors of 30 g/GJ and 150 g/GJ respectively for various flue gas O_2 concentrations (Nussbaumer, 2011).

TABLE 3.4
PM and NO_x Concentrations in mg/m³ for Excess O_2, Excess Air and λ Corresponding to RHI Pollution Emissions Standards

O_2 concentration in flue	0%	6%	10%	11%	12%
Excess air	0%	28.6%	47.6%	52.4%	57.1%
Stoichiometric ratio (λ)	1.0	1.29	1.48	1.52	1.57
PM	119	85	62	56	51
NO_x	593	423	311	282	254

Source: Defra. 2011–2012. *Conversion of Biomass Boiler Emissions Concentration Data for Comparison with the Renewable Heat Incentive Emission Criteria (RHI).* Defra Report. Governement, Great Britain, www.gov.uk/decc; Ofgem. 2012. *UK Solid and Gaseous Biomass and Biogas Carbon Calculator.* https://www.ofgem.gov.uk/publications-and-updates/uk-solid-and-gaseous-biomass-carbon-calculator [Accessed June 30, 2014]; Ofgem. 2013a. *Renewable Heat Incentive Guidance Volume One: Eligibility and How to Apply.* https://www.ofgem.gov.uk/ofgem-publications/83036/rhiguidancevolone1.pdf [Accessed June 30, 2014]; Ofgem. 2013b. *Renewable Heat Incentive Guidance: Non-domestic Scheme Metering Placement Examples.* https://www.ofgem.gov.uk/publications-and-updates/renewable-heat-incentive-guidance-non-domestic-scheme-metering-placement-examples [Accessed June 30, 2014].

3.3.7.4 Carbon Monoxide (CO)

Carbon monoxide (CO) is created by the incomplete combustion of carbon and is well known to be lethally toxic. Less well known is that sustained subclinical exposure (which produces no symptoms) can result in incremental damage to the brain and other organs. Chronic, subclinical exposure to CO causes cumulative damage to oxygen-using organs in the body, hence the World Health Organisation (WHO) recommends a 24 hour exposure limit of 7 mg/m^3 (6 ppm). Subclinical exposure is that which produces no symptoms, that is, <50 ppm in the case of CO. Biomass pellets in the period up to 8 weeks after manufacture will give off gas producing CO. CO concentrations at lethal levels have been measured in inadequately ventilated pellet stores and the holds of bulk carrier ships transporting wood pellets. The concentration of CO may also be such as to form an explosive atmosphere. Hence, ATEX regulations apply to any electrical equipment installed in pellet stores. ATEX derives its name from the French title of the 94/9/EC directive *"Appareils destinés à être utilisés en ATmosphères EXplosives."* The ATEX as an EU directive finds its US equivalent under the hazardous-location (HAZLOC) standard. This standard given by the Occupational Safety and Health Administration defines and classifies hazardous locations such as explosive atmospheres.

3.3.7.5 Benzene and PAH

Benzene is created by the incomplete combustion of the hydrocarbon vapors in the oxidation phase (see Figure 3.5). It is produced during pyrolytic decomposition and gasification and causes cancer in both humans and animals. PAH are potent atmospheric pollutants consisting of fused aromatic rings and are chemically related to benzene. They, too, are the product of incomplete combustion in the oxidation phase. Pyro(a)benzene is particularly dangerous and is the component in tobacco smoke that causes lung cancer.

3.3.8 THE EFFECTS OF BURNING OUT-OF-SPECIFICATION FUELS

3.3.8.1 Too Wet

If wet fuel is not dried sufficiently by the boiler (because the fuel moisture content is outside the fuel tolerance range of the boiler) incomplete gasification and oxidation will occur and black smoke will be produced. In addition, the tars released at combustion Stage 2 will gradually coat the heat exchanger surfaces resulting in reduced heat exchange efficiency and the eventual failure of the boiler. Tar accumulation is also one reason why many manufacturers recommend minimum running periods for their boilers to ensure that combustion chambers and heat exchangers reach full working temperature to drive off the heavy volatiles deposited during the heat-up phase. The energy used to evaporate the moisture is not available to the appliance user.

3.3.8.2 Too Dry

If the fuel is too dry for the boiler, grate (Stage 2) and oxidation zone (Stage 4) temperatures can be too high, resulting in the formation of slag and a higher concentration of NO_2. This latter issue can be addressed by installing flue gas recirculation to primary air. This allows dry fuel to be used in a boiler designed to burn wet fuel by maintaining the primary gas flow rate while reducing its O_2 content, thus reducing the gasification rates (Stage 3) on the grate.

3.3.8.3 Safety Considerations

As with all fuels some excess air above that required for stoichiometric combustion is required to obtain full combustion of the fuel. Failure to obtain full combustion, by provision of inadequate air, will result in a significant presence carbon monoxide in the flue gases. This is a major boiler efficiency loss in the calorific value that it carries away but, more importantly, presents a significant explosion risk in that the lower explosive limit in the flue gases can be reached very quickly after load is removed from a boiler. The Combustion Engineering Association and Carbon Trust's guide (CEA, 2011) provides comprehensive guidance on all aspects of health and safety related to the design and operation of biomass systems and should be referred to as the definitive text on this subject.

Participants in the nondomestic RHI also have ongoing obligations, including to ensure that the fuel used is in accordance with that specified on a valid RHI emissions certificate for their plant. The use of other fuels, or the use of fuels at a moisture content exceeding the certificated moisture content, will result in noncompliance, which would be subject to sanctions.

3.4 ENVIRONMENTAL ASPECTS OF HEAT PRODUCTION BASED ON BIOFUELS

Bioenergy system operation is generally characterized as environmentally friendly and with minimal pollution production. The precise emission formation depends on the boiler and any additional devices' design, quality of applied fuel, its ultimate composition, and further cooperative factors. During burning of different fuels, a whole range of solid, gaseous, or liquid potential polluting matter in different amounts may originate. The most significant of them being CO_2, CO, solid pollution mater (SPM) of different sizes, that is, flying ash and soot, water vapor, SO_x, NO_x, OC (organic compounds), and trace elements.

3.4.1 Solid Pollution Matter

Pollution is the introduction of contaminants into the natural environment that causes adverse changes. Pollution can take the form of chemical substances or energy, such as noise, heat, or light. Pollutants, the components of pollution, can be either foreign substances or energies, or naturally occurring contaminants. Pollution is often classed as point source or nonpoint source pollution.

Of these, PM (and its intermediate products) and NOx are considered to be the most relevant emissions when considering biomass combustion. PM is considered to be one of the most significant pollutants produced from biomass combustion.

PM measured in the ambient air is a combination of primary aerosols, directly emitted from both natural and anthropogenic sources, and secondary aerosols, formed in the atmosphere from the conversion of other gaseous compounds such as SO_2 and ammonia (NH_3).

The size of PM varies, with two categories commonly identified when analyzing the impacts of PM on human health and the environment; PM10 and PM2.5. PM10, or coarse particles, includes all particles that have an aerodynamic diameter that is ≤10 μm, while PM2.5, or fine particles, includes all particles that have an aerodynamic diameter that is ≤2.5 μm. Estimates indicate that >90% of PM emissions from the efficient combustion of wood fuel fall within the PM10 category, while >75% fall within the PM2.5 category.

Along with the increasing levels of interest in PM2.5 in recent years, there is also an increasing focus on the portion of PM known as "black carbon," which are light-absorbing, fine soot-like particles that are released from the incomplete combustion of fossil fuels. Black carbon is present in the ultrafine fraction of PM (PM0.1) and is known to be a significant component of diesel soot, a substance that the WHO has identified as being carcinogenic.

3.4.1.1 Why Burning Wood Threatens the Health of Europeans

Alongside PM, oxides of nitrogen (NO_x) are pollutants that are of highest concern when considering emissions from biomass combustion. NO_x emissions relevant to biomass combustion include nitric oxide (NO), nitrogen dioxide (NO_2), and nitrous oxide (N_2O). N_2O is generally less commonly produced from the combustion process than NO or NO_2.

NO_2 is an irritant in the lungs that is known to have direct negative respiratory impacts on human health.

NO_x emissions can have significant adverse impacts on the environment. N_2O, while not commonly produced from biomass combustion, is a greenhouse gas that contributes directly to the impacts of climate change. The various gaseous forms of NO_x can react with SO_2 and other substances to form acid rain, which can be damaging to vegetation and buildings, as well as contributing directly to eutrophication.

These deleterious substances cause, among other less serious effects, greenhouse effects, acid rain, and disturbance of the ozone layer.

Table 3.5 illustrates typical emissions and achieved values for bioenergy systems by woodchip firing, which are significant especially from the environment protection point of view.

How far this corresponds to the truth is verified by the experimentally way of analysis of emission measurement results carried out at the chosen boiler with solid biofuels—woodchips. Measurements were carried out to prove via observance concerning emission norm limits within the framework a compliance delivery by a local authority of air protection during starting operation of middle size heat source regards air pollution (Table 3.6) (Viglasky, 2001).

One important aspect in commercialization of biofuels and bioenergy systems is to reduce gaseous emissions to acceptable levels, meaning that they comply with current and future legislation.

If the burning process within your boiler is faulty or the system has not been set up correctly, then the emissions (due to the incomplete combustion of the fuel) can produce carbon monoxide (CO), benzene, and volatile organic gases, among other toxic substances that can be carcinogenic, so please ensure your boiler is serviced well and is operating inside the manufacturer's guidelines at all times.

TABLE 3.5
Typical Emission Values in Connection with Woodchip Firing

Main Pollution Emission	Unit	Typical Value	Typical Variation
SO_x as SO_2	$g \cdot GJ^{-1}$	15.0	5–30
NO_x as NO_2	$g \cdot GJ^{-1}$	90.0	40–140
Dust, multicyclone	$mg \cdot m_r^{-3}$	300.0	200–400
Dust, flue-gas condensation	$mg \cdot m_r^{-3}$	50.0	20–90
CO_2	$kg \cdot GJ^{-1}$	102.3	
Condensate	$L \cdot (MWh)^{-1}$	–	150–250

Source: Defra. 2011–2012. *Conversion of Biomass Boiler Emissions Concentration Data for Comparison with the Renewable Heat Incentive Emission Criteria (RHI).* Defra Report. Governement, Great Britain, www.gov.uk/decc; Ofgem. 2013a. *Renewable Heat Incentive Guidance Volume One: Eligibility and How to Apply.* https://www.ofgem.gov.uk/ofgem-publications/83036/rhiguidancevolone1.pdf [Accessed June 30, 2014].

Note: The figures vary in practice, even beyond typical variations listed within emission limits.

TABLE 3.6
Increased Moisture Content of Fuel Leads to Pollution Emissions Higher than All Limits

Chosen Pollution Emission	Unit	Fuel-1/mix Woodchips 60% + Sawdust 40%, Moisture Content 51%	Fuel-2/mix Woodchips 60% + Sawdust 40%, Moisture Content 41%	0.2–2.0 MWth Limit
Dust—SPM	mg/m_{nr}^3	436–492	223–259	250
CO	mg/m_{nr}^3	4729–9794	303.9–360.3	850
CO_2		Woodchips as fuel is considered CO_2-neutral.		
SO_2		Many analyses of sulfur content in fuel-woodchips show values that they are below the laboratory equipment limits of detection.		
$NO_x = NO + NO_2$	mg/m_{nr}^3	154–205	296–385	650
TOC	mg/m_{nr}^3	75–778	3–17	50

Source: Viglasky J. 2001. In: *Proceedings of Inter. Conference "Renewable Energy Sources on the Verge of the XXIst Century".* PL, Warsaw, 10–11th December 2001, pp. 281–285.

3.4.2 OTHER POLLUTANT EMISSIONS AND WORKING CONDITIONS

In addition to particles, SO_2, NO_x, and CO, flue gases may contain other pollutants, such as PAH, dioxins, hydrogen chloride (HCl), and so on. PAH is a joint designation for a range of chemical compounds consisting of carbon and hydrogen. It occurs with incomplete burning. Some PAHs are noxious (some even cancer-causing) and should therefore be avoided. Since 1985, several investigations have been carried out all showing that there is a close connection between the formation of PAH and CO. Low CO content and low PAH content go together (Kona, 1995). Alongside emissions, which are connected with work safety and health protection, noise is also ranked, which must comply with the required exterior and interior noise limits. Finally, we must not neglect fire protection and so-called in-plant safety.

3.4.3 Conclusions

Practical experiences point out that not every installation of modern bioenergy systems is perfect and it is not possible to automatically classify among these, those which are environmentally friendly in comparison with conventional energy systems based on fossil fuels. It is important to mention that often owing to operators and their working discipline and not technology systems, which have been proved by our measurements.

3.5 CASE STUDY: A POTENTIAL FEEDSTOCK RESOURCE

3.5.1 Amaranth (*Amarantus* L.) as Raw Material Resource for Biofuels Production

Biomass, generally, has been taken as the only indigenous renewable energy resource capable of displacing large amounts of solid, liquid, and gaseous fossil fuels. As a widely dispersed, naturally occurring carbon resource, biomass is a logical choice as a raw material for the production of a broad range of fossil fuel substitutes.

3.5.1.1 Introduction

Amaranth is a plant species with the C4 photosynthetic pathway. These plant species are distinguished by a significantly high dry matter (DM) yield potential and lower quality in comparison with plant species with the C3 photosynthetic pathway (Viglasky et al., 2009a,b).

Amaranth is distinctive. It is known for its significantly high yield as well as quality. Some of its numerous species are becoming an increasingly important resource for healthy food (the seeds have high nutritional value); the unprocessed biomass is used primarily as fodder in many countries, but especially by Central America Indians, who were the original cultivators. The utilization of biomass as a renewable energy source for some advanced bioenergy systems indicates, however, that biomass can be an excellent and perspective biofuel. It has become clear that current knowledge of biomass physical properties, especially of new energy crops or plants cultivated in the polluted fields, is not sufficient for further optimization of industrial energy plants or domestic bioenergy units. For this reason, it is necessary to explore biomass, especially new energy plant species as fed into the bioenergy system. Without this, there can be no guarantee that the biofuel will perform satisfactorily under operating conditions. It is clear that our research on Amaranth as a potential energy crop has not included a definitive taxonomic study but it could be an important key link in future environmental protection (Viglasky et al., 2009b).

Objectives of experimental activities:

- To study physical properties including thermal behaviour of Amaranth phytomass as fuel;
- To determine:
  - Higher heating value of Amaranth phytomass
  - Ash content and its chemical composition.

3.5.1.2 Material and Methods

3.5.1.2.1 Crop Material

The crop samples of Amaranth for chemical analyses and bomb calorimetry were obtained from experimental sites of Slovak University of Agriculture (SUA) in Nitra, Slovakia. The actual and potential yield of Amaranth is still being investigated for Slovak climate and soil conditions. Yields of Amaranth biomass are in the range of 50–260 tons of green matter (GM) (or 10–60 tons of DM) per hectare and growing season, depending on chosen plant species and location of cultivation, inclusive soil condition.

3.5.1.2.2 Methods Used for the Characterization of Amaranth-Phytomass

Solid fuels: Amaranth phytomass dry matter energy value (combustion heat) was determined by means of adiabatic instrument IKA C4000 (Analysentechnik Heitersheim. IKA Works GmbH & Co. KG in Staufen, Germany). Gross calorific value and heat capacity of the calorimetric system were determined by a software program C-402, according to the DIN 51900 standard.

Biofuels: Determination of ash content and its chemical composition, and analyses of basic properties of waste-combustion residues, were made according to the Statute of Slovak Government No. 15/1996 Digest about the treatment of waste.

3.5.1.2.3 Experimental Procedures

Thermo-oxidation reactions from the given samples of Amaranth phytomass were carried out under laboratory conditions at controlled burning temperature by means of air oxygen. The surplus that remained after oxidation samples were gravimetrically fixed and adapted for analytic fixing of the chosen elements. The analysis consisted in the preparation of water lye with deionized water followed by the process of elution with solid matter in 3% solution of nitric acid, purified for spectral analysis. Eluates were analytically prepared and individual elements were fixed by GBC 932AB Plus AAS (AAS–ICP) methods. Mercury was fixed by mercury analyzer. The results showed that the methods are suitable for the determination of selected heavy metal contents in vegetables and grains, for example, Amaranth.

3.5.1.3 Results

Different crops are likely to have a similar calorific value per unit weight of DM (Grimm and Strehler, 1987). In general, moisture content (MC), ash content (AC), as well as growing parameters (e.g., location, different fertilization treatments, nutrient balance, ley, year, and others) mainly effect calorific value of the phytomass as a potential raw material for biofuel production.

3.5.1.3.1 Calorific Value of Amaranthus Cruentus: Giganteus

The lower and higher heating values were determined according to DIN 51900. Data on proximate and ultimate compositions and heating values are given on a dry basis.

HHV of Amaranth DM varies from 15.5 to 17.0 MJ·kg^{-1}, LHV is range 13–14 MJ·kg^{-1} at 10% MC. The analysis results confirm Amaranth phytomass to be a low-grade fuel in comparison with coal.

A summary of the results obtained in laboratory analyses of Amaranth phytomass as a fuel is presented in Table 3.7. Thermal conversion quality is determined by many factors, including water content at harvest, and ash, alkali, Cl, and N content. Limited data available show that as a low input C4 plant with a low water requirement, Amaranth has high ash and alkali content in comparison with a typical wood crop and a typical C3 crop such as grass species.

3.5.1.3.2 Ash Analyses of Amaranth

Data regarding ash content with its chemical components are among the most important for thorough analysis and classification of potential biomass as a fuel.

3.5.1.3.3 Ash Content

Another aspect of biofuel quality is mineral content. Silicon and other mineral contents are important in that they affect quantity and quality of ash; therefore, these values should be clearly defined.

TABLE 3.7
Results of Lab-Analyses of Amaranth Cruentus "Giganteus" Phytomass

Measured Value	Fuel Type	Unit	Amaranth Matter, 2001 Inflorescence and Leaves	Stems	Average	Amaranth, DM 2000 Stems
MC as received	Wt-r	%	11.80–19.70	12.80–20.10		14.70–22.40
Ave.	Wt-r	%	15.75	16.45	16.10	18.60
MC in analytical sample	W-a	%	6.94	6.18	6.56	6.05
Ash content	A-d	%	15.40	10.17	12.79	13.03
Sulfur content	St-d	%	0.57	0.11	0.34	0.18
Carbon content	C-d	%	41.00	43.50	42.25	40.00
Hydrogen content	H-d	%	5.70	6.20	5.95	5.10
Nitrogen content	N-d	%	2.50	0.80	1.65	1.20
Higher heating value Analytical sample	HHV Qs-d	MJ/kg MJ/kg	15.72	16.61	16.17	15.48
Lower heating value Dry matter, DM	LHV Qi-d	MJ/kg MJ/kg	14.48	15.26	14.87	14.35
"As received," Ave. MC	Qi-r	MJ/kg	11.81	12.34	12.08	11.23

Source: Viglasky J. et al. 2009b. *Agronomy Research*, Estonia. No. 2, vol. 7, pp. 865–873, ISSN 1406-894X.

This parameter connected with Amaranth is not common in literature and the published values are considerably different (varying from 5%–22%). This difference is significant and, if true, it is necessary to explore it by conducting further experiments.

Experimental findings of ash content in Amaranth phytomass: The fixed state of Amaranth ash was carried out on five weighed air dried samples, with material weighing from 1–5 grams, annealing at 520°C, for 48 hours. The solid matter remaining after annealing—ash—was determined by gravimetric method; the gained value was 19.09% ($S_x = 0.18$).

3.5.1.3.4 Chemical Composition of Ash

Ash from our experiments was used for fixing state of individual elements, while particular attention was paid to ascertaining the presence of heavy metal, and aimed at the following groups of elements:

1. Mercury, thallium, cadmium;
2. Arsenic, nickel, chromium, cobalt
3. Lead, copper, manganese.

These data are important in view of the proposal, respectively recommended working parameters of the technological equipment, which should use phytomass of Amaranth as feedstock and then as substrate for biofuel production.

Results of the analyses were numbered according to their original solid surplus matter, which remained after the thermo-oxidation reaction and are listed in Table 3.8 as well as plotted by a column graph in Figure 3.8.

3.5.1.4 Discussions

The analyses of Amaranth samples were directed to products that arise during annealing and remain in the experimental system as solid material. The experiment itself was carried out under laboratory conditions and should be the basis for realization of larger experiments in the quarter-operating capacity. At this planned experiment, a fully-fledged analysis should take place, which would be directed also to the study of gas production, which occurs during the process of thermal oxidation.

Analyses of solid matter after thermo-oxidation were made. In the case of the utilization of parts of Amaranth for technical purposes in production of heat energy, occurrence of solid parts can be expected. These, from the point of view of evaluation of technological processes affecting environmental protection, are part of solid emissions and also solid waste, which belong within the category of dangerous waste (Anonymous, 1991, 1996a,b) in waste management.

The realized measurements and analyses showed that, in solid matter after thermo-oxidation, compound elements are found that have unfavorable physiological effects in relationship to the health of human organisms. From among the elements with cancerous characteristics, cadmium (Cd) was found in solid matter (in ash) in concentrations of 1 mg·kg^{-1}; furthermore, chromium (Cr^{VI}), cobalt (Co), and nickel (Ni) under the laws regarding the atmosphere (Anonymous, 1991) and its strict

TABLE 3.8
Representation of Chosen Elements in Water Lye and Lye of 3% Nitric Acid from Solid Matter After Sample Annealing at Temperatures 800–850°C Calculation for 1 kg of Solid Residue—Ash [mg·kg^{-1}]

No.	Element	Element Concentration in Solid Surplus From Water Lye	From Lye with 3% NO$_3$	Sum [mg·kg^{-1}]	Emission Limit [mg·m^{-3}]
1.	F	<0.05	<0.05	<0.05	10.0
2.	Pb	<0.10	1.25	1.25	5.0
3.	Cd	<0.20	1.00	1.00	0.2
4.	Cr	<0.02	2.50	2.50	5.0 (no Cr^{VI})
5.	Co	0.50	5.05	5.05	2.0
6.	Cu	<0.30	11.28	11.28	5.0
7.	Ni	<0.10	4.90	4.90	2.0
8.	Zn	<0.10	34.30	34.30	5.0
9.	As	<3.00	<3.00	<3.00	2.0
10.	Sb	<0.50	<0.50	<0.50	5.0
11.	Sn	<1.00	<1.00	<1.00	5.0
12.	Se	<5.00	<5.00	<5.00	2.0
13.	Hg	0.0134	0.005	0.0184	0.2
14.	B	1.00	8.25	9.25	–
15.	Conductivity	5.21 µS·cm^{-1}	–	–	–
16.	pH	11.13	–	–	–

Source: Blaho J., Viglasky J. 1997. In: *Proceedings of the Conference Biologization of Plant Productions VII.* Slovak University of Agriculture. Nitra, 1996/7, pp. 5–7.

claim in G SR No. 92/96 (Anonymous, 1996a,b) belong to Group 1, that is, polluted materials with cancerous activity. As with Cd, their content is limited by burning at 0.2 mg·m^{-3}, which is also true for Cr, Co, and Ni at 2 mg·m^{-3}.

Another noticeable and dangerous element that was determined was mercury (Hg); its concentration was found at 13.4 µg·kg^{-1}, which is unfavorable; additionally, cobalt and mercury were found in parts and in water soluble form. Further observed elements that are regulated by law where they are found in the atmosphere and in waste materials are copper, lead, zinc, and chromium (Cr^{III}). These elements, as emissions, belong to Group 3, under Group 2 polluted materials; their emission limit is fixed at 5 mg·m^{-3} in relief gas. The most common concentration reached up to 50 mg·m^{-3} in ash in of the given Amaranth samples, which were cultivated in soil with higher contamination level, but it is still within limit of contamination content for soil in Slovakia.

The results of the analyses, aimed at fixing concentrations of chosen heavy metals in solid matter after thermo-oxidation of Amaranth phytomass, proved that this plant has

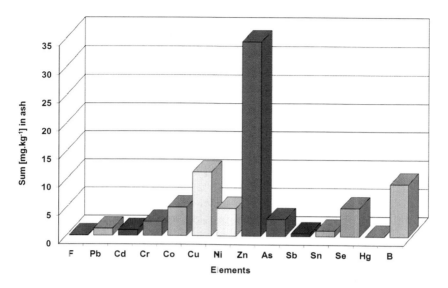

FIGURE 3.8 Representation of chosen elements in water lye and lye of 3% nitric acid from solid matter after sample annealing at temperatures 800–850°C calculation for 1 kg of solid residues—ash [mg·kg^{-1}]. (From Blaho J., Viglasky J. 1997. In: *Proceedings of the Conference Biologization of Plant Productions VII*. Slovak University of Agriculture. Nitra, 1996/7, pp. 5–7.)

a tendency to absorb some heavy metals from the surrounding soil in which it grows. Owing to their physiological activity, heavy metals are dangerous for living organisms.

These results were confirmed by independent measuring carried out at the Department of Gardening of the Agronomy Faculty, Slovak University of Agriculture in Nitra, Slovakia. The content of heavy metals was observed under two pedological conditions: in garden soil and in contaminated soil from the region of Rudnan, Slovakia (Kona, 1995). Cultivation was compared also with green lettuce for eating. These experiments and results showed that Amaranth can be listed among plants that have a greater ability to accumulate up to: 103 times more Pb, 240 times more Cd, and 5.9 times more Hg, than Lettuce (bot: *Lactuca* or *Lactuca sativa*). However, the presence of contamination in seeds was low and was within the norm in all tests. This is important, especially when considering the use of Amaranth seeds in the food industry. The ability of Amaranth to accumulate heavy metals has not been evaluated in other Slovak and foreign literature.

Equal success can be evaluated for growing Amaranth in regions with burnt fuel containing oxides of sulfur. Tests with Amaranth were carried out on land near a chemical factory, where ammonium sulfate is produced. Their results showed that the growing of Amaranth is suitable under these conditions (Veresova and Hoffmanova, 1995).

3.5.1.5 Conclusions
- Amaranth could prove to be a very attractive biomass-phytomass source because of its high yield under marginal conditions.

- The Amaranth agroenvironmental system is a key link in the sustainable production of agriculture. It will play an important role as a raw material source for industrial biofuel production as well as for environmental protection in this century.
- Energy generated from Amaranth-based biofuel has the potential to reduce greenhouse gas (CO_2) emissions and to decrease dependence on diminishing supplies of fossil fuels.
- Thermal conversion quality is determined by many factors, including water content at harvest, and ash, alkali, Cl, and N content.
- HHV of Amaranth DM varies from 15.5 to 17.0 MJ·kg^{-1}; LHV is range 13–14 MJ·kg^{-1} at 10% MC.
- Limited data available show that as a low input C_4 plant with a low water requirement, Amaranth has high ash and alkali content in comparison with a typical wood crop and a typical C_3 crop such as grass species.
- Ash content of investigated Amaranth species varies from 12% to 22%; in reality, it is influenced by the soil in which the plant was grown or cultivated.
- At present, economic viability is still uncertain, as is the case for all biomass crops.
- It will be necessary to exploit the multifunctional uses of Amaranth crop species to increase the value per area of land and/or per ton of biomass-phytomass.
- Many successful applications in food production, in industrial as well as the energy sector of Amaranth show promise, although further research still remains to be carried out.

These are the main reasons for increasing interest in exploring the phytomass quality of different Amaranth species, especially giant Amaranth (*Amarantus* L.), and to receive their definitive taxonomic studies. Giant Amaranth species for example, *Amaranthus Australis L.* or *Amaranthus Cruentus L.*(it was confirmed by cultivation of 30 years in Slovakia) are one of the main crops being considered as a source of raw material for solid biomass-based production processes to acquire one of its main products, that is, high quality biofuel.

3.6 ENVIRONMENTAL ASPECTS OF BIOMASS GASIFICATION

3.6.1 Introduction

Biomass,* with regard to its availability and possibility of new technology use, seems to be the most important renewable energy source from an economical, energy, and political point of view, considering quality of biofuel—level of MC by lowering 10% resulting in 50% decreased emissions as mentioned in Table 3.6, for example CO, NOx,

---

* Biomass is defined as all forms of plant-derived matter (terrestrial and aquatic) other than that which has been fossilized. This includes dedicated agricultural and forest products, for example, firewood (15% from harvested wood), sugarcane, rape seed, etc., agricultural and forestry residues, animal wastes—both dung and abattoir-derived—municipal solid waste (MSW) including sewage, and peat where its use can be demonstrated to be renewable.

TOC, in most countries. In terms of size of resources, there is the potential to produce at least 50% of Europe's total energy requirement, from purpose-grown biomass using agricultural land no longer required for food, and from wastes and residues from agriculture, commerce, and consumers (Kaltschmitt and Dinkelbach, 1997).

3.6.1.1 Biomass Power System's Environmental Characterization

Power production from biomass and its effect on the environment extend beyond the power plant gate to incorporate the full fuel cycle of production/procurement, harvesting, processing, and delivery. These operations are analogous to the mining, drilling, refining, and delivery of fossil fuels for power production. Figure 3.9 depicts current and future options for various biomass power operations (Viglasky, 1998).

For the power producer, biomass fuels can be produced and harvested as energy crops from a dedicated feedstock supply system (DFSS), or purchased as urban wood waste or residues from the wood products industry (e.g., mill shavings). Such feedstocks may be chipped or shredded on the fuel supplier's site prior to transportation to the generating station. Once at the plant, the fuel may undergo further processing prior to combustion for power generation. Today's systems utilize wood residues and wastes in a direct-fired Rankine cycle. Future systems will use DFSS energy crops with an intermediate thermochemical process (e.g., gasification, pyrolysis) to produce fuel gas or biocrude oil which can be fired in high-efficiency, gas turbine-based power cycles (e.g., Integrated gasification combined cycle [IGCC]; HATCH – Innovation Company dealing with design, engineering, procurement, and construction management of renewable bio-systems, Canada).

In both near and long-term, it will be the acceptability of environmental discharges and impacts from the entire production-generation fuel cycle which will determine

1. Production
- Woody energy crops
- Herbaceous energy crops
- Wood products industry residues (untreated)
- Urban wood waste (treated)

2. Harvesting
- Woody energy crops
- Herbaceous energy crops

3. Field Preprocessing
- Chipping
- Baling

4. Transportation
- Truck
- Rail
- Barge

5. Storage
- Open
- Covered

6. Processing
- Drying
- Chipping / Hogging
- Pelletizing
- Separation
- Classifying
- Screening
- Gasification
- Pyrolysis

7. Power Generation
- Combustion boilers
- Fluidized bed combustion boilers (FBCs)
- Whole tree burner
- Simple cycle gas turbine
- Steam-injected gas turbine (STIG)
- Combined cycles (CHP)
- Fuel cells
- Advanced power cycles utilizing gasification technologies (IGCC - Integrated Gasification Combined Cycles, HATC - Hot Air Turbine Cycles).

FIGURE 3.9 Technology options.

the success of biomass power. It can be assumed that within this decade biomass will mostly be burnt. However, at the beginning of the third millennium it is more than likely to find biomass as the base raw material for production of high-grade fuels. The most promising and intensive research and development works throughout the world are aimed at the improvement of the biomass gasification process (van Ree, 1994).

3.6.2 Methods of Systematic Environmental Assessment, Applied to Biomass Gasification Technologies

An analytical framework that breaks down the environmental assessment problem into steps is a useful aid for communicating the complexity of the problem and for characterizing the coverage of existing data and literature.

The operation of a gasification plant can result in occupational health and safety hazards unless adequate and effective preventive measures are taken and continuously enforced. A gasification system comprises:

- Fuel storage, handling, and feeding systems
- The gasifier, gas cooling, and cleaning equipment
- Utilization of the gas.

Each part of the plant creates specific occupational, health, and safety hazards. Table 3.9 describes the main environmental concerns and major hazards associated with the operation of a gasification system.

TABLE 3.9
Environmental Aspects of Gasification Systems

Process Activity	Fuel Preparation	Feeding System	Gasifier	Gas Cleaning	Gas Utilization
Environmental Concerns					
Dust	•	•		•	
Noise	•	•	•	•	•
Odor	•		•	•	
Wastewater				•	•
Tar				•	•
Fly ash				•	
Exhaust gases					•
Hazards					
Fire	•	•	•	•	•
Dust explosion	•	•	•		
Mechanical hazard	•	•	•		•
Gas poisoning		•	•	•	•
Skin burns			•	•	•
Gas explosion			•	•	•
Gas leaks			•	•	•

This section examines the sources of environmental concerns and describes the measures that have to be taken to limit the environmental impact; it considers only in-plant environmental factors concerning gas, liquid and solid emissions and wastes. The major hazards are then discussed and safety guidelines are provided for the appropriate operation of gasifiers systems. Although there is probably nothing unusual in the problems and plant requirements, an overview of these will be helpful in plant specification and evaluation.

3.6.2.1 Environmental Aspects of Gasification Plant Operations

In comparison to fossil fueled energy generation systems, biomass gasifiers are relatively benign in their environmental emissions producing no sulfur oxides, low levels of particulate matter, and if consuming biomass produced on a sustainable basis, no net increase in global CO_2 levels.

Dust: Dust is generated during preparation, storage and handling of the feedstock, feeding, and fly ash removal by particulate collection equipment; all of these present particular problems when the solids are dry and friable. Dust generation creates several possible problems, including:

- Formation of explosive mixtures with air (a primary explosion can render the dust airborne, causing secondary explosions which can be devastating);
- Inhalation causing lung damage;
- Eye and skin irritation;
- Pollution odors from, or smouldering and ignition of, layers of combustible dust;
- Dust settlement on all exposed horizontal surfaces, leading to safety problems for personnel in routine operations, as well as increased maintenance and aesthetic detraction;
- Increased friction and wear of mechanical equipment caused by dust deposition, increasing costs and reducing reliability, both increasing the potential for accidents.

Preventive measures include:

- Minimization of solids handling and avoidance of rough handling to minimize attrition of fuel particles and suspension of dust;
- Complete enclosure of all solids handling, particularly conveying equipment at the discharge points;
- Installation of suction hoods and gas cleaning equipment to control localized dust sources, for example, mills and screens;
- Maintenance of an under pressure in enclosed environments to prevent the spreading of dust into adjacent premises, again with suitable gas cleaning equipment.

Solid particles also arise in the product gas, such as cinders, fly ash, filter dust, charcoal, fluid bed inertness, and catalyst fines. Since such sources are localized, they are in principle easier to control. Fine materials such as fly ash may need to be wetted to prevent re-entrainment during handling and disposal. Carbon formed by secondary

Environmental Impacts of Biofuel-Fired Small Boilers and Gasifiers

cracking or incomplete gasification can also form explosive mixtures with air, but it is usually contained in appropriate vessels. Charcoal from biomass can be pyrophoric and needs to be adequately cooled before discharge and storage if arising in significant amounts. Some types of gasifiers may produce hot particles, as a consequence of malfunction or equipment faults, which may ignite flammable materials and cause a fire.

From an occupational health viewpoint, dust particles may be classified by:

1. Size: particles >5 μm are arrested by wet hairs in the nostrils. Those <0.2 μm do not settle in the lungs and are breathed out again. Hence, the intermediate size range is the most dangerous.
2. Shape and composition: some materials are known to cause lung damage, for example asbestos (asbestosis) and silica (fibrillose). The latter may arise from fluid bed materials.

Dust originating from fly-ash removal may be toxic owing to adsorption of chemicals on to the dust particles. Several compounds with carcinogenic properties, such as benz[a]anthracene and benzo[a]pyrene are adsorbed on to the dust particles. They are dangerous to human health after inhalation and skin contact and/or after accumulation in the food chain. Dust particles may adsorb nonpolar organic compounds up to 40% of their weight, possibly higher for soot and carbon black with their very high specific surface areas. The dispersion of gasifier's dust may lead to air and food contamination.

Wastewater and condensates: The present trend is to produce a clean tar- and ash-free gas that may be directly burned in engine or turbine, and there is a positive effort to design and develop systems that do not produce a liquid effluent, because of the potential problems of treatment and disposal. However, if wastewater and condensates are produced during gas cooling and wet gas cleaning, they will require treatment. The condensate is known to contain, for example, acetic acid, phenols, and many other oxygenated organic compounds that may be soluble or insoluble in water. There is a risk of water pollution and adverse health effects from the suspended tars and soluble organic compounds. The condensate and wastewater consist mainly of water and can be divided into an aqueous, that is, water-soluble fraction, consisting of tars and oils. However, separation is not always simple, since wood tar tends to emulsify in the aqueous phase. The insoluble fraction consists mainly of ash and particulates, tars, phenolic compounds, and light oils.

The tars in particular, as well as the condensates, are toxic and require careful evaluation of their occupational and safety aspects. Little research has been carried out to determine the mutagenic and carcinogenic effects of biomass tar, but research on coal tar has confirmed the above reservations. It would be prudent to assume that some of the tar components may be carcinogenic. High-temperature gasifier operation can increase these problems, since the mutagenic and carcinogenic effects are related to the presence of polycyclic aromatics and their relative concentrations increase as process temperature increases. Direct contact between skin and tars or condensate, clothing and complete instruction.

Tars present an insignificant fire hazard, as their flash points are comparatively high. Tar disposal has not been examined, but it is usually assumed that it may be recycled to the gasifier or incinerated. Other disposal options are unlikely to be acceptable. No work is known to have been carried out on other uses for the tar, such as recovery of chemicals and direct applications such as road tar, although chemicals from biomass via flash pyrolysis are of major interest. There are too few substantial biomass gasification plants currently operating from which tars could be recovered, and all of the current interest is focused on production of clean gas through tar cracking rather than on removal and recovery.

Since tars are such a potential problem in wastewater and in their own right, every effort should be made to reduce their environmental impact (Viglasky and Langova, 2003): by cracking tars during or after gasification, applying hot gas clean-up and so avoiding wet gas scrubbing, and reducing gasification temperatures to limit the production of refractory polycyclic tars. Pressurized gasifiers are assumed to operate with hot gas clean-up, no wastewater, and no tar production. Atmospheric pressure gasifiers are more likely to include a wet scrubbing system, particularly if an engine is specified for power generation.

Wastewater treatment is usually assumed to be relatively simple and low cost, although there is remarkably little information on treatment methods or costs. The design of the wastewater treatment plant would be expected to rely partly on chemical treatment, such as solvent extraction for high concentrations of phenolic compounds, with incineration of recovered organics, and orthodox biological treatment of the dilute, low-BOD (biological oxygen demand) effluent. Although tar separation and recycling to the gasifier for thermal destruction have been proposed, no examples of application have been found.

Fly ash and char: Fly ash and char present similar problems to those caused by dust as described above. There is an additional risk of fire, which dictates that fly ash and char should be stored moist. Disposal of this wetted mixture presents its own environmental problems. The solids need to be separated from the water in a water treatment facility. Extracted water will be contaminated and may require further treatment before discharge, using orthodox water treatment technology. There are no known special problems. The solid fraction should be considered an industrial waste and discharged accordingly to licensed landfill sites.

Odor problems: Odors may arise because of the degradation of organic matter (e.g., in refuse or sewage gasification), the occurrence of even minute gas leaks, the handling and storage of tar, wastewater, fly ash, and other by-products. Wood tar has a strong, characteristic and persistent odor, even in minute concentrations. The smell of coal tar is somewhat aromatic owing to the presence of naphthalene, anthracene, and phenanthrene. Tar derived from lignocellulosic feedstocks is more pungent.

When sulfur- or nitrogen-containing feedstocks are used, the producer gas also contains odorous gases, such as H_2S, COS or NH_3. The tar and wastewater may be contaminated by even more strongly smelling organic sulfur and nitrogen compounds, although this is unlikely to be a problem with biomass-derived products, owing to the very low levels of sulfur and nitrogen. Some waste materials such as sewage sludge may cause such problems.

Noise: Noise is produced whenever a mechanical part or an engine or motor is in operation. Particular plant areas where noise levels are likely to be significant are reception, storage and handling equipment, the feeding systems, the compressors, the gas motor or turbine.

Environmental Impacts of Biofuel-Fired Small Boilers and Gasifiers

The effects on humans of prolonged exposure to noise are well documented. Adequate measures must be taken to minimize noise, for example, by using sound- and vibration-absorbent materials between supports or acoustic enclosure. Operators are also required to be provided with ear protection plugs.

3.6.2.2 Hazards of Gasifier Plant Operation

Gasification technologies and operation of systems concerned are not risk free, although they are taken as considerably less detrimental than conventional fossil fuel-fired power systems.

Combustible gases and vapors: A flammable gas is combustible only within a certain range of concentrations, bounded by the LEL (lower explosion limit) and the UEL (upper explosion limit). Below the LEL, the mixture is too lean to sustain combustion. Above the UEL, the reaction stops because of a deficiency in oxygen. In both cases, the generation of heat becomes too slow to give rise to the characteristic acceleration in reaction rates which marks the start of an explosion. The range between the LEL and UEL values depends on the reactivity of the flammable compound or mixture. It widens when the flammable gas or the combustion air is preheated or under pressure. Some data are given in Table 3.10 (Directive 2009/125/EC).

Explosive mixtures could arise in two situations:

1. Air leaks into the gasifier plant as a consequence of a reduction in operating pressure. Reduced pressure may arise owing to rapid cooling, condensation of vapor such as water, chimney effects, or the suction of an induced-draft fan or of an engine.
2. Fuel gas leaks out of the gasifier plant into a confined space, thus building up a substantial concentration in an enclosed space. A source of ignition is necessary for an explosion, so explosion-proof, flameproof or spark proof motors would be specified in any such areas. In addition, such an atmosphere is likely to present a lethal toxicity hazard from carbon monoxide, so suitable detectors should be fitted.

TABLE 3.10
Some LEL and UEL Values and Self-Ignition Temperatures in Air

Gas	LEL [vol. %]	UEL [vol. %]	Self-Ignition Temperature [°C]
Hydrogen, H_2	4.0	76.0	400
Carbon monoxide, CO	12.5	74.0	[a]
Methane, CH_4	4.6	14.2	540
Ethane, C_2H_6	3.0	12.5	515
Ethylene (ethene), C_2H_4	3.1	32.0	490
Propane, C_3H_8	2.2	9.5	450

Source: Knoef H.A.M., Stassen H.E. 1997. Small-Scale Biomass Gasifier Monitoring—Findings and Gasifier Performance. BGMP Report No.1129/Volume I&II, University of Twente—TEI/BTG B.V., NL-Enschede. Prepared for The World Bank and UNDP. 1995–7, 54/233 p.

[a] The value is highly dependent on the presence of traces of moisture

When a flammable mixture of gas and air is formed, an explosion may occur if the mixture is ignited. Ignition may occur as a result of static electricity, sparking equipment such as motors, or contact with a hot surface. In view of the wide explosion limits of the main components of producer gas—hydrogen and carbon monoxide—the accidental formation of explosive gas mixtures should be prevented. Mixtures of producer gas with oxygen-enriched air or pure oxygen have a higher UEL than do mixtures with air. The LEL does not change significantly. This means that oxygen gasifiers present an even higher explosion risk, such as when oxygen breaks through the fuel layer or there is a perturbation in the fuel supply.

Combustible dusts: Combustible solids such as wood, flour, and coal in very small particles can also form explosive mixtures with air within certain concentration limits, usually ranging from 20 to 50 mg·m^{-3} for LEL and 2 to 6 g·m^{-3} for UEL. Numerous carbohydrate materials, including starch, sugar, and wood flour, have given rise to extremely destructive explosions.

Fire risks: The main fire risks in gasifier systems are associated with fuel storage, combustible dusts formed in fuel comminution, fuel drying (in forced-draft conditions a fire is likely to expand quickly), ignition procedure (especially for moving-bed gasifiers), and the product gas.

There are also the usual risks associated with any construction involving a thermal unit. Local rules and guidelines should be followed for construction and materials of buildings. Adequate means for firefighting should be provided and the gasifier operators should be well acquainted with their existence, location, and operating instructions. Such fires can be avoided by proper procedures and proper layout of the plant.

Carbon monoxide poisoning: Carbon monoxide is a major constituent of producer gas and is by far the most common cause of gas poisoning. It is particularly insidious owing to its lack of color or smell. The accepted threshold limit value (TLV) is 50 ppm CO (0.005 vol.%), although concentration and exposure are closely linked. There is extensive documentation available on the effects, treatment, and controls laid down by the relevant statutory authorities. Since carbon monoxide is an odorless, colorless gas, it may be detected only through instrumentation, and personal detectors are necessary when working in confined spaces. All operating personnel should be aware of the hazards presented by the gas. The best way of avoiding the risk of CO poisoning is to build the gas generator in the open with the minimum of containment and with adequate ventilation, particularly where gases may collect.

Other toxic compounds: It is well established that extremely toxic dioxins and furans (polychlorinated dibenzodioxins [PCDDs], or simply dioxins, are a group of polyhalogenated organic compounds that are significant environmental pollutants; and polychlorinated dibenzofurans [PCDFs] are a family of organic compounds with one or several of the hydrogens in the dibenzofuran structure replaced by chlorines) are formed during most combustion and gasification processes when some chlorine (Cl) is present. Under normal operating conditions the concentrations of these compounds in wood-fired units are extremely small, although pesticide-treated wood and waste materials can provide the source of Cl necessary for their formation. A close-coupled IGCC (integrated gasification combined cycle) system normally provides satisfactory operating conditions for thermal destruction of these compounds and it is generally believed that they do not present any problem, though very few data exist to support this.

Other hazards: Other hazards include skin burns, mechanical hazards, and electrical hazards. Some of the surfaces of the gasifier, the cyclones, the gas lines, the engine, and its exhaust may get sufficiently hot during operation to create the hazard of skin burns. For permanent stationary installations, such surfaces should be insulated to protect the operators and also to reduce heat losses. Covers or rails should be installed to keep personnel at a safe distance. All equipment with moving parts such as blowers, fans, screw conveyors, front-end loaders, and pretreatment and feeding equipment present a hazard from moving parts and should be suitably protected. All electric appliances provide the potential for electric shocks, and suitable precautions need to be taken, using standard procedures and equipment specifications. Any work on elevated equipment involves the hazard of possible falls. Similarly there is some hazard from falling objects, tripping on hot equipment, or slipping on oil-stained floors. Unauthorized or untrained personnel should be prevented from entering the plant, to reduce the risk of injury through improper use of equipment or facilities. For all "normal" hazards that may arise in a process plant there are well-documented and statutory requirements.

3.6.3 Examples of Environmentally Advanced Biomass Gasification-to-Electricity Systems

3.6.3.1 Technology Overview

Gasification combined cycles (IGCC, including variants such as the STIG or steam-injection gas turbine) are discussed in this part as the basic configuration for environmentally advanced, high-efficiency biomass conversion power system with larger-scale capacity. The data available characterizing environmental impacts of this conversion technology are very limited and results presented are used to suggest trends and possible concerns.

Process description: Gasification combined cycles increase the efficiency of power generation by firing gas turbines at temperatures up to 1260°C and capturing the gas turbine exhaust heat in a steam Rankine cycle for overall power cycle efficiencies near 50%. Combined cycles are now widely used in natural gas-fired systems for intermediate or base loads. The additional subsystem of the biomass-fired power system is the gasification step for converting solid biomass fuels to combustion turbine-quality fuel gases. By close coupling of the gasification and power cycle equipment, thermal energy of the hot fuel gas is put to work in the power cycle. The primary unit processes include:

- Gasification including fuel feed, metering and flow control, the gasifier reactor, booster air compressors and piping, blast steam piping and controls.
- Fuel gas cleaning including hot gas clean-up systems and "quench and scrub" designs.
- Gas turbine including combustors, power turbines and compressors, exhaust ducting, and power generator.
- Heat recovery steam generator (HRSG).
- Steam turbine including power turbines, generator.
- Cooling tower and exhaust stack.

Gasification is a two-step, endothermic process where a solid fuel (biomass or coal) is thermochemically converted into a low- or medium-caloric gas. In the first reaction, pyrolysis, the volatile components of the fuel are vaporized at temperatures below 600°C by a set of complex reactions. Included in the volatile vapors are hydrocarbon gases, hydrogen, CO, CO_2, tar, and water vapor. Because biomass fuels tend to have more volatile components (70%–86% on a dry basis) than coal (30%), pyrolysis plays a larger role in biomass gasification than in coal gasification. Char (fixed carbon) and ash are the by-products of pyrolysis that are not vaporized. The second process, char conversion, involves the gasification and/or combustion of the carbon that remains after pyrolysis. In this reaction, a portion of the char is burned to provide the heat required for pyrolysis and for gasification of any remaining char. Converting solid biomass into a gas suitable for gas turbine operation creates the potential to integrate biomass gasifiers with the simple cycle gas turbine and its variations (combined cycle—IGCC, STIG, etc.). Close coupling of gasification and the power system increases overall conversion efficiency by utilizing both the thermal and chemical energy of hot product gases to fuel the power cycle. Combined cycles, with their high efficiency and low emissions characteristics, are a prime choice for biomass gasification systems. An alternative to hot gas clean-up is a quench and scrub process that removes particulates and metals prior to combustion. This approach results in an efficiency penalty caused by cooling the hot fuel gas. It also increases wastewater flows from the system. The impacts of scrubber wastewater on plant wastewater discharges have not been evaluated. On a smaller scale, STIGs offer a variant of the combined steam and gas cycle configuration. This configuration reduces system complexity by using a single turbine through which both gas and steam are expanded for power. Unlike the combined cycle, steam expanded in the turbine is not captured, condensed, and returned to the steam generator.

Environmental impacts: Combined gas and steam cycles, along with their variants (e.g., IGCC, STIG, and humid air turbine), offer power cycle efficiencies approaching 50%* (6800 kJ/kWh) (Consonni and Larson, 1994). The efficiency gains of the combined cycle directly reduce the levels of environmental discharges and other impacts per unit electrical output in all aspects of biomass power production, from harvesting through stack pollution emissions. The second environmental advantage for the IGCC configuration is also its most difficult technical development problem. To meet the fuel gas specifications of the gas turbine, the raw fuel gas must be cleaned of several contaminants. This cleaning process for the turbine coincidentally removes many of the potential environmentally harmful emissions.

The primary environmental disadvantage of the biomass-fired IGCC is the additional NO_x arising from fuel-bound nitrogen in this high temperature, efficient thermodynamic cycle. The total NO_x emissions (thermal and fuel-bound) will always exceed the NO_x produced by natural gas firing of the same equipment at the same turbine inlet temperature and with comparable NO_x controls. Current regulatory standards for combustion turbines are based on natural gas and oil firing. A special concern for the STIG cycle is the relatively large amounts of demineralized water

---

* All gas turbine efficiencies in this section are based on operation with natural gas unless otherwise noted. This is a consistent standard of comparison for combustion turbines.

Environmental Impacts of Biofuel-Fired Small Boilers and Gasifiers

used in the open cycle. Water is primarily a conservation issue rather than an environmental issue, although disposal of water treatment sludge can be a concern.

3.6.3.2 Environmental Discharges and Controls

Figure 3.10 lists chemicals/materials generated in biomass IGCC systems which, if present in sufficient concentrations in certain waste streams, can trip regulatory limits for environmental protection. These materials either directly enter the power conversion system in the fuel and in other chemicals used in operations or are generated in the power generation process (primarily combustion). They tend to be regulated in terms of their concentrations in the plant waste streams shown at the right of the diagram. In the paragraphs that follow, the materials of possible concern are discussed in terms of how they are generated, factors affecting their generation, and their relative impacts on the plant waste streams. The discussion of the waste streams follows, with an evaluation of the potential for waste stream constituents to trip restrictive regulatory limits.

Primary discharges of concern for the gasification combined cycle are air emissions (especially NO_x), and solid wastes, particularly gasifier-generated ash. For IGCC,

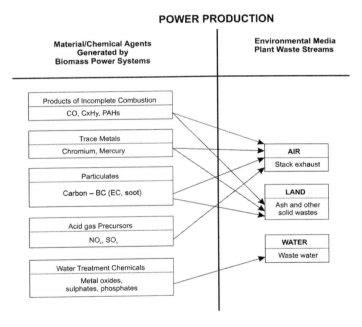

FIGURE 3.10 Evaluated materials for environmental impacts. PAHs—polycyclic aromatic hydrocarbons. Particulates: Carbon—BC (EC, soot); solid biofuels combustion is related to three basic types of primary particles, which are summarized as "salts," "soot," and Condensable Organic Compounds "COC." Organic matter (OM) can be either primary or secondary, the latter part deriving from the oxidation of VOCs; organic material in the atmosphere may either be biogenic or anthropogenic. Organic matter influences the atmospheric radiation field by both scattering and absorption. PM is considered to be one of the most significant pollutants produced from biomass combustion.

water discharge concerns are comparable to those faced by modern coal-fired boilers including discharges from ash cooling and transport, boiler blow down, and water treatment processes. Other potential air pollutants such as particulates, alkali metals, and hazardous trace elements tend to be removed prior to the gas turbine to protect the hot sections of the turbine from erosion and corrosion. Thus, environmental controls for these pollutants serve a dual purpose. Turbine fuel specifications are generally sufficiently stringent concerning these pollutants to require controls that also meet current environmental criteria. This situation could change in the future, especially concerning air pollutants.

Possible environmental control processes include the following:

- Cyclone or ceramic filter particulate control at the gasifier outlet
- Gas stream cooling and removal of condensable with particulates
- Staged combustion at the combustor, or urea/ammonia injection in the exhaust stream with or without catalysts for NO_x control
- Conventional ash treatment and disposal systems
- Use of sorbents to remove contaminants in fluidized-bed gasifiers.

Gasification combined cycles for biomass are a developmental technology. Existing data on discharges are largely from experimental or process development equipment. In the case of recent biomass gasification tests at General Electric (GE) facilities, the Process Development Unit (PDU) is one developed for coal gasification, (Furman et al., 1992). Fuel gas was flared rather than used to fire the gas turbine simulator at the facility. Although the results of these tests provide better guidance than combustion models, they do not yet represent a data set for a full scale plant under typical operation conditions.

Nitrogen oxides: The constituents of primary concern for the gasification combined cycle are the nitrogen oxides. Sources include both FBN (fuel-bound nitrogen), and nitrogen in the oxidant, which will be air unless oxygen-blown gasifiers are used. Based on the composition of the fuel gas produced by the GE fixed-bed gasifier, GE estimates of fuel-bound and thermal NO_x generated in a conventional gas turbine combustor are shown in Table 3.11.

These preliminary estimates indicate that FBN-generated NO_x will be the predominant source of NO_x emissions by a factor of 10 or more. Compared to natural gas-fired combustion turbine emissions with no FBN sources, biomass will generate significant levels of NO_x unless NO_x controls are applied either pre- or post-combustion. Table 3.11 also shows anticipated removal efficiencies for NO_x controls in use with other combustion systems.

The higher the combustion temperature at which the fuel gas is burned, the more thermal NO_x is produced. Therefore, thermal NO_x production can be reduced by lowering the peak temperatures during combustion. The most common way of doing this is through steam injection (the technique also used for power augmentation in the STIG). Steam (or water) injection is capable of reducing NO_x from natural gas-fired turbines to below 25 ppmvd (parts per million by volume, dry basis) computed at 15% exhaust gas oxygen.

TABLE 3.11
NO$_x$ Emissions for an Integrated Gasification Combined Cycle (IGCC)

Fuel Gas	Gasified Mill Waste
Turbine inlet temperature (°C)	1175–1180
FBNO$_x$[a]	200–300 ppmv
Thermal NO$_x$[a]	25 ppmv
Total NO$_x$[a]	225–325 ppmv
Control technology	Selective catalytic reduction (SCR)
Removal efficiency	90%
Exhaust emissions	22–32 ppmv

Source: Directive 2009/125/EC. Commission Regulation (EU) 2015/1185 of 28 April 2015 implementing Directive 2009/125/EC of the European Parliament and of the Council with regard to ecodesign requirements for solid fuel local space heaters, that will be valid from January 1st, 2022. Also labeling of these devices will be introduced then.

[a] Preliminary estimates based on fuel gas composition from gasifier tests.

Unlike thermal NO$_x$, FBNO$_x$ (fuel-bound NO$_x$) cannot be controlled by reducing flame temperature. Although the FBN in biomass fuels is less than in coal-derived gases, levels of FBNO$_x$ emitted from gases may exceed emission regulations. One commercially available process which can help eliminate both thermal and FBNO$_x$ is selective catalytic reduction (SCR). In this process, ammonia, which is diluted with air or water, is injected into the flue gas stream as it enters the catalyst cells. As the combustion gas passes through the cells, the nitrogen oxides are converted into nitrogen and water vapor. This process has been proven to remove up to 90% of the nitrogen oxides from flue gas streams, (Valentiny, 1991).

Particulates: Particulates in a gas turbine present erosion problems for the turbine blades. Although there are more particulates in product gases from the fluidized-bed gasifier than from the fixed-bed system, these particles tend to be larger, and therefore easier to remove. For these larger particles, cyclones are an effective means of removal. The cyclone removal process is less effective in capturing the lighter and smaller portions of the wood particulates. For these smaller particles, barrier filters offer more complete particulate removal, but have not been proven for integrated gasification gas turbine applications. Filtering systems being developed that show promise are silicon carbide candle and ceramic filters. The removal efficiency of these systems is well in excess of 85%–99%; however, their success will ultimately depend on their survival in gas turbine systems.

Trace metals: Trace metals in the fuel will be concentrated in particulates (as in the FBC or fluidized-bed combustor) with similar concentrations for fluidized-bed gasifiers. Trace elements such as alkali metals found in the raw fuel gas pose a special problem. Potassium and sodium in the feedstock are of primary concern because they accelerate the hot corrosion and cementing of particulates on turbine blades. Preliminary calculations suggest that in the gasifier operating temperature ranges, alkali levels will be greater than is acceptable for gasifier/gas turbine systems.

Stack emissions: The rules governing air emissions specifically for the biomass IGCC have yet to be written. Thus, the regulatory limits discussed in this section are those developed for conventional utility plants and, with some exceptions, serve simply as guides to the possible requirements that will be imposed on the technology. As explained previously, there are no measured data for biomass IGCC emissions, only rough engineering estimates. This fact further complicates the process of analyzing the capability of the IGCC technology to meet regulatory requirements.

Solid waste stream: Ash discharges and other solid wastes (tramp materials in the fuels and sludge from make-up water treatment and cooling towers) are expected to be about the same as the FBC boiler on a basis of volume of discharge per kJ of fuel input. However, on a unit output basis, the higher efficiency of the IGCC results in lower discharges per kW·h produced.

Water discharges: The levels of contaminants in water discharges for the IGCC plant are expected to be comparable to those of conventional boilers. FBC boilers are applicable to the IGCC where the major source of water discharges is the heat recovery steam generator and steam turbine loop.

Water supply issues: Raw water requirements for a combined cycle (CC) unit are approximately 11–15 L/min/MW (including steam injection for NO_x control), significantly less than the approximately 38 L/min/MW required for a traditional Rankine steam cycle. The difference is attributed to the power produced by the gas turbine; except for steam injection for NO_x control (14%–18% of a CC's requirement), the gas turbine requires no significant water supply. Nearly all of the required supply is make-up water for the conventional cooling tower associated with the steam turbine bottoming cycle. For the CC, remaining water requirements (fire system, demineralization regeneration water, etc.) and conservation methods (dry cooling, etc.) are identical to those for the steam cycle.

3.7 DISCUSSION

The IGCC system is in an early developmental stage for biomass. There is no existing experimental system as there is for coal at Coolwater, California, from which early environmental monitoring results are available. Thus nearly all of the preliminary findings presented above for air emissions will need to be verified and NO_x should receive priority. In the interim, laboratory simulations of gasifier operations feeding a typical combustion turbine combustor would provide important indicators of the potential problems and indicate the need for research on emissions controls.

To develop a bioenergy environmental strategy requires some understanding of both the value that we, as a society, place on the environment and what the potential environmental impacts of bioenergy systems are. The value we place on the environment is not yet fully realized in our current market system, although attempts have been made to value environmental externalities and include these values in, for example, new electric power generation. Unfortunately, the linkages between the environmental emissions and actual damages are difficult to quantify.

3.7.1 POTENTIAL APPLICATIONS FOR GASIFICATION

Gasification is an efficient process to obtain valuable products from biomass with several potential applications, which has received increasing attention over recent decades.

Further development of gasification technology requires innovative and economical gasification methods with high efficiencies, and various conventional mechanisms of biomass gasification as well as new technologies. In fact, the increasing attention on renewable resources is driven by climate change owing to GHG emissions caused by the widespread utilization of conventional fossil fuels, while biomass gasification is considered as a potentially sustainable and environmentally-friendly technology. Nevertheless, social and environmental aspects should also be taken into account when designing such facilities, to guarantee the sustainable use of biomass, and considering various feedstocks.

3.7.1.1 Small-Scale Downdraft Gasification Systems

Since the middle of the 1980s, the Technical University in Zvolen (the former University of Forestry and Wood Technology was renamed as the Technical University in Zvolen, from December 17, 1991) has carried out research work in the field of the gasification of biomass and exploitation of generator gas as biofuel or chemicals for several applications.

3.7.1.2 Woody Generator Gas as an Alternative: Renewable Energy Carrier

Artificial drying of lumber is one of the major components of heat consumption in wood industry. It is generally known as having high energy ratio, that is, up to 60%–70% of the total energy used for the production of lumber is consumed mainly for drying lumber drying. Wood drying, as any other goods drying, needs a large quantity of energy. Therefore, we were looking for an alternative energy carrier suitable substitute gas to natural gas, and we decided to design a gasifier of wood gas with calorific value varying from 4.5 to 6.1 $MJ.nm^{-3}$. The pilot gasification plant was projected for a downdraft gasifier fired with wood chips with a fuel effect of 1.0 MW, see the Figure 3.11. We only have to remind ourselves that in a dryer of 50 m^3 of pine, for example, we can evacuate, that is to say transform from liquid state to the gaseous state, from 20 to 25 tons of water and in a dryer of 100 m^3 of oak, approximately 50 tons. This represents, according to the outside conditions and temperatures, about 28,000 kWh and in the latter case 65,000 kWh. Therefore, we have striven to find effective and less consuming drying systems as well as to look for some new, renewable energy resources for wood during processes and to emphasize their ecological reliability.

Therefore, in order to be able to indicate and determine ways in which to improve operation efficiency, decrease overall costs of drying, achieve satisfactory quality of dried lumber, and at the same time implement the technology in environmentally-friendly manner, it is necessary to specify both in a complex way and in detail the currently very topical process of lumber drying.

One of the possibilities to save fossil fuels spent for lumber drying is the use of generator gas produced as a result of woody biomass—wood residue or energy

FIGURE 3.11 Experimental Downdraft gasifier of the Technical University in Zvolen, TUZVO, Slovakia. The gasification group of TUZVO has developed and constructed this pilot gasification plant, mainly for long-term testing of the gasifier, and testing of essential (material) components. The main advantage of a downdraft gasifier is the production of a gas with a low tar content (<100 mg tar.Nm-3), this type of gasifiers is used in power production applications in a range from 80 up to 500 kWe. (From Viglasky J., Langova N., 2003. Mathematical Model for Efficiency Tar-Removal from Generator Gas I. In: *Acta Facultatis Technicae Zvolen* VI. No. 1, Technical University in Zvolen, - ISBN 80-228-1256-2. pp. 113–120.)

woodchip gasification—as an alternative fuel for flue-gas-based lumber-drying kilns. In economical and energy/policy sense, biomass, owing to its availability and flexibility in terms of usage of new technology, seems to be the most important and at the same time in our environment, the most promising renewable energy resource. Energy extraction from biomass is compatible with the carbon dioxide cycle in nature and it does not increase the concentration of this gas in the air, hence it does not worsen the "greenhouse effect."

3.7.1.3 Direct-Fired Heating System of Wood Drying Kilns

Dryers with direct-fired heating systems are heated by combustible gases, which are introduced directly in the drying kiln. This principle is not new; it has worked with many advantages and disadvantages for a long time in many countries around the world. First, dryers with direct-fired systems were featured by the fact that the ashes, carried by the hot gases, form a deposit on the lumber and give to it a gray color. This settled ash has a poor appearance disadvantage if the lumber is processed or sold immediately after the drying process. Moreover, some of the ash remains in

Environmental Impacts of Biofuel-Fired Small Boilers and Gasifiers

the hot air pipes and obstruct them, so it is necessary to clean them regularly. Thus, these old types of dryers are destined to disappear and therefore should be replaced. Newly designed direct-fired heating systems should solve such problems, as follows:

1. The poor appearance of the wood is a commercial problem.
2. The defect rate owing to drying is generally higher than in dryers fed by a steam or hot water boiler, because the regulating of the wood burners and the effective distribution of hot air in the dryer it is difficult to achieve and requires constant adjustments.
3. The fire risk is higher than in other dryer types, because burning ashes can fall on the lumber and cause a fire.
4. Electric consumption is very high owing to the fans blowing hot air, therefore is recommended to find a solution, which would help to decrease electricity request.
5. Because of these problems, direct-fired heating systems at lumber dryers in many countries have been abandoned. However, the principle of direct-fired heating, because of its high reduction of investment, compared with the setting with a boiler, remains interesting for many sawmills and wood processing plants from the viewpoint of energy supplies and costs.
6. New idea would change economics of this system.
7. Operator of this system should take in account all details to find a correct solution.

3.7.1.3.1 Flue-Gas as Drying Environment

Improvement of combustion of liquid and gaseous fuels enables the wider utilization of flue-gas for creating a drying environment in materials drying technologies. It has been proved experimentally that many materials originally dried by air could be also be dried with flue-gas, without any risk of damage. Transformation of biomass into generator gas and its subsequent use as an alternative fuel for flue-gas-based drying kilns provides us with an option to save not only fossil fuels, but also the environment by using gasification technologies. This technology enables us to use effectively also biological raw materials (such as biomass from forestry and agriculture), which are often considered to be environmentally hazardous. The usage of flue-gas directly in technologies for materials drying offers several advantages:

1. In order to use the drying kiln, a boiler house does not need to be operating.
2. No distribution piping and heating bodies are needed.
3. Equipment is much simpler.
4. Capital and operational costs per output unit of drying chamber are substantially lower.

The direct use of flue-gas in drying technology at the same time more efficiently uses the inner energetic value of the fuel, since direct drying with flue-gas does not produce heat losses, which cannot be avoided, for example, in internal or external heat carrier distribution piping, etc. As an example, we can demonstrate a comparison of heat energy from the basic fuel, which is necessary for the production of 1 GJ of heat used directly in the drying device and we will compare three different ways of drying carrier preparation:

1. Standard method of drying carrier preparation, which assumes that efficiency of the steam boiler is 80%, heat losses in distribution piping is a minimum of 5%, heat losses in the condenser are 15%, and service efficiency is 90%.
2. The direct use of flue-gas in drying technology with combustion equipment efficiency is approximately 93%.
3. The direct use of flue-gas from generator gas in the creation of drying media, which can assume that gasification efficiency is 75% and combustion chamber efficiency is 95%.

The results of effectiveness of drying media preparation implemented using the aforementioned methods are in the following Table 3.12.

The use of combustion of solid fuels (including wooden residues) for direct lumber-drying processes is nowadays unrealistic for various reasons:

1. Difficulties with perfect flue-dust free combustion.
2. Problems with combustion process regulation.
3. Unresolved automated regulation of ovens, etc.

Because of these reasons, the only feasible option for advanced flue-gas dryers are noble fuels; however, these have been recently absent from the market and their prices increase continuously. One option to resolve the aforementioned problems is the gasification of wood with subsequent combustion of the produced generator gas, thus creating a drying carrier. In addition to fuel saving, this method of preparation of drying carrier in lumber companies could resolve the issue of the utilization of woody residues, which is still often taken to waste dumps and burned outside.

The aforementioned approach also covers environmental issues by using the woody residues on the one hand, and by producing flue-gas from a gas generator (originating from wood) combustion, which does not contain any sulfuric compounds and therefore does not pollute the air, on the other hand.

A small experimental direct-fired lumber-drying kiln system including wood gas generator was designed and installed at the Technical University in Zvolen. By direct-fired heating, we must understand that hot combustion gases, which are introduced directly in the drying kiln heat dryers, are going (after a mixing with the air of

TABLE 3.12
Effectiveness of Drying Media Preparation

Mode	Calculation of Heat Energy from Fuel Needed for the Production of 1 Gj in Heat for Drying	Results GJ in fuel/GJ in heat
a.	$(\eta_B \cdot \eta_P \cdot \eta_S \cdot \eta_{SE})^{-1} = (0.80 \times 0.95 \times 0.85 \times 0.90)^{-1}$	1.720
b.	$(\eta_{CE})^{-1} = (0.93)^{-1}$	1.705
c.	$(\eta_{GG} \cdot \eta_{CCH})^{-1} = (0.75 \times 0.95)^{-1}$	1.403

Environmental Impacts of Biofuel-Fired Small Boilers and Gasifiers

the dryer) directly onto the wood. The gas combustion, with no pollution, has the characteristic of producing water vapor in large quantities.

Elements taken into account to define the drying quality include several quality controls that have been made regarding the timber-drying kiln with settings equipped with direct-fired gas heating. There are controls on the homogeneity of temperature and air speed, and on the quality of the timber wood, such as surface shakes, end shakes, knots, color of wood, final moisture homogeneity, and drying time. The experimental results obtained from this unit have shown that the drying process prepared in such a manner has no negative influence on the quality of the dried material—timber. The quality is comparable or even superior to that obtained by the conventional kiln drying method. During the experiment, a volume of 0.5 m^3 of spruce and fir timber of 32-mm thickness with the initial absolute moisture content of 51% was dried to a moisture content 29%. The average consumption of generator wood gas with a heating value of 5.45 $MJ \cdot m^{-3}$ was obtained during the drying process, 1.7 $m^3 \cdot h^{-1}$, which when converted to solid fuel represents 0.8 kg of wood residues with an absolute moisture content of 12.5%. From the energetic viewpoint, the use of generator gas for preparation of the drying medium represents a saving on the use of fossil fuels.

3.7.1.3.2 *Environmental Aspects*

The combustion of such a poor gas as a generator for woody gas produces water (145 $g \cdot m^{-3}$ of smoke), but in smaller quantities than natural gas. The composition of the smoke gases after the combustion (in volume percent) is as follows: H_2—0%; O_2—6.99%; N_2—65.4%; CO—0%; CO_2—11.33%; CH_4—0%; C_2H_2 and C_2H_4—0%; and H_2O—9.3%.

The aim of this contribution was to show its possibilities by using woody gas as a production supply of heat for the lumber-drying process.

Combustion gases as a drying medium for lumber dryers and other tasks were being researched as part of the main task of utilization of nonconventional and secondary energy sources for lumber drying and dryers.

3.7.1.3.3 *Conclusions*

One of the possibilities of how to replace, at least partly, conventional energy sources of heat in a drying process is bioenergy. Biomass could be one of the most important renewable energy sources in the Slovak Republic, as well as in other European countries.

Biomass thermochemical conversion such as gasification could offer an attractive solution: direct heating as well as "direct-indirect heating" for lumber-drying kilns. The use of diluted combustion gases in lieu of heated air as a drying-medium appeared satisfactory for the lumber-drying processes and tested lumber, as this could further enhance the economics of biomass-firing a production drier. The improvement of the drying quality is real. From the energetic viewpoint, the use of generator gas in lieu of a conventional heating system (e.g., hot water) for preparation of the drying medium in a lumber-drying kiln represents saving on the use of fossil fuels, which converted to a specific fuel represents a value of 0.33 kg for each 1 kg of evaporated water.

The results mentioned above show that some improvements could still be found before arriving at a satisfactory use of woody gas in all cases. Important technological progress could only have been made by adding the research of the direct-fired heating to those made to improve, among others, biomass gasifier designs, the drying kinetics, etc. At the present point of development, can be summed up:

1. Gas energy is easy to use, causes very little pollution, and is cheap in most countries.
2. Reduced investment in an energy system for dryers compared with other energy types (e.g., traditional boilers fired with wood residues or current technologies of dehumidifying by heat pumps).

Nevertheless, principles of "direct-fired" as well as "direct-indirect" heating systems, because of the high reduction on investment, compared with a setting with a boiler, remain interesting.

3.7.2 Future Trends

The EU wants to double the use of biomass by 2020 and also to increase the share of CHP in power generation. The design, construction and operation data of a 75 kW two-stage Viking gasifier is presented in Figure 3.12.

TK Energi A/S has developed, constructed, and tested a three-stage gasifier with dry gas cleaning. In this concept, the pyrolysis, oxidation, and reduction zones are separated. This enables optimization of each step individual and test results showed a raw gas, which is almost free of tars. The concept is quite similar to the design developed at the Laboratory for Bioenergy in Lyngby DTU, Denmark (Ahrenfeldt et al., 2006; Bentzen et al., 2004; Brandt et al., 2000; Henriksen et al., 2006; Hofmann et al., 2007; Iversen et al., 2006; Mudgal et al., 2009).

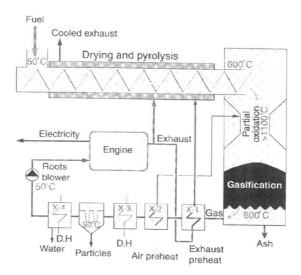

FIGURE 3.12 Flow sheet of the Viking gasifier. (Henriksen U. et al. 2006. *Energy*, vol. 31, issue 10–11, pp. 1542–1553.)

Using ash beneficially, ash valorization, or high moisture content, still poses various practical problems, although in principle, these have been worked on for more than 30 years.

Large-scale commercial torrefaction is still waiting for higher CO_2 prices within the EU Emissions Trading Systems. Large-scale pellet production is already a reality. The first commercial biomass gasifiers and pyrolyzers are in operation and their effect on biomass usage is awaited. There is currently debate about biomass sustainability and how much of it there is (ECF, 2008; Directive, 2009/125/EC).

The argument is not about the carbon in biomass, which is renewable. The argument is about the magnitude of the economic, societal, and climatic changes that increased use of biomass may cause. Biomass burning is still seen as carbon neutral by the Intergovernmental Panel on Climate Change (IPCC, 2011).

3.8 CONCLUSIONS

No energy producing technology is completely benign to the environment, but environmental impacts, if properly valued and managed, can be mitigated with minimal adverse impact on economic growth. Indeed, potential growth opportunities exist for new technologies that convert energy efficiently and at competitive costs.

Biomass-derived energy appears poised to assume an important role in the development of a sustainable global energy system. Significant environmental benefits can be obtained by using biomass fuels in gasification systems, especially IGCC, although some uncertainties still exist. Sulfur dioxide and carbon dioxide production will be far lower for biomass-fired power systems than for fossil fuel combustion and conversion systems. Emissions of potential air toxic from gasification of biomass and wastes will require further characterization, but will probably be less problematic than the air toxics emissions from fossil fuels.

The gasification system has to be designed to meet all local environmental and safety requirements. If it is operated correctly and no accidents occur, then the environmental impact will normally be acceptable. Safety design is a major consideration and there is a range of precautions that have to be included and provided for, which will be defined internationally, nationally, and locally. The general factors are outlined above but there may be additional specific requirements in respect of the following:

- Dust from feed handling
- Dust explosions
- Gas explosions
- Carbon monoxide poisoning
- Tars and wastewater management
- Solids disposal, and noise levels.

All of these factors can be adequately managed through good design and operating practice. Information contained here should be taken as the author's interpretations and not as authoritative guidance.

REFERENCES

Ahrenfeldt J., Henriksen U., Jensen T.K., Gøbel B., Wiese L., Kather A., Egsgaard H. 2006. Validation of a Continuous Combined Heat and Power (CHP) Operation of a Two-Stage Biomass Gasifier. *Energy & Fuels*, vol. 20, issue 6, pp. 2672–2680.

Alakangas E., Erkkilä A., Oravainen H., Rubik A., English M., Corbella L., Floc´h-Laizet C., Pennequin J. 2008. *Efficient and Environmentally Friendly Biomass Heating: Firewood Production and Use in Fireplaces and Stoves*. Technical Research Centre of Finland (VTT), Espoo, Finland.

Anonymous. 1991. The Act No. 309/1991 digest, about the air protection against polluting matters. Slovak Office of Standards, Metrology and Testing (the UNMS SR, previously also the SOSMT) in Bratislava, www.unms.sk, Czech and Slovak Federative Republic.

Anonymous. 1996a. The statute of the government of the SR No. 15/1996 digest about waste treatment. Slovak Office of Standards, Metrology and Testing (the UNMS SR, previously also the SOSMT) in Bratislava, www.unms.sk, Slovakia or Slovak Republic.

Anonymous. 1996b. The statute of the government of the SR No. 92/1996 digest about the air protection against polluting matters. Slovak Office of Standards, Metrology and Testing (the UNMS SR, previously also the SOSMT) in Bratislava, www.unms.sk, Slovakia or Slovak Republic.

Bentzen J.D., Hummelshøj R.M., Henriksen U., Gøbel B., Ahrenfelt J., Elmegaard B. 2004. Upscale of the Two-Stage Gasification Process. In: van Swaaij W.P.M., Fjällström T., Helm P., Grassi A. (eds.), *Proceedings of 2. World Conference and Technology Exhibition on Biomass for Energy and Industry*. Rome, Italy.

Blaho J., Viglasky J. 1997. "*Charakteristika odpadu pri spaľovaní láskavca*" (Properties of Ashes from Burned Amaranth). In: *Proceedings of the Conference Biologization of Plant Productions VII.*, Slovak University of Agriculture. Nitra, 1996/7, pp. 5–7.

Brandt P., Larsen E., Henriksen U. 2000. High Tar Reduction in a Two-Stage Gasifier. *Energy & Fuels*, vol. 14, issue 4, pp. 816–819.

CEA. 2011. *Health and safety in biomass systems: Design and operation guide*. The Combustion Engineering Association and Carbon Trust.

Consonni S., Larson E.D. 1994. Biomass-Gasifier/Aeroderivative Gas Turbine Combined Cycles, Part A: Technologies and Performance Modelling. In: *Paper Presented at Cogen Turbo Power'94*, Portland, Oregon, 25–27 October 1994.

DECC. 2014, *UK Timber Standard for Heat and Electricity*. Governement, Great Britain, www.gov.uk/decc

Defra. 2011–2012. *Conversion of Biomass Boiler Emissions Concentration Data for Comparison with the Renewable Heat Incentive Emission Criteria (RHI)*. Defra Report. Governement, Great Britain, www.gov.uk/decc.

DGS, Ecofys. 2005. The German Solar Energy Society (DGS). First published by James & James (Science Publishers) Ltd in the UK and USA in 2005, ISBN 1-84407-132-4.

Dias J., Costa M., Azevedo J.L.T. 2004. Test of a Small Domestic Boiler Using Different Pellets. *Biomass and Bioenergy*, vol. 27, pp. 531–539.

Directive 2009/125/EC. Commission Regulation (EU) 2015/1185 of 28 April 2015 implementing Directive 2009/125/EC of the European Parliament and of the Council with regard to ecodesign requirements for solid fuel local space heaters, that will be valid from January 1st, 2022.

EC. 2010. *Biomass Sustainability. Report from the Commission to the Council and the European Parliament on sustainability requirements for the use of solid and gaseous biomass sources in electricity*, heating and cooling SEC(2010) 65 final SEC(2010) 66 final, Brussels, Belgium.

ECF. 2008. European Climate Foundation, A low-carbon society for prosperity and energy security. https://europeanclimate.org/mission/vision/

Fiedler F. 2004. The State of the Art of Small-Scale Pellet-Based Heating Systems and Relevant Regulations in Sweden, Austria and Germany. *Renewable and Sustainable Energy Reviews*, vol. 8, pp. 201–221.

Furman A.H. et al. 1992. *Biomass Gasification Pilot Plant Study*. General Electric, Schenectady, New York, July 1992.

Grimm A., Strehler A. 1987. "Harvest and Compaction of Annual Energy Crops for Heat Generation". In: *Producing Agricultural Biomass for Energy—Report and Proceedings of CNRE (European Co-operative Network on Rural Energy) Workshop.* CNRE Bulletin Number 17, 97–102. FAO, Rome, 129 p.

Henriksen U., Ahrenfeldt J., Jensen T.K., Gøbel B., Bentzen J.D., Hindsgaul C., Sørensen L.H. 2006. The Design, Construction and Operation of a 75 kW Twostage Gasifier. *Energy*, vol. 31, issue 10–11, pp. 1542–1553.

Hofmann Ph., Schweiger A., Fryda L., Panopoulos K.D., Hohenwarter U., Bentzen J.D., Ouweltjes J.P., Ahrenfeldt J., Henriksen U., Kakaras E. 2007. High Temperature Electrolyte Supported Ni-GDC/YSZ/LSM SOFC Operation on Two-Stage Viking Gasifier Product Gas. *Journal of Power Sources*, vol. 173, issue 1, pp. 357–366.

IPCC. 2011. Summary for Policymakers. In: Edenhofer O., Pichs-Madruga R., Sokona Y., Seyboth K., Matschoss P., Kadner S., Zwickel T., Eickemeier P., Hansen G., Schlömer S., von Stechow C. (eds.), *IPCC Special Report on RES and Climate Change Mitigation.* Cambridge University Press, Cambridge, UK and New York, NY, USA. https://www.uncclearn.org/sites/default/files/inventory/ipcc15.pdf

Iversen H.L., Henriksen U., Ahrenfeldt J., Bentzen J.D. 2006. *D25 Performance Characteristics of SOFC Membranes at Two Stage Gasifier (Confidential).* Technical report from the EU project BioCellus (Biomass Fuel Cell Utility System), 6th Framework Programme, Contract No: 502759. 2006.

Kaltschmitt M., Dinkelbach L. 1997. Biomass for Energy in Europe—Status and Prospects. In: *Proceedings of the International Conference "Gasification and Pyrolysis of Biomass—State of Art and Future Prospects"*, 9–11 April 1997, pp. 7–23.

Kasimir D.I., Nemestothy P. (eds.). 2008. *EVA Fuel Cost Calculator.* Austrian Energy Agency, Wien.

Keller R. 1994. *Primärmassnahmen zur NOx Minderung.* Ph.D. Thesis: Laboratorium für Energiesysteme, ETH, Zürich.

Knoef H.A.M., Stassen H.E. 1997. *Small-Scale Biomass Gasifier Monitoring—Findings and Gasifier Performance.* BGMP Report No.1129/Volume I&II, University of Twente—TEI/BTG B.V., NL-Enschede. Prepared for The World Bank and UNDP. 1995–7, 54/233 p.

Kona J. 1995. Accumulation of Heavy Metals by Amaranth. In: *Proceedings from the Conference "Biologization of a plant production VI."*, SUA, Nitra, pp. 120–123.

Madsen A.M., Mårttensson L., Schneider T., Larsson L. 2004. Microbial dustiness and particle release of different biofuels. *Annals of Occupational Hygiene*, vol. 48, issue 4, pp. 327–338.

Møller and Esbensen, 2005. Representative sampling of wood chips – a contribution to fulfil the Kyoto Protokol. *Proceedings from the 2nd World Conference on Sampling and Blending*, Queensland, Australia, pp. 205–209.

Mudgal S. Turunen L. Stewart R. Woodfield M. Kubica K. Kubica R. Preparatory Studies for Eco-design Requirements of EuPs (II); Task 6 Technical analysis of BATs; December 2009; https://www.eceee.org/static/media/uploads/site-2/ecodesign/products/solid-fuel-small-combustion-installations/bio-eup-lot-15-task6-final.pdf

Nussbaumer Th. 1997. Primary and Secondary Measures for NO_x Reduction in Biomass Combustion. In: Bridgwater A.V., Boocock D.G.B. (eds.), *Developments in Thermochemical Biomass Conversion.* Blackie Academic and Professional, London.

Nussbaumer Th. 2011. *Characterisation of Particles from Wood Combustion with Respect to Health Relevance and Electrostatic Precipitation.* University of Applied Sciences, Lucerne.

Ofgem. 2012. *UK Solid and Gaseous Biomass and Biogas Carbon Calculator.* https://www.ofgem.gov.uk/publications-and-updates/uk-solid-and-gaseous-biomass-carbon-calculator [Accessed June 30, 2014].

Ofgem. 2013a. *Renewable Heat Incentive Guidance Volume One: Eligibility and How to Apply.* https://www.ofgem.gov.uk/ofgem-publications/83036/rhiguidancevolone1.pdf [Accessed June 30, 2014].

Ofgem. 2013b. *Renewable Heat Incentive Guidance: Non-Domestic Scheme Metering Placement Examples.* https://www.ofgem.gov.uk/publications-and-updates/renewable-heat-incentive-guidance-non-domestic-scheme-metering-placement-examples [Accessed June 30, 2014].

PAS 111. 2012. Specification for the requirements and test methods for processing waste wood. British Standards Institution, UK. ISBN 9780580696435, 40 p. www.bsigroup.com.

Quaak P. et al. 1995. *Comparison of Biomass Combustion and Gasification Systems.* Report No. 1289, University of Twente—TEI/BTG Biomass Technology Group B.V., NL-Enschede. Prepared for The World Bank. 1995, 96 p.

Topsoe Fuel Cell. 2010. Topsoe Fuel Cell A/S. Business Areas, Distributed Generation. 2010. http://www.topsoefuelcell.com/business_areas/dg.aspx [Accessed April 2010].

Valentiny M. 1991. Combined-Cycle Plants: Burning Cleaner and Saving Fuel. *Mechanical Engineering*, vol. 113, issue 9, September 1991, pp. 46–50.

van Ree R. 1994: *Biomass Gasification—A "New" Technology to Produce Renewable Power. An Inventory of Technologies.* Netherlands Energy Research Foundation ECN, ECN-CX--94-057, Petten.

Veresova A. and Hoffmanova Z. 1995. The Evaluation of an Experimental Growing of Amaranth. In: *"Biologization of a Plant Production VI"*. SAU, Nitra, Slovakia, pp. 172–180.

Viglasky J. 1998. Biomass as Fuel for a Gas Producer–Engine System. In: *Traktori i pogonske mašine = Tractors and Power Machines: časopis Jugoslovenskog društva za pogonske mašine, traktore i održavanje.* YU ISSN 0354-9496. - Vol. III, no. 3, YU-Novi Sad, pp. 104–111.

Viglasky J. 2001. Demonstration Project of Wood Fired Heating Plant in Zvolen. In: *Proceedings of Inter. Conference "Renewable Energy Sources on the Verge of the XXIst Century"*. PL, Warsaw, 10–11th December 2001, pp. 281–285.

Viglasky J. 2002. Straw as a Fuel and its Characteristics. In: *Acta Facultatis Technicae Zvolen V. No. 1.* Technical University in Zvolen, ISBN 80-228-1256-2. pp. 141–150.

Viglasky J., Langova N., 2003. Mathematical Model for Efficiency Tar-Removal from Generator Gas I. In: *Acta Facultatis Technicae Zvolen VI. No. 1.* Technical University in Zvolen, ISBN 80-228-1256-2. pp. 113–120.

Viglasky J. Langova N. 2005. Industrial Wood Waste or Strategy Energy Raw Material. In: *Bioenergy 2005. Inter. Bioenergy in Wood Industry Conference and Exhibition.* Jyväskylä, pp. 285–287.

Viglasky J., Langova N. 2008. Successful biogas case studies within Slovakia. In: *Proceedings of the Central European biomass conference 16th–19th January 2008 in Graz, Austria.*

Viglasky J. et al. 2009a. The Development Philosophy of Biomass-to-Energy in the Slovak Republic. In: *Forum ifl: vom Landwirt zum Energiewirt—die Landwirtschaft Südosteuropas zwischen Euphorie und Skepsis. Heft 10/ed. Elke Knappe.* Leibniz-Institut für Länderkunde, Leipzig, 2009. ISBN 978-3-86082-066-7. pp. 37–64.

Viglasky J. et al. 2009b. Amaranth (Amaranthus L.) is a Potential Source of Raw Material for Biofuels Production. In: *Agronomy Research*, Estonia. No. 2, vol. 7, pp. 865–873, ISSN 1406-894X.

Viglasky J. et al. 2011. Brown Coal Substitution by Wood Chips at the Large-Scale District CHP Plant. In: *Proceedings of the Central European Biomass Conference 26th–29th January 2011 in Graz, Austria.*

Viglasky J. et al. 2015. *Heating, Ventilation and Air Conditioning. University Textbook.* Technical University, Zvolen, 1st edition, 248 p. ISBN 978-80-228-2748-5.

4 Emissions of Pollutants from Biomass Combustion
Relevant Regulatory Measures and Abatement Techniques

Robert Kubica

CONTENTS

4.1 Introduction 99
4.2 State of the Art and Legal Background Regarding SCIs Fueled
 with Solid Fuels Including Biomass 101
    4.2.1 Appliances 101
    4.2.2 Fuels 103
        4.2.2.1 Fossil Fuels 104
        4.2.2.2 Solid Biofuels 105
        4.2.2.3 Non-Woody Biomass: Straw as Fuel 105
        4.2.2.4 Fuel Properties 106
        4.2.2.5 Pollutants Produced by Combustion of Fuels 107
        4.2.2.6 Emissions Factors 112
    4.2.3 Requirements and Relevant Standards 112
        4.2.3.1 Boilers 113
        4.2.3.2 Space Heaters 114
    4.2.4 Testing Standards 116
4.3 Emission Abatements Measures 118
    4.3.1 Primary Measures for Emission Abatement 120
    4.3.2 Secondary Measures for Emission Abatement 121
4.4 Summary 122
References 123

4.1 INTRODUCTION

One of the most important challenges of an economy in transition, along with mitigating climate change, is air quality improvement. The second is of crucial importance with regards to the direct adverse health effects. Yet a derived effect is the negative influence on the economy, by so-called external costs. The issue of poor, insufficient air quality can only be solved by reducing emissions from both transport and residential sectors. The discussion presented here is focused on reducing emissions of pollutants into the air, from small combustion appliances equipped with short emitters, usually with a height below 40 m, used in residential sectors. These emissions are commonly known as "short stack emissions" or "residential emissions" and result from the production of useful heat in individual households, predominantly based on solid fuel combustion.

In Poland, the residential sector is one of the major sources of emissions of particulate matter (PM) and persistent organic pollutants (POPs). In 2016, in Poland, the shares of emissions from the residential sector, in the total annual emissions of carbon monoxide (CO), total suspended particles (TSP), PM10, PM2.5, black carbon (BC), polyaromatic hydrocarbons (PAH), and polychlorinated dibenzo(p) dioxins (PCDDs)/polychlorinated dibenzo(p)furans (PCDFs) were respectively 61.3%, 44.2%, 45.3%, 48.2% 25.1%, 88.0%, and 52.3% [1]. In 2015, PAH emissions into the air, estimated on the basis of emissions of four indicator compounds, that is, (benzo(a)pyrene, benzo(b)fluoranthene, benzo(k)fluoranthene, and indeno(1,2,3-cd)pyrene, amounted to 128.8 mg, while PM10 and PM2.5 amounted to 117,406.4, and 70,184.2 mg, respectively. According to the data provided by the European Environment Agency (EEA), such a large load of pollutants emitted into atmosphere meant that Poland's southern voivodeships, but also the valley of the Po river in Italy, were placed on the list of so-called "hot spots." These are the areas with the highest air concentrations of PM10, PM2.5, and benzo (a)pyrene (BaP), exceeding the limit values set by relevant legal acts. In the latest EEA report [2], Poland is regrettably placed in last place when it comes to achieving the annual BaP reference level, and with regards to PM10 emissions in the penultimate last place, just behind Bulgaria.

Over the years, the share of the pollutant emissions from the residential sector, in the total national emissions, has not changed. It has not been reduced, as this predominantly depends on the amount of fuel consumed, as well as the quality of fuels used and the technical condition of the heating devices. Two important national documents, adopted in 2015, that is, the National Air Protection Plan [3] and the so-called "anti-smog resolutions" have become the basic nontechnical, regulatory options for air quality improvement. At the level of territorial self-government, the voivodeship and municipal levels, they allow implementation, among others, of specific technical measures for the reduction of residential emissions.

Despite numerous undertakings carried out recently in Poland, to improve air quality, though improving, it does not yet meet the standards set out in the directive on ambient air quality and cleaner air for Europe CAFE, Directive 2008/50/EC of the European Parliament and of the Council of 21 May 2008 on ambient air quality and cleaner air for Europe [4]. The directive specifies the admissible values of PM concentrations, in particular subfractions PM10 and PM2.5, and substances related to them, including benzo(a)pyrene, the metabolites of which are mutagenic and highly carcinogenic. As mentioned, the source of the problem are high loads of hazardous pollutants derived from the combustion of solid fuels by individual households [2,4]. The true reason is the use of outdated heating devices, boilers, and stoves fueled by "bad practice" with low quality fuels, both mineral and biomass, as well as incineration/coincineration of municipal waste [4].

Emissions of Pollutants from Biomass Combustion 101

4.2 STATE OF THE ART AND LEGAL BACKGROUND REGARDING SCIS FUELED WITH SOLID FUELS INCLUDING BIOMASS

The main source of particulate matter emissions (PM and subfractions PM10, PM2.5, and PM1) into the atmosphere are so-called small combustion installations (SCIs) fueled with solid fuels, including biomass [5]. These are common sources employed by individual households to produce hot water and provide central heating. They are an important source of useful heat in residential, commercial agriculture, and forestry sectors, thus they are the predominant source of emissions.

4.2.1 APPLIANCES

The variety of fuels used, as well as heat exchange mechanisms and combustion techniques employed, makes the selection of appliances very broad. SCIs may include [6]:

Direct heat sources, space heaters:

- Stoves, manually and automatically fueled
- Fireplaces, opened and closed
- Cooking stoves

Indirect heat sources:

- Boilers, manually and automatically fueled—In general, a combustion appliance is specifically designed for use with given type of fuel, either a gaseous, liquid, mineral, wood, or other fuel such as straw. Appliances designed for one fuel type will not operate satisfactorily with another fuel type. Traditional open fireplaces and stoves can be reasonably tolerant of variation in fuel type and quality through more user intervention, but modern automatic appliances require more consistent fuels. Use of an inappropriate fuel quality (different moisture or size), or just the wrong fuel type will generally lead to combustion problems. This has a huge impact on emissions and appliance efficiency; many significant pollutants are indeed products of incomplete combustion [6].
- For manually fueled appliances (e.g., for domestic use), typical fuels are wood logs, coal, manufactured solid fuels, and non-woody biomass such as straw. Note that, for example, an appliance designed for wood is not generally suitable for straw and vice versa. Yet, there are also multifuel enclosed appliances that can be fueled with, for example, manufactured fuels, briquettes, or wood logs.
- Automatic-fueled appliances are designed to be fueled with specific fuel types and usually a manufacturer will define the limits of key criteria (e.g., moisture, fuel size, ash). Fuel size is of particular relevance for reliable fuel transfer and combustion process, to some extent. Smaller automatic appliances, for example, domestic appliances below 25 kW, tend to be designed to use manufactured fuel such as wood pellets, woodchips, and manufactured

mineral fuel briquettes. Woodchip is used by small and large wood fueled appliances, but is more common for nondomestic appliances. So it is with straw, which is used both as manufactured, in bales, or as bulk fuel in bigger appliances, usually operating in combination with heat sinks (buffer tanks).

The optimum process of fuel combustion, with regards to energy efficiency and environmental impact, is determined by the fulfillment of the so-called 3T principle (temperature, turbulence, time) [6], that is, maintaining the appropriate ratio of the amount of air to the load of fuel combusted, ensuring its complete combustion and ensuring homogeneity of the mixture of oxidizer-air/fuel (homogeneity of the mixture of incomplete combustion volatile products with oxidizer/oxygen combustion air), adequately high combustion temperature, and as much as necessary combustion reaction time, ensuring the presence of reagents in the oxidation zone. Using solid fuels in SCIs employs fixed bed combustion (layer combustion) technology. This technology can be implemented according to three different techniques (organization of the combustion process): countercurrent flow, cocurrent flow, and crosscurrent flow. It should be noted that the organization of the combustion process, especially in terms of the appropriate air/fuel ratio, is different for solid biofuels and coal. Biomass is characterized by almost twice as big a content of volatile parts than coal fuels, while the oxygen content in biomass is far higher (see Table 4.1).

In traditional boilers and space heaters, that is, stoves, fireplaces, and kitchen ovens with an outdated design, a countercurrent flow regimen is used. It is combustion in the entire volume of the bed. These traditional boilers and space heaters operate in the natural draft systems, and are fed manually, periodically. They neither employ distribution of combustion air into primary and secondary fractions, nor the control over the combustion air and fuel introduced to the furnace. As a result, these appliances have low energy efficiency and high emission of pollutants—the products of incomplete combustion. The basic fuels in these boilers are coal (e.g., nut, cube, or bean), wood logs, briquettes, and sometimes coke.

Modern boilers are stoked manually, and operate based on the crosscurrent combustion technique (best available technology, BAT) [6,7]. They are also fed manually, but have controlled and forced air supply to the combustion zone, as well as air distribution into primary and secondary fractions. Appliances using this technique are often called: boilers with semi-automatic or gravity feeding of fuel into the combustion chamber. The use of appropriate construction solutions and ceramic materials ensures high energy efficiency, over 87% at nominal power. Relevant emission target values, specified by the most challenging standards are also met (see Section 4.2.3 in Chapter 4, Requirements and relevant standards).

The group of boilers with a cocurrent arrangement of the combustion process includes solid biofuel appliances, which are manually stoked, for example, with chipped wood or wood logs (gasification boilers). In this particular case, the furnace is equipped with a strictly separated precombustion (gasification) chamber. Inside the precombustion chamber, under oxygen shortage, the degassing/gasification process takes place. The pyrolysis gas produced undergoes a successive combustion in the main chamber of the furnace. Within this group of manually fueled appliances, one can find BAT solutions, that is, wood log gasifying boilers, reaching efficiencies of 92% [7].

TABLE 4.1
Physical and Chemical Properties of Coal and Biomass

Composition and Properties	Symbol	Unit	Biomass	Coal
Carbon	C^{daf}	%	44–51	75–85
Hydrogen	H^{daf}	%	5.5–7	4.8–5.5
Oxygen	O_d^{daf}	%	41–50	8.8–10
Nitrogen	N_d^{daf}	%	0.1–0.8	1.4–2.3
Sulfur	S_t^d	%	0.01–0.9	0.3–1.5
Chlorine	Cl_t^d	%	0.01–0.7	0.04–0.4
Volatile matter	V^{daf}	%	65–80	35–42
Ash content	A^d	%	1.5–8	5–10
Calorific value	Q_s^a	MJ/kg	16–20	21–32
Main Components of Ash				
SiO_2	–	%	26.0–54.0	18.0–52.3
Al_2O_3	–	%	1.8–9.5	10.7–33.5
CaO	–	%	6.8–41.7	2.9–25.0
Na_2O	–	%	0.4–0.7	0.7–3.8
K_2O	–	%	6.4–14.3	0.8–2.9
P_2O_5	–	%	0.9–9.6	0.4–4.1

Source: Kubica K. et al. 2007. Small combustion installations: Techniques, emissions and measures for emission reduction, EUR 23214 EN–2007, ISBN 978-92-79-08203-0, ISSN 1018-5593, pp. 25–27; (accessed February 21, 2019), http://publications.jrc.ec.europa.eu/repository/bitstream/JRC42208/reqno_jrc42208_final%20version%5B2%5D.pdf.

The most modern heating appliances, based on cocurrent technology, fulfill the optimal combustion principles, 3T. These are boilers with precisely controlled, continuous mechanical fuel and air supply, as well as air staging into primary and secondary fractions. This group of boilers includes coal-fired boilers with a combustion retort or underfeed burner, fired with high-grade coal of appropriate grain size and high quality parameters. Automatic boilers can be also stoked automatically with wood pellets (pellet boilers) or woodchips (usually with a capacity above 50 kW). Modern boilers with automatic fossil fuel feeding (retort burner) and solid biofuel (pellet burner), of BAT type, are characterized by high energy efficiency, above 90% and 95%, respectively [7].

4.2.2 Fuels

Different economy sectors including power production, heat production, and transport use certain energy carriers. The world's production of energy is based on resources such as fossil, nonrenewable fuels, that is, coal, oil, and gas. However it is actively and rapidly being redirected toward the renewable energy sources such as solar, wind, water, and geothermal energy, as well as biomass. Selected sectors of the economy, namely residential, commercial, institutional, and agricultural, consume energy predominantly in the form of useful heat. The heat is produced by SCIs using a wide variety of fossil fuels, including gas, oil, hard and lignite coals, but increasingly renewables such as wood and straw. In general, gas and oil are the predominant types of fuel used by SCIs. Yet in many countries, in particular those with economies in transition, Central and Eastern European Countries (CEEC), fossil and biomass solid fuels are the main type of fuel used for domestic and commercial heating. In some regions such as Scandinavia, biomass fuels are predominant. Currently, it is observed that the use of biomass for production of heat and energy in SCIs, especially up to 500 kW, has been growing across Europe, also owing to newly built strategies promoting a zero CO_2 economy.

The emission inventory guidebook by European Monitoring and Evaluation Programme (EMEP)/Corinair [5], which provides guidance for the compilation of emission inventories for countries reporting under the UN Convention on Long Range Transboundary Air Pollutants (CLRTAP), defines solid fuel in the context of small-scale combustion activities into two subcategories:

- Mineral fuels: Hard coal, brown coal, patent fuels, brown coal briquettes, coke, peat.
- Solid biomass fuels, which comprise wood: Lumps, logs, chips, pellets, and charcoal, as well as wood wastes (e.g., bark, sawdust), charcoal, and agricultural wastes used as fuels (straw, corncobs, etc.).

4.2.2.1 Fossil Fuels

Taking the physical-chemical properties of fuels into consideration, fossil fuels can be divided into three categories: solid, liquid, and gas fuels.

Gas fuels that can be used in small combustion installation include gasses distributed to the municipal economy by communal gas piping, such as natural gas with a gross heating value (GHV) of between 16 and 31 MJ/m^3, a mixture of propane and butane with air, having a GHV of about 22 MJ/m^3, artificial gases and its mixture with propane and butane, with a calorific value between 15 and 17,5 MJ/m^3, and liquid petroleum gas (LPG) [7].

Liquid fuels such as kerosene, gas oil (gas/diesel oil), residual oil, residual fuel oil (which has a GHV between 35 and 42 MJ/kg) are used in SCIs [7].

Solid fuels refer to solid fossil fuels, upgraded solid derived fuel, briquettes, patent fuels, solid biofuels, and any solid combustible organic material containing mainly carbon, hydrogen, and oxygen, and minor proportions of nitrogen, sulfur, chlorine, and mineral matter (profitably less than in coal) [6]. The solid fossil fuels subgroup includes hard coal, lignite coal, brown coal, and peat. SCIs use mainly hard coal, lignite coal, peat, and upgraded solid derived fuel. Different types of hard coal with a GHV greater than 17,435 kJ/kg on ash-free basis are used in SCIs, including: hard coal (GHV > 23,865 kJ/kg), subbituminous coal (17,435 kJ/kg < GHV < 23,865 kJ/kg), and anthracite. Peat, which is applied in SCIs, is referred to as combustible organic material such as lignite materials of low coalification degree (peat-like material) with moisture contents (air-dried material 20%–25%), of lower heating values (Hu) 9.5 MJ/kg. The residential sector has still been applying solid upgraded coal fuels from hard/subbituminous coal and anthracite, such as smokeless fuels, and briquettes with or without binder, with calorific values (GHW) usually below 27 MJ/kg [7].

4.2.2.2 Solid Biofuels

Solid biomass fuels are called solid biofuels, and are produced from different sources of organic material. In accordance with the description given by relevant standards, general biomass materials include the following [7]:

- Products from agriculture and forestry (e.g., short rotation willow coppice, miscanthus, switchgrass, straw, etc.)
- Vegetable waste from agriculture and forestry (e.g., lump firewood, pine spruce, beech, poplar and willow, bark and sawdust, cereals and straw from agriculture, etc.)
- Vegetable waste from the food processing industry
- Wood waste, with the exception of wood waste which may contain halogenated organic compounds or heavy metals as a result of treatment with wood preservatives or coating, and which includes in particular such wood waste originated from construction and demolition waste (e.g., demolition wood, fiberboard residues, railway sleepers, etc.)
- Fibrous vegetable waste from virgin pulp production and from production of paper from pulp, if it is coincinerated at the place of production and the heat generated is recovered
- Cork waste.

The range of fuels produced from biomass includes a variety of different forms, such as logs, pieces, chips, briquettes, pellets, sawdust, etc. Manufactured biofuels are made from pulverized biomass with or without pressing aids, usually with a cylindrical form, and a random length typically 5–30 mm. Biomass briquettes and pellets are formed by press or pellet process, with or without a binder, containing at least 90% of fine biomass material, with nitrogen, sulfur, chlorine, and mineral matter contents profitably less than in best quality raw biomass, with a calorific value of between 16 and 20 MJ/kg (daf). SCIs also use a temperature treated wood such as charcoal.

4.2.2.3 Non-Woody Biomass: Straw as Fuel

Straw, with its diverse physical properties and composition, is a very challenging type of fuel. The combustion process is accompanied by corrosion risk and low temperatures of ash fusion. Studies on the process of straw combustion indicate that the biggest amount of corrosive deposits are formed in the convective part of the boiler heat exchangers [8,9]. The phenomenon of slagging and the formation of deposits is predominantly caused by alkali species, namely chlorine and potassium. Straw, especially when "yellow", may contain large quantities of the slagging precursors. Fly ash produced by the combustion of straw may contain more than 20% of unburnt carbon, but also significant shares of potassium and chlorine compounds [10]. Almost 40% of the chlorine condenses toward KCl.

Significant amounts of alkali metals in the straw leads to a low sintering temperature for ashes, ranging between 700 and 900°C. The softening point of the ash is usually below 1000°C, and the melting point is below 1200°C [11]. These characteristic points of ash are the reason for slagging, that is, fuel agglomeration in the bed during the combustion process. In the case of straw, separate, viscous, and partially melted particles of ash tend to agglomerate. Unlike wood ash, agglomeration of straw ash is caused by the viscous layer of alkaline calcium silicate, formed in the particles [12]. Owing to the physicochemical properties of straw, its use for energy production is problematic, both in large industrial boilers (power plants) and in SCIs. Therefore, it is of crucial importance to design and build the installations compliant with the requirements relevant for straw combustion boilers. In large industrial plants, straw can by cofired with coal. In smaller units, such as SCIs, it is used in the form of pellets, briquettes, or chaff. It is often used by medium- and small-scale water boilers in agriculture, forestry, and residential sectors. In such small appliances, water boilers in particular, the risk of high temperature corrosion is limited, yet all the problems associated with bed agglomeration and deposits formation on heat exchange surfaces are in place [13].

Straw leaching and aging is the most common technique to diminish the amount of chlorine and alkali metals in the material. This process may be carried out in the field. After the harvest of cereals, straw is left exposed to the rain. It may also be carried out by technical means, such as a special leaching plant [14]. Rainwater may leach as much as 80% of potassium and chlorine form the straw [15]. Through this process, so-called "gray straw" is obtained, which has lower amounts of mineral matter and improved energy parameters.

Using additives is another way to mitigate the risks of creating corrosive deposits in straw combustion units. Compounds with high melting points, which may also react with the constituents of the ash, are usually used. Additives not only increase the characteristic temperatures of the ash, but they also reduce the amount of potassium chloride by reaction with mineral matter [16].

There is no legal act at the EU level that covers straw combustion. Non-woody biomass is excluded from the majority of SCIs' relevant requirements. However, in some countries such as the UK, regulatory measures are in place. In the UK, newly built systems installations fueled with biomass including straw must meet admissible emission requirements with regards to PM and NO_x, as set by the Domestic Renewable Heat Incentive [17]. Systems must not exceed the emission limit values of 30 g/GJ for PM and 150 g/GJ for NO_x. Similar requirements are being considered in countries such as Czech Republic and Estonia.

4.2.2.4 Fuel Properties

Emission of pollutants strongly depends on the fuel and the combustion technologies, as well as on operational practices and maintenance. In particular, it refers to emissions of these pollutants that are products of incomplete combustion of fuel, for instance, CO, TSP, PM2.5, PM10, volatile organic compounds (VOCs), and POPs. In general, combustion of liquid and gaseous fuels runs under similar conditions to those in SCIs and in industrial combustion plants, so the emissions are comparable. Opposed to that, technologies for solid fuels combustion vary widely owing to very different fuel properties and technical means employed [18]. Solid biofuels (biomass) roughly consist of the same elements as coal, but their content is different (Table 4.1).

For instance, the average C, H, and O contents are 45% and 80%, 6% and 5%, 45% and 10%, respectively in biomass and coal. For this reason, the volatile matter content in biomass fuels is about two times higher in comparison to coal and its combustion process with optimum energetic and ecological efficiency requires special conditions of low-intensity combustion. However, the content of elements such as sulfur, nitrogen, chlorine, and mineral matter (containing heavy metals) in biomass is lower than in coal, it is, respectively 0.01%–0.9% and 0.3%–3.0%, 0.011%–0.04 (but also 0.7%) and 0.04%–0.4%, 1%–8% and 3%–10% and 1%–8%. In the case where the optimum conditions of the biomass combustion process are achieved, a decrease in emission of pollutants is observed. But, for the combustion process of poor quality, usually carried out in small appliances (fireplaces, stoves, old design small boilers) the emission factors of pollutants from both types of solid fuels, biomass and coal (as well as peat, solid alternative fuels, and refused derived fuels), are many times higher than the factors for the same pollutants in the case of power plants. The chemical composition of liquid fuels is more uniform. The average for carbon content is between 84% and 87%, hydrogen amounts to about 13%, but the content of sulfur ranges between 0.2% and 3% and depends on its content in crude oil. Good quality light oil contains less than 0.3% sulfur, but some residue oil may contain about 3% sulfur.

4.2.2.5 Pollutants Produced by Combustion of Fuels

In any type of combustion process, a range of typical pollutants is usually formed, but their properties and amount differs depending on the fuel and appliance type, and on the operational mode [19,20]. The emissions that can be of concern in the context of small-scale combustion are:

- Particulate matter and subfractions PM_{10}, $PM_{2.5}$ and $PM_{1.0}$
- Black carbon
- Carbon monoxide
- Sulfur dioxide
- Nitrogen oxides
- Organic gaseous carbon
- Volatile organic compounds
- Polycyclic aromatic hydrocarbons
- Polychlorinated dibenzo(*p*)dioxin and furan
- Ammonia
- Heavy metals.

The above listed pollutants and relevant formation processes have been defined as outlined in the following sections [5].

4.2.2.5.1 Carbon Monoxide

Carbon monoxide (CO) is found in the gaseous combustion by-products of all carbonaceous fuels. It is an intermediate product of the combustion process and final by-product in substoichiometric conditions. CO is the most important intermediate product of fuel conversion to CO_2; it is oxidized to CO_2 under appropriate temperature and oxygen availability. Thus, CO can be considered as a good indicator of combustion quality. The mechanisms of CO formation, thermal NO, VOCs, and PAHs, are in general similarly influenced by combustion conditions. The emissions are dependent on the excess air ratio as well as the combustion temperature and residence time of the combustion products in the reaction zone.

4.2.2.5.2 Aerosols and Particulate Matter

Aerosols are defined as a suspension of particles and droplets in the size range between 0.001 and 100 μm in a surrounding gas phase. The total mass of particles and droplets is indicated as PM or TSP. Since fine particles, that is, smaller than 10 μm, are only partially precipitated in the upper respiratory tract, they can be inhaled and transported into human lungs. Hence, the particle fraction PM10 (particulate matter <10 μm able to stay suspended in the air) is commonly used for the definition of emission limits (ambient air concentrations). The number indicates the aerodynamic particle diameter in μm according to a separation efficiency of 50% in the sampling system. Further, PM2.5 and PM1.0 fractions are also used for the fact that PM2.5 fraction is easily inhalable while PM1.0 can reach the human blood system through the lungs.

PM10 and its subfractions are of major concern for human health, and biomass combustion mainly leads to particulate emissions smaller than 10 μm. As it has been shown in many investigations, the main fraction of PM in flue gas is even smaller than 1 μm [7]. As a consequence, most investigations focus on submicron particles.

PM in flue gases from combustion of fuels, in particular of solid fuels, might be described as carbon, smoke, soot, stack solid, or fly ash. PM matter can be split between three groups of fuel combustion products.

The first group of PM is formed via gaseous phase combustion or pyrolysis because of the incomplete combustion of fuels (products of incomplete combustion or PICs) [18]. Soot and organic carbon particles (OC) are formed during combustion as well as from gaseous precursors through nucleation and condensation processes (secondary organic carbon). These precursors occur as a product of chemical radicals' reactions in the presence of hydrogen and oxygenated species within a flame. Condensed heavy hydrocarbons (tar substances) are an important, and in some cases, the main contributor to the total level of particle emissions in small-scale solid fuels combustion appliances such as fireplaces, stoves, and old design boilers.

The second and third groups of particulate may contain ash particles or cenospheres that are largely produced from mineral matter in the fuel; they contain oxides and salts (S and Cl) of Ca, Mg, Si, Fe, K, Na, P, and heavy metals, and unburned carbon form from incomplete combustion of carbonaceous material (black carbon or elemental carbon (BC); this is called carbon-in-ash (or loss on ignition) [21].

PM emissions from SCIs are typically combined with PICs associated and/or adsorbed onto particulate. Size distribution depends on combustion conditions. Optimization of the solid fuel combustion process, by continuous process control employing automatic fuel feed and air staging, leads to a decrease of TSP emissions and to a change of PM distribution [21]. Several studies have shown that the particulate emissions from modern residential biomass combustion technologies predominantly produce submicron particles (<1 μm) and the proportion [22] of the mass concentration of particles larger than 1 μm is normally <10% for SCIs.

Aerosols from incomplete combustion of, for example, biomass can be a result of incomplete combustion such as soot BC [23], PAHs, unburnt carbon, and of unburnt biomass fragments. In simple combustion systems and/or under poor combustion conditions, the mass fraction of unburnt particles can reach more than 90% of the total particle mass, while by good combustion it can be limited to less than 1%. Hence, one of the main aims is to design and operate combustion devices that enable almost complete combustion.

Aerosols are formed by different mechanisms, depending on the fuel composition and combustion process run. The pathways of formation, as well as the fuel type, influence the composition and physical properties of products formed.

Coarse and super coarse particle formation (coarse, 1–10 μm; super coarse >10 μm) are predominantly produced by mechanical entrainment of combustion by-products. The mechanism pathway includes:

- Fuel combustion
- Char formation agglomeration and melting of low volatile species (Ca, Fe, Si, Ti)
- Formation of bottom ash, for example, $K_2Ca_2(SO_4)_3$
- Entrainment and ejection of coarse fly ash and unburnt char particles.

Fine particles, including soot, are mainly formed by condensation and/or nucleation. The mechanism pathway includes:
For inorganic aerosol:

- Vaporization of inorganic matter (K, Na, S, Cl, Zn)
- Sulfating and oxidation
- Nucleation/condensation of alkali sulfates.

For organic aerosol and soot:

- Vaporization and condensation of vapors of volatile organic compounds
- Release of pyrolysis gases
- PAH formation
- PAH polymerization, nucleation
- Formation of primary soot particles, agglomeration
- Oxidation and partial soot burnout.

4.2.2.5.3 Nitrogen Oxides

"Oxides of nitrogen," recalculated and expressed as NO_2, which is a general convention for reporting these emissions, are the sum of nitric oxide (NO) emissions (>90% of the NO_x emissions) and nitrogen dioxide (NO_2, typically <10% of the NO_x) emissions [6]. Nitrogen emissions are the result of the partial oxidation of fuel nitrogen. The emissions of NO_x increase with increasing nitrogen content in the fuel, as well as with increasing excess air ratio accompanied by higher combustion temperatures. Nitrogen content in fuels vary both among and within fuel types: coals contain nitrogen mainly in N-organic form (0.5% to 2.9% daf, average about 1.4%). Biomass fuels contain nitrogen in N-organic form (0.05% to 0.8% daf); for coke the N-content is between 0.6% and 1.55% (daf); and for peat between 0.7% and 4.4% (daf). Additional NO_x may be formed from nitrogen in the air under high temperature conditions, as "thermal NO" and as "prompt NO." Thermal and prompt NO are generated in the flames—high temperature zones, surrounding individual particles, through free radical reactions. Nitrogen in the air starts to react with O-radicals and forms NO at temperatures above approximately 1300°C. The amount of pollutants formed depends on O_2 concentration and residence time. However, the combustion temperatures in SCIs, in general, are far lower than 1300°C and hence thermal NO_x formation is not usually important (especially for fireplaces and insert appliances). However, the majority of thermal NO_x is formed in the post-flame gases, therefore with the development of small boiler design, the significance of such emissions is increasing.

4.2.2.5.4 Nonmethane Volatile Organic Compounds

Nonmethane volatile organic compounds (NMVOCs) are the name given to a large variety of chemically different species, for example, benzene, ethanol, formaldehyde, cyclohexane, 1,1,1-trichloroethane, or acetone. NMVOCs are VOCs, but with methane excluded. They are intermediates, a by-product in the process of thermal conversion of carbonaceous matter into CO_2 and H_2O. As for CO, emissions of NMVOCs are a result of insufficient temperatures and residence time in the oxidation zone, as well as a shortage in oxygen availability. Emissions of VOCs are lower for bigger combustion installations, predominantly owing to application of advanced and well-controlled combustion techniques.

4.2.2.5.5 Organic Gaseous Carbon

Organic gaseous carbon (OGC) is defined in EN303-5 and is essentially equivalent to a VOC emission.

4.2.2.5.6 Polychlorinated Dibenzo(p)dioxin and Furan

The emissions of dioxins and furans are highly dependent on the conditions under which cooling of the combustion and exhaust gases is carried out. Carbon, chlorine, a catalyst, and oxygen excess are necessary for the formation of PCDD/Fs. Coal-fired stoves in particular have been reported to release very high levels of PCDD/Fs when using certain kinds of coal [24]. The emission of PCDD/Fs is significantly increased when plastic waste is cocombusted in residential appliances or when contaminated/treated wood is used. The emissions of PCDD/Fs can be reduced by introduction of advanced combustion techniques of solid fuels [25].

4.2.2.5.7 Polycyclic Aromatic Hydrocarbons

Emissions of PAHs result from incomplete (intermediate) conversion of fuels. As for CO and VOCs, emissions of PAHs depend on the organization of the combustion process, particularly on the temperature (too low temperature favorably increases their emission), the residence time in the reaction zone, and the availability of oxygen [26]. It has been reported [6] that coal stoves and old type boilers (hand-fueled) emit several times higher amounts of PAH in comparison to new design boilers (capacity below 50 kW), such as boilers with semi-automatic feeding. Technologies of cocombustion of coal and biomass that can be applied in commercial/institutional and in industrial SCIs lead to reduction of PAHs as well as TSP, VOCs, and CO emissions [27].

4.2.2.5.8 Ammonia

Small amounts of ammonia (NH_3) may be emitted because of the incomplete combustion of solid fuels containing nitrogen. This occurs at very low combustion temperatures (e.g., in fireplaces, stoves, and old-design boilers). NH_3 emissions generally can be reduced by primary measures aiming to reduce products of incomplete combustion and increase energy efficiency.

4.2.2.5.9 Sulfur Dioxide

The emission of sulfur dioxide (SO_2) depends on the fuel sulfur content. For coal, sulfur content normally varies between 0.1% and 1.5% (daf) (up to an extreme value of 10%) and for biomass, it varies between 0.01% and 0.9%. Both the coal and biomass sulfur content may vary significantly depending on their origin. Sulfur content of biomass also depends on the type of biomass. In biomass fuels, sulfur appears mainly as organic and salts sulfur. In coal, sulfur can be found in three forms:

- Inorganic—mainly pyritic sulfur (FeS_2), and sulfur salts.
- Organic—as sulfides, disulfides, and cyclic compounds mainly tiophene compounds.
- Elemental sulfur—pyritic and organic sulfur are a major part of the sulfur in coal; both types are responsible for SO_x formation (SO_2>95%, and SO_3 <5% is formed at lower temperatures).

Moreover, for coals, the content of calcium carbonate is also relevant, due to its capacity to absorb the SO_2 generated.

For biomass fuels, the average sulfur retention in ash can be taken as zero. For coal fuels, a default value of 0.1 can be taken in the absence of real data. Precleaning, pretreatment processes of raw coals, and improvement of their quality can achieve a reduction of SO_2 emissions [6].

4.2.2.5.10 Heavy Metals

Most heavy metals (HMs) (As, Cd, Cr, Cu, Hg, Ni, Pb, Se, and Zn) are usually released as compounds associated and/or adsorbed with particles (e.g., sulfides, chlorides, or organic compounds) [28]. During the combustion of coal and biomass, particles undergo complex changes, which lead to vaporization of volatile elements. Some metals (Hg, Se, As, and Pb) are at least partially present in the vapor phase. The rate of volatilization of heavy metal compounds depends on technology characteristics (e.g., type of boiler; combustion temperature) and on fuel characteristics (e.g., their metal content, and the fraction of inorganic species, such as chlorine, calcium, etc.).

Less volatile elements tend to condensate onto the surface of smaller particles in the exhaust gases. Higher emissions of Cd and Zn have been observed in biomass in comparison to coal.

The chemical form of the mercury emitted may depend in particular on the presence of chlorine compounds. The nature of the combustion appliance used and any associated abatement equipment will also have an effect [28]. Mercury emitted from SCIs occurs in elemental form (elemental mercury vapor Hg^0), reactive gaseous form (reactive gaseous mercury, RGM) and total particulate form (total particulate mercury, TPM) [29]. The distribution of particular species of emitted mercury from SCIs is different to the distribution observed in large-scale combustion.

Contamination of biomass fuels, such as treated or painted wood, may cause significantly higher amounts of heavy metals emitted (e.g., Cr, As). Emissions of heavy metals can be reduced by secondary emission reduction measures, with the exception of Hg, As, Cd, and Pb, which have a significant gaseous phase component.

4.2.2.6 Emissions Factors

An emission factor relates the rate of release of a pollutant to the rate of an activity. For solid fuel combustion, emission factors are typically expressed as a mass of pollutant emitted per mass of fuel burned (e.g., g/kg) or, as used in this chapter, per unit of energy input (e.g., g/GJ). In this study, emission factors have been derived from concentration data provided by manufacturers to enable a comparison with published and experimentally derived values.

By combining fuel and appliance the emission factors, defining environmental impacts can be derived (see Table 4.2).

4.2.3 REQUIREMENTS AND RELEVANT STANDARDS

The main barrier to achieve the required air quality standards was a lack of nationwide legislation for boilers fired with solid fuels, that would cover:

- The emission standards for heating devices operated in individual households, combustion installations below 1MW;
- The quality of solid fuels used in individual households in the municipal sector, as well as;
- The control and supervision system concerning both the condition of installations and the market of the solid fuels utilized in these installations.

The National Air Protection Program strongly emphasizes the need to resolve these barriers [3]. In April 2015, after more than 8 years of work, the European Commission adopted two regulations implementing the ErP (Energy related Products) Directive, concerning solid fuel combustion installations with nominal output up to 0.5 MW (500 kW), treated as standard products available on the market; one of these was Commission Regulation (EU) 2015/1189 of 28 April 2015 implementing Directive 2009/125/EC of the European Parliament and of the Council with regard to ecodesign requirements for solid fuel boilers [31]. This regulation entered into force on August 10, 2015 and has direct effect in Member States, thus not requires transposition into

TABLE 4.2
Emission Factors of Key Pollutants for Selected Appliances

Fuel		Appliance	Emission Factor, g/GJ, for BaP mg/GJ					
			TSP	CO	OGC	NO_x	SO_2	BaP
Wooden biomass	Wood logs dry aged	Room heater (closed fireplace, cooking oven)	840	5250	630	60	20	130
	Wood pellet	Room heater, pellet stove	85	530	20	95	20	55
	Wood logs dry aged	Boiler, old design	530	4200	370	80	12	130
	Wood logs and briquettes, dry aged	Manually fed gasifying boiler	120	2850	290	80	10	50
	Wood pellet	Automatic boiler	45	470	20	115	11	19
Agricultural biomass, straw	Bale	Manual boiler	250	3000	250	150	50	80
	Straw pellet and briquette,	Automatic boiler (including cigar boiler)	70	300	20	150	20	12
Coal	Pea coal	Boiler, old design	540	12,500	260	160	675	390
	Pea coal	Automatic boiler new design	22	500	18	190	600	15

Source: Stanek, W. et al. 2018. Wielowariantowa analiza eliminowania przestarzałych, niskoefektywnych energetycznie i wysokoemisyjnych źródeł wytwarzania energii użytkowej ze spalania węgla w indywidualnych gospodarstwach domowych; (accessed February 21, 2019) http://ios.edu.pl/wp-content/uploads/2018/02/WIELOWARIANTOWA-ANALIZA.pdf.

national law. The resulting criteria for ecodesign will apply from January 1, 2020. EU Member States can implement national legislation earlier, before 2020.

4.2.3.1 Boilers

Commission Regulation (EU) 2015/1189 introduces the concept of seasonal space heating emissions (Es) and seasonal space heating energy efficiency (η_s) relevant for solid fuel boilers. Seasonal space heating emissions are the weighted average determined for the emission of a given pollutant, at boiler operation with rated heat output at minimal load, at active mode of operation (continuous operation). Automatically stoked solid fuel boilers can be operated at 30% of the rated heat output in continuous mode, whereas manually stoked solid fuel boilers, at 50% of the rated heat output in continuous mode. The seasonal space heating energy efficiency (η_s) of automatically stoked solid fuel boilers is the weighted average energy efficiency of the boiler, determined similarly as in the case of pollutants, taking into account losses of electricity consumption for the boiler's own needs (fuel supply, control, etc.) and factors related to temperature regulators. The method of calculating individual criteria indicators is detailed in Annex III to the above mentioned Regulation. Table 4.3 presents requirements for boilers with a rated power below 0.5 MW, including emission limit values (ELVs) in accordance with Commission Regulation (EU) 2015/1189.

Emission factors corresponding to ELVs for the key pollutants set by the ecodesign directive are given in Table 4.4.

The Regulation of the Minister of Development and Finance of 1 August 2017 on the requirements for solid fuel fired boilers—fossil and biogenic (effective from October 1, 2017)—contains provisions specifying the limits as for total suspended particles (TSP), carbon monoxide (CO) and organic pollutants (OGC) [32], as summarized in Table 4.5.

The Regulation of the Minister of Development and Finance sets out no minimum energy efficiency requirements for solid fuel boilers.

4.2.3.2 Space Heaters

As in case of boilers, local space heaters must comply with relevant regulations, that is, Commission Regulation (EU) 2015/1185 of 24 April 2015 implementing Directive 2009/125/EC of the European Parliament and of the Council with regard to ecodesign requirements for solid fuel local space heaters [33]. From January 1, 2022, solid fuel local space heaters shall comply with the requirements as listed in Table 4.6.

TABLE 4.3

Energy and Emission Requirements for Solid Fuel Boilers with Heat Output ≤500 kW, Consistent with Commission Regulation (EU) 2015/1189

Fuel Type	Seasonal Energy Efficiency, η_s %	Seasonal Emissions, E_s (mg/m³ at 10% O_2)[d]			
		TSP mg/m³	OGC mg/m³	CO mg/m³	NO$_x$ mg/m³
Valid from 2020[a]					
Automatic Fuel Feeding					
Biomass	75[b]; 77[c]	40	20	500	200
Fossil	75[b]; 77[c]	40	20	500	350
Manual Fuel Feeding					
Biomass	75[b]; 77[c]	60	30	700	200
Fossil	75[b]; 77[c]	60	30	700	350

Source: Commission Regulation (EU) 2015/1189 of 28 April 2015 implementing Directive 2009/125/EC of the European Parliament and of the Council with regard to ecodesign requirements for solid fuel boilers.

[a] EU Member States can implement domestic legislation earlier, before 2020.
[b] For ≤20 kW boilers, indicated only for rated power.
[c] For >20 kW boilers.
[d] Related to dry exhaust gases under standard conditions, 0°C.

TABLE 4.4
Ecodesign ELVs and Corresponding Emission Factors for Solid Fuel Boilers ≤500 kW Output[a]

		Concentration mg/m³ at 10% O_2 Dry and STP (0°C, 101.3 kPa)[a]				Emission Factor g/GJ Net Heat Input[b]			
Stoking	Fuel	PM[c]	CO	OGC	NO_x	PM[c]	CO	OGC	NO_x
Auto	Biogenic	40	500	20	200	19.4	243	9.7	97
Manual		60	700	30	200	29.1	340	14.6	97
Auto	Fossil	40	500	20	350	19.8	247	9.9	173
Manual		60	700	30	350	29.6	346	14.8	173

[a] Commission Regulation (EU) 2015/1189 of 28 April 2015 implementing Directive 2009/125/EC of the European Parliament and of the Council with regard to ecodesign requirements for solid fuel boilers.

[b] Conversion from EMEP/EEA air pollutant emission inventory guidebook; 1A4 Small combustion 2016—July 2017 https://www.eea.europa.eu/publications/emep-eea-guidebook-2016 (Conversion from concentrations and emission factors assume a stoichiometric specific flue gas volume of 253 m³/GJ net fuel input for biomass and 258 m³/GJ net fuel input for bituminous coal).

[c] PM emission limits based on filterable material only.

TABLE 4.5
Emission Limit Values Based on Regulation of the Minister of Development and Finance

	ELVs[a] (mg/m³ at Reference Level of 10% O_2)[b]		
Fuel Feed	CO	OGC	TSP
Manual	700	30	60
Automatic	500	20	40

Source: The Regulation of the Minister of Development and Finance on requirements for solid fuel boilers of nominal heat output of up to 500 kW, August 1, 2017.

[a] The fact that the emission limit values are met is confirmed taking into account the PN-EN 303-5:2012 standard.

[b] The emission limit values are expressed in mg/m³ of waste gases, at 0°C and 1013 mbar pressure, standardized to a dry flue gas basis.

TABLE 4.6
Seasonal Space Heating Energy Efficiency and Seasonal Space Heating ELVs for Biogenic Local Space Heaters (from January 1, 2022)

		Seasonal Emissions Limits, E_s (in mg/m³ at 13% O_2)[a]				
Local Space Heaters	Seasonal Space Heating Energy Efficiency, η_s [%]	PM (mg/m³)[b]	(g/kg)	CO (mg/m³)	OGC (mg/m³)	NO_x (mg/m³)
Open fronted	30	50	6[c]	2000	120	200
Closed fronted	65	40	5[c]; 2.4[d]	1500	120	200
Pellet stoves	79	20	2.5[c]; 1.2[d]	300	60	200
Cookers	65	40	5[c]; 2.4[d]	1500	120	200

Source: Commission Regulation (EU) 2015/1185 of 28 April 2015 implementing Directive 2009/125/EC of the European Parliament and of the Council with regard to ecodesign requirements for solid fuel local space heaters.

[a] Referred to dry exit flue gas, 0°C, 1013 mbar.
[b] PM measurement by sampling a partial dry flue gas sample over a heated filter. PM measurement as measured in the combustion products of the appliance shall be carried out while the product is providing its nominal output and if appropriate at part load.
[c] PM measurement by sampling, over the full burn cycle, a partial flue gas sample, using natural draft, from a diluted flue gas using a full flow dilution tunnel and a filter at ambient temperature.
[d] PM measurement by sampling, over a 30 minute period, a partial flue gas sample, using a fixed flue draft at 12 Pa, from a diluted flue gas using a full flow dilution tunnel and a filter at ambient temperature or an electrostatic precipitator.

4.2.4 Testing Standards

The CEN Technical Committees 57 and 295 cover heating boilers and residential solid fuel appliances, respectively. Current EN standards are listed in Table 4.7. A number of these standards cover testing boilers with a rated heat output of up to 0.5 MW to ensure those available on Poland's and the EU's market are tested in compliance with the PN-EN 303-5:2012 standard [34] (Table 4.8). This standard classifies boilers into three classes—3, 4, and 5 with gradual tightening of emission limits. Compliance with this standard is not obligatory. The boiler efficiency is calculated based on NCV (net calorific value). The boilers classified as the highest class, that is, class 5, must have a thermal efficiency of not less than:

$$\eta k\ [\%] = 87 + \log Q$$

in the heat power range up to 100 kW and 89% for more powerful boilers (Q—nominal boiler power in kW). Thus, the efficiency of the boiler with a nominal power of 10 kW cannot be less than 88.0%, and that of a 30-kW boiler not less than 88.5%. The efficiency of class 5 boilers is at least 30% higher when compared to those in class 1.

TABLE 4.7
EN Standards for Solid Fuel SCIs

Technical Committee	Standard Reference	Title
TC 57 Central heating boilers	EN 303-5:2012 (under development)	Heating boilers—Part 5: Heating boilers for solid fuels, hand and automatically stocked, nominal heat output of up to 500 kW—Terminology, requirements, testing and marking
	EN 15270:2007	Pellet burners for small heating boilers—Definitions, requirements, testing, marking (applies to burners applied to nonintegral boiler)
	EN 16510-1:2018	Residential solid fuel burning appliances. General requirements and test methods
		Part 2-1: Room heaters;
		Part 2-2: Inset appliances including open fires;
	EN 16510 (under development)	Part 2-3: Cookers;
		Part 2-4: Independent boilers—Nominal heat output up to 50 kW
TC 295 Residential solid fuel burning appliances	EN 14785:2006	Residential space heating appliances fired by wood pellets—Requirements and test methods
	EN 15250:2007	Slow heat release appliances fired by solid fuel—Requirements and test methods
	EN 15281:2010	Multi-firing sauna stoves fired by natural wood logs. Requirements and test methods
	EN 15544:2009	One off tiled/mortared stoves—Dimensioning

TABLE 4.8
Limit Values of Pollutant Emissions from Solid Fuel Boilers with Rated Power Less Than/or Equal to 0.5 MW, According to the Product Standard EN 303-5: 2012

		ELVs (mg/m³)[a]								
		CO			OGC[b]			TSP		
		Class			Class			Class		
Fuel Type	Rated Heat Output, kW	3	4	5	3	4	5	3	4	5
Biofuel	≤50	3000			100			150		
	>50 to 150	2500			80			150		
	>150 to 500	1200	1000	500	80	30	20	150	60	40
Fossil fuel	≥50	3000			100			125		
	>50 do 150	2500			80			125		
	>150 do 500	1200			80			125		

Source: Testing Standard EN 303-5: 2012 Heating boilers—Part 5: Heating boilers for solid fuels, hand and automatically stocked, nominal heat output of up to 500 kW—Terminology, requirements, testing and marking.

[a] Referred to dry exhaust gases under standard conditions, 0°C, 1013 mbar at 10% O_2, emissions at 6% oxygen content.
[b] The content of organic carbon, given as the content of C (carbon) in dry exhaust gases.

It must be stressed again that TSP values arising from solid fuel combustion differ significantly according to the measurement method used, that is:

- Gravimetric method, in stack by The Association of German Engineers (VDI)
- Gravimetric method with dilution tunnel (Norwegian)

While less frequently used methods deemed to have significant potential are:

- Electrostatic precipitation
- Particles count.

Currently, research is being carried out to compare the PM measurements obtained with different test methods. Intense work is also ongoing to develop a new unified measurement method across Europe [35]. The dilution tunnel method results in significantly higher observed TSP values than the gravimetric method. This may be caused by the presence of primary, condensable aerosols under dilution conditions (produced by the condensation of gaseous precursors). This may also be caused by the less convenient measurement conditions in the direct measurement method—lower linear velocities of flue gases, cause nonuniform concentration distribution in the stack, and greater losses of particles.

Solid biofuels included in the EN 303-5 standard are categorized as follows:

- Log wood with "as used" moisture content M25, according to EN-ISO 17225-5
- Chipped wood ≤M35 (wood chipped by machine, usually up to a maximum length of 15 cm) with moisture content from M15 to M35, according EN-ISO 17225-4
- Chipped wood >M35 according to EN ISO 17225-4
- Wood pellets according to EN ISO 17225-2
- Wood briquettes according to EN ISO 17225-3
- Sawdust with moisture content ≤M50
- Non-woody biomass, such as straw, miscanthus, reeds, kernels, and grains according to EN ISO 17225-6.

Technical specifications, requirements, and testing methods, as well as relevant definitions for biogenic fuels are provided by a set of EN standards as listed in Table 4.9.

4.3 EMISSION ABATEMENTS MEASURES

To improve the environmental performance of SCIs, a broad set of technical means is available and used; these include, in particular:

- Primary options, which are focused on the improvement of the combustion process, for example, by means of rearrangement of the combustion process, fuel substitution, clean heat sources; and
- Secondary options, which include emission abatement techniques, such as catalysts, PM removal devices [22], etc.

TABLE 4.9
Standards Related to Biogenic Fuels

Standard Reference	Title
EN/TS 14588:2004	Solid biofuels—Terminology, definitions and descriptions
EN/TS 14774-1:2004	Solid biofuels—Methods for determination of moisture content—Oven dry method—Part 1: Total moisture—Reference method
EN/TS 14774-2:2004	Solid biofuels—Methods for the determination of moisture content—Oven dry method—Part 2: Total moisture—Simplified method
EN/TS 14774-3:2004	Solid biofuels—Methods for the determination of moisture content—Oven dry method—Part 3: Moisture in general analysis sample
EN/TS 14775:2004	Solid biofuels—Method for the determination of ash content
EN/TS 14778-1:2005	Solid biofuels—Sampling—Part 1: Methods for sampling
EN/TS 14778-2:2005	Solid biofuels—Sampling—Part 2: Methods for sampling particulate material transported in lorries
EN/TS 14779:2005	Solid biofuels—Sampling—Methods for preparing sampling plans and sampling certificates
EN/TS 14780:2005	Solid biofuels—Methods for sample preparation
EN/TS 14918:2005	Solid Biofuels—Method for the determination of calorific value
EN/TS 14961:2005	Solid biofuels—Fuel specifications and classes
EN/TS 15103:2005	Solid biofuels—Methods for the determination of bulk density
EN/TS 15104:2005	Solid biofuels—Determination of total content of carbon, hydrogen and nitrogen—Instrumental methods
EN/TS 15148:2005	Solid biofuels—Method for the determination of the content of volatile matter
EN/TS 15149-1:2006	Solid biofuels—Methods for the determination of particle size distribution—Part 1: Oscillating screen method using sieve apertures of 3.15 mm and above
EN/TS 15149-2:2006	Solid biofuels—Methods for the determination of particle size distribution—Part 2: Vibrating screen method using sieve apertures of 3.15 mm and below
EN/TS 15149-3:2006	Solid biofuels—Methods for the determination of particle size distribution—Part 3: Rotary screen method
EN/TS 15150:2005	Solid biofuels—Methods for the determination of particle density
EN/TS 15210-1:2005	Solid biofuels—Methods for the determination of mechanical durability of pellets and briquettes—Part 1: Pellets
EN/TS 15210-2:2005	Solid biofuels—Methods for the determination of mechanical durability of pellets and briquettes—Part 2: Briquettes
EN/TS 15234:2006	Solid biofuels—Fuel quality assurance
EN/TS 15289:2006	Solid Biofuels—Determination of total content of sulfur and chlorine
EN/TS 15290:2006	Solid Biofuels—Determination of major elements
EN/TS 15296:2006	Solid Biofuels—Calculation of analyses to different bases
EN ISO 16968:2015	Solid Biofuels—Determination of minor elements
ISO 17225-5:2014	Solid biofuels—Fuel specifications and classes

Along with the above mentioned options, one cannot neglect the really significant impact of user interference, which must, by all means, follow best practice requirements. A schematic diagram presents pathways for the improvement of environmental performance of SCIs fueled with solid fuels (see Figure 4.1).

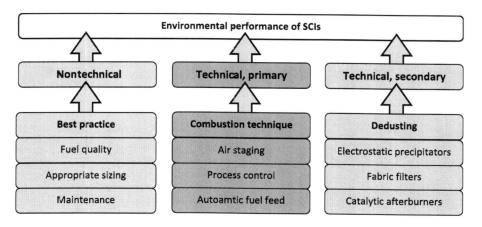

FIGURE 4.1 Possible pathways to improve SCIs' environmental performance.

4.3.1 Primary Measures for Emission Abatement

The set of primary technical measures employed to improve the combustion process includes the following:

- Appropriate air staging forms the basis for modern low-emission combustion concepts in all capacity ranges:
  - Primary combustion air is supplied directly to the fuel bed;
  - Secondary combustion air is injected into the burnout zone;
  - Tertiary air (biomass).
- The overall excess air ratio should be kept as low as possible in order to achieve high efficiencies.
- Sufficient residence time for gas phase burnout forms another requirement for low emission combustion.
- High combustion temperatures strongly support a complete gas phase burnout. Therefore, a proper insulation of combustion chamber must be assured.
- Application of an advanced control system that supports the implementation of the air staging strategies based on oxygen and/or CO content in the flue, as well as the flue temperature (lambda probes, CO-probes, flue temperature).

The primary measures are supplemented and enhanced by a set of "good practice" routines:

- Correct boiler and stove dimensioning in order to minimize stop and go operation at boilers and partial load operation at stoves.
- Appropriate user training and information for stove users as an important primary measure to avoid operating errors.

TABLE 4.10
Comparison of Pollutant Emissions by Old Appliances and BAT Solutions, g/GJ

Appliance		Fuel	TSP	CO	OGC	NO$_x$	SO$_2$	BaP[a]
Roomheater (closed fire place, cooking stove)	Old design	Wood logs, dry aged	840	5250	630	60	20	130
		Wood logs, low quality, moist	1680	10,500	3150	60	45	740
	Ecodesign	Wood logs, dry aged	25	950	125	75	n.d	13
		Wood logs, low quality, moist	31	1150	150	92	n.d.	15
Pellet stove	Old design	Pellet, optimum quality	85	530	20	95	20	55
		Pellet, low quality	260	2620	115	100	55	105
	Ecodesign	Pellet, optimum quality	11	158	32	105	n.d	5
		Pellet, low quality	12	177	35	118	n.d.	6

Source: Kubica, K. 2018. Clean combustion of solid fuels. Polish small capacity boilers in terms of Commission Regulation (EU) 2015/1189 Ecodesign. COP 24 KATOWICE. (accessed February 21, 2019) http://www.pie.pl/materialy/_upload/COP24_2018/Prezent_KK_11_12_2018/Clean_Combustion.pdf.

[a] mg/GJ.

Replacement of obsolete sources fueled with solid fuels by modern, low-emission sources employing all the relevant primary abatement measures, allows the achievement of a significant ecological effect. The typical degree of reduction achievable for biogenic fuels is shown in Table 4.10.

As given in Table 4.10, reductions may reach 90% and more, especially in the case of TSP, CO, OGC, and BaP, which are all the products of incomplete combustion.

4.3.2 Secondary Measures for Emission Abatement

Where further improvement is desirable, secondary measures for emission abatement should be undertaken. These measures, besides dedusting, include others such as denitrification, desulfurization, and OGC reduction—catalytic afterburning (post-combustion). However, both technical feasibilities and the costs should meet the specific requirements of the market, which is mainly in the domestic, commercial, and agriculture sectors.

Catalysts/afterburning for the reduction of CO and OGC emissions meant for small stoves were under development. Owing to the high flue gas temperatures required for catalytic oxidation, these devices are typically not available during start-up where the highest emissions usually occur. The pressure drop of a catalytic converter may negatively influence the combustion behavior of natural draft systems. Susceptibility to catalyst poisoning was also an issue, therefore this particular technique is not commonly applied; however, appropriate combustion process can almost completely reduce these pollutants.

In recent years, there has been a major focus on secondary measures for PM emission abatement. Different separation mechanisms are employed by apparatus commonly used in technical practice:

- Gravitational force; settling chambers
- Inertia force; baffled inertial separators and cyclone type dedusters
- Filtration; fabric filters (FF), ceramic filters, packed bed filters
- Electrostatic precipitation; electrostatic precipitators (ESPs)
- Wet separators; scrubbers, wet cyclones, etc.

Again, well-known emission abatement techniques developed for industrial, large combustion plants are usually not suitable for smaller plants. Furthermore, significantly improved SCIs tend to emit very fine particles. So, to meet the emission limit values specified by relevant requirements, only highly efficient solutions can be used, namely FFs and ESPs. Although they are technically feasible in general, the economic and maintenance issues require significant improvement before they can be implemented by SCIs. Some recent developments prove that simple ESPs applied by SCIs can reach efficiencies of up to 90%. This performance is offered under moderate capital and operating costs.

The situation is again different in the case of straw combustion, where coarse unburnt particles can be emitted from the appliance. In such cases, a first, preliminary dedusting stage can employ less efficient apparatus such as cyclone-type dedusters. Of course, it must be followed by a second dedusting stage if the relevant ELVs are not met [7].

4.4 SUMMARY

The discussion presented herein was focused on the energy and environmental performance of SCIs fueled with solid fuels, including biomass. First, the relevance of the problem was revealed, that is, the environmental and health impacts were discussed, in brief. Next, the definitions and technicalities, relevant for solid fuel SCIs, were provided, including:

- The types of appliances used
- Combustion techniques employed
- Solid fuel grades and their properties
- Qualitative and quantitative characteristics of pollutants' emissions resulting from solid fuel combustion, including emission factors.

Up to date information on relevant emission limit values, set by EU regulatory measures and testing standards used for different groups of appliances and fuels were also provided. All the above information is followed by a brief discussion of emission abatement measures, both primary and secondary, that allow for effective reduction of SCIs' environmental impacts.

REFERENCES

1. Poland's informative inventory report 2018, Warszawa, February 2018, (accessed February 21, 2019), http://www.kobize.pl/uploads/materialy/materialy_do_pobrania/krajowa_inwentaryzacja_emisji/IIR_2018_POL.pdf
2. Air quality in Europe 2016, report, EEA Report No 12/2018, (accessed February 21, 2019), https://www.eea.europa.eu//publications/air-quality-in-europe-2018
3. National Air Protection Plan, (Polish: Krajowy program ochrony powietrza do roku 2020, z perspektywą do 2030), Polish Ministry of Environment, (accessed February 21, 2019), https://archiwum.mos.gov.pl/fileadmin/user_upload/mos/srodowisko/lesnictwo/KPOP_do_roku_2020.pdf
4. Directive 2008/50/EC of the European Parliament and of The Council of 21 May 2008 on ambient air quality and cleaner air for Europe, (accessed February 21, 2019), https://eur-lex.europa.eu/legal-content/EN/TXT/HTML/?uri=CELEX:32008L0050&from=EN]
5. EMEP/EEA air pollutant emission inventory guidebook – 2016, 1.A.4 Small combustion 2016, (accessed February 21, 2019), http://www.eea.europa.eu/publications/emep-eea-guidebook-2016
6. Kubica, K., Paradiz, B., and P. Dilara. 2007. Small combustion installations: Techniques, emissions and measures for emission reduction, EUR 23214 EN–2007, ISBN 978-92-79-08203-0, ISSN 1018-5593, pp. 25–27; (accessed February 21, 2019), http://publications.jrc.ec.europa.eu/repository/bitstream/JRC42208/reqno_jrc42208_final%20version%5B2%5D.pdf
7. Mudgal, S., Turunen, L., Stewart, R., Woodfield, M., Kubica, K., and R. Kubica. 2009. Preparatory Studies for Eco-design Requirements of EuPs (II), (accessed February 21, 2019), https://www.eceee.org/static/media/uploads/site-2/ecodesign/products/solid-fuel-small-combustion-installations/bio-eup-lot-15-task6-final.pdf
8. Olanders, B. and B.M. Steenari. 1995. Characterization of ashes from wood and straw. *Biomass and Bioenergy* 8(2): 105–115.
9. Zeuthen, F.J., Livbjerg, H., Glarborg, P., and F. Frandsen. 2007. *The formation of aerosol particles during combustion of biomass and waste*, Final Report, Technical University of Denmark, Lyngby, (accessed February 21, 2019), https://core.ac.uk/download/pdf/13677500.pdf
10. Sander, B. 1997. Properties of Danish biofuels and the requirements for power production. *Biomass and Bioenergy* 12(3): 177–183.
11. Luan, C., You, C., and D. Zhang. 2014. Composition and sintering characteristics of ashes from co-firing of coal and biomass in a laboratory-scale drop tube furnace. *Energy* 69: 562–570.
12. Jensen, P.A., Stenholm, M., and P. Hald. 1997. Deposition investigation in straw-fired boilers. *Energy & Fuels* 11(5): 1048–1055.
13. Lund, H., Möller, B., Mathiesen, B.V., and A. Dyrelund. 2010. The role of district heating in future renewable energy systems. *Energy* 35(3): 1381–1390.
14. Kubica, K., Jewiarz, M., Kubica, R., and A. Szlęk. 2016. Straw combustion: Pilot and laboratory studies on a straw-fired grate boiler. *Energy and Fuels* 30(6).
15. Jenkins, B.M., Bakker, R.R., and J.B. Wei. 1996. On the properties of washed straw. *Biomass and Bioenergy* 10(4): 177–200.
16. Zeuthen, J.H., Jensen, P.A., Jensen, J.P., and H. Livbjerg. 2007. Aerosol formation during the combustion of straw with addition of sorbents. *Energy & Fuels* 21(2): 699–709.
17. RHI emission certificate, (accessed February 21, 2019), https://www.ofgem.gov.uk/key-term-explained/emission-certificate-rhi
18. Van Loo, S. and J. Koppejan. 2008. *The Handbook of Biomass Combustion and Co-firing*. London, Sterling, VA: Earthscan Ltd ISBN 978-1-84407-249-1.

19. Ross, A.B., Jones, J.M., Chaiklangmuang, S. et al. 2002. Measurements and prediction of the emission of pollutants from the combustion of coal and biomass in a fixed bed *Fuel* 81: 571–582.
20. Mitchell, E.J.S., Lea-Langton, A.R., Jones, J.M., Williams, A., Layden, P., and R. Johnson. 2016. The impact of fuel properties on the emissions from the combustion of biomass and other solid fuels in a fixed bed domestic stove. *Fuel Processing Technology* 142: 115–123.
21. Kupiainen, K. and Z. Klimont. 2007. Primary emissions of fine carbonaceous particles in Europe. *Atmospheric Environment* 41: 2156–2170.
22. Boman, Ch., Nordin, A., Boström, D., and M. Öhman. 2004. Characterization of inorganic particulate matter from residential combustion of pelletized biomass fuels. *Energy & Fuels* 18: 338–348.
23. Fitzpatrick, E.M., Jones, J.M., and M. Pourkashanian. 2009. The mechanism of the formation of soot and other pollutants during the co-firing of coal and pine wood in a fixed-bed combustor. *Fuel* 88: 2409–2417.
24. Quass, U., Fermann, M., and G. Bröker. 2000. The European Dioxin Emission Inventory - Stage II "Desktop studies and case studies"; Final Report 31.21. *North Rhine Westphalia State Environment Agency* 2: 115–120.
25. Pfeiffer, F., Struschka, M., Baumbach, G., Hagenmaier, H., and K.R.G. Hein. 2000. PCDD/PCDF emissions from small firing systems in households. *Chemosphere* 40: 225–232.
26. Ross, A.B., Bartle, K.D., Hall, S. et al. 2011. Formation and emission of polycyclic aromatic hydrocarbon soot precursors during coal combustion. *Journal of the Energy Institute* 84: 220–226.
27. Kubica, K. 2004. *Combustion and Co-Combustion of Solid Fuels*. Chapter in Management of energy in the town. ISBN 83-86492-26-0. Polish Academy of Science, Łódź: 102–140.
28. Pye, S., Jones, G., Stewart, R. et al. 2005. Costs and environmental effectiveness of options for reducing mercury emissions to air from small-scale combustion installations. AEAT/ED48706/Final report v2, (accessed February 21, 2019) http://ec.europa.eu/environment/chemicals/mercury/pdf/sci_final_report.pdf
29. Hlawiczka, S., Kubica, K., and U. Zielonka. 2003. Partitioning factor of mercury during coal combustion in low capacity domestic heating appliances. *The Science of the Total Environment*, Elsevier, 312: 261–265.
30. Stanek, W., Kubica, R., Plis, W., Bogacz, W., and A. Falecki. 2018. Wielowariantowa analiza eliminowania przestarzałych niskoefektywnych energetycznie i wysokoemisyjnych źródeł wytwarzania energii użytkowej ze spalania węgla w indywidualnych gospodarstwach domowych (Multi-variant analysis of eliminating old, inefficent and high emissive sources producing useful heat by combustion of coal, in individual households), (accessed February 21, 2019) http://ios.edu.pl/wp-content/uploads/2018/02/WIELOWARIANTOWA-ANALIZA.pdf
31. Commission Regulation (EU) 2015/1189 of 28 April 2015 implementing Directive 2009/125/EC of the European Parliament and of the Council with regard to ecodesign requirements for solid fuel boilers.
32. The Regulation of the Minister of Development and Finance on requirements for solid fuel boilers of nominal heat output of up to 500 kW, August 1, 2017.
33. Commission Regulation (EU) 2015/1185 of 28 April 2015 implementing Directive 2009/125/EC of the European Parliament and of the Council with regard to ecodesign requirements for solid fuel local space heaters.
34. Testing Standard EN 303-5: 2012. Heating boilers - Part 5: Heating boilers for solid fuels, hand and automatically stocked, nominal heat output of up to 500 kW - Terminology, requirements, testing and marking.

35. Schön, C. and H. Hartmann. 2018. Status of PM emission measurement methods and new developments Technical report, IEA Bioenergy, (accessed February 21, 2019), http://task32.ieabioenergy.com/wp-content/uploads/2018/09/IEA-Paper_PM_determination.pdf
36. Kubica, K. 2018. Clean combustion of solid fuels. Polish small capacity boilers in terms of Commission Regulation (EU) 2015/1189 Ecodesign. COP 24 KATOWICE. (accessed February 21, 2019) http://www.pie.pl/materialy/_upload/COP24_2018/Prezent_KK_11_12_2018/Clean_Combustion.pdf

5 Monitoring and Modeling of Pollutant Emissions from Small-Scale Biomass Combustion

Janusz Zyśk

CONTENTS

5.1 Introduction 127
5.2 Air Quality.. 128
5.3 The Impact of Emissions from the Residential Fuel Burning Sector
    on Air Quality... 129
5.4 Emissions from the Residential Sector... 133
5.5 The Share of Biomass Use in the Residential Sector.................................... 134
5.6 Estimation of Emissions from Small-Scale Biomass Combustion............... 135
5.7 Models of Small-Scale Biomass Combustion... 136
5.8 Measurements of Pollutants from Biomass in Ambient Air........................ 138
5.9 Modeling of Atmospheric Dispersion.. 138
5.10 Conclusions.. 140
References... 140

5.1 INTRODUCTION

Improving the air quality is one of the most crucial and difficult challenges faced by the majority of countries in the world today. Even in developed countries, which are leaders in the field of emission reduction and emission standards, still a large part of the population is exposed to concentrations of pollutants, such as particulate matter, benzo(a)pyrene, ozone, and nitrogen oxide, that exceed these standards. One of the most important sources of emissions that directly affect local air quality is the residential sector. People use fuels at home to heat a living space. Often, poor quality fuel is burned in old, ineffective, and uncontrolled devices that cause high emissions. Emissions from such sources are released at low altitude, so they spread over a relatively small area and significantly contribute to local pollution concentrations. Biomass is the most important, or one of the most important, sources of heating in many regions of the world. Determination of the amount of emissions from these sources has a key impact on taking action to reduce pollution and improve air quality, and consequently to reduce their exposure and impact on human health. The aim of this chapter is to indicate the current state and review of scientific achievements in the area of monitoring and modeling of pollutant emissions from small-scale biomass combustion. In this chapter, air quality worldwide, the impact of polluted air on human health, and the contribution of emissions from the residential sector to air pollution concentrations will be presented. Estimation of the amounts of emissions from this sector and the share of biomass combustion in the residential sector will also be determined. Then, models developed for assessment of the emissions from biomass combustion and the possibilities of modeling their transport in the atmosphere will be discussed. An attempt is also made to determine the method for measuring pollutants from biomass in ambient air. Reference is made both to the data and experiences of developed and developing countries.

(*General remark*. In this chapter, the word "household" is used for emissions realized in homes (often from open fires) and that affect directly indoor air quality. The words "domestic" or "residential" links to emissions from human residences (houses), but those that are released directly into the air through chimneys.)

5.2 AIR QUALITY

The impact of air pollution on human health has been the subject of numerous studies conducted in many places around the world for many years. The obtained results confirm that exposure of the population to concentrations of air pollutants have negative health effects, especially for respiratory, cardiovascular, and nervous systems. In addition, air pollution is linked to an increased risk of cancer. Children and the elderly are particularly exposed to these effects. The most important harmful substance is particulate matter, especially the smallest particles with diameters of 2–3 μm and smaller, which are so-called PM2.5 (for comparison, the thickness of a human hair is approximately 50 μm). While larger particulate matter fractions are retained in the upper respiratory tract (nose, sinuses, pharynx, or larynx), small fractions can be deposited in the pulmonary alveolus and next enter the circulatory system and various organs of the human body. In addition, the chemical composition of dust is also extremely important. Particulate matter consists of heavy metals and polycyclic aromatic hydrocarbons (PAHs), including benzo(a)pyrene (B(a)P).

The health effects mentioned above cause an increased mortality rate and shorten life expectancy. According to World Health Organisation, 23% of deaths globally are caused by poor environment quality (WHO, 2018). In 2016, air pollution was linked to 7 million deaths, which was 12% of the global mortality. This number includes deaths owing to ambient air pollution and indoor (household) air pollution. The mortality rates (per 100,000 population) owing to poor air quality (ambient and household) were estimated in Africa at 180.2, in South-East Asia at 160.4, in the Eastern Mediterranean at 123.7, in the Western Pacific at 103.1, in Europe at 36.3, and in the Americas at 30.6. The WHO estimates show that the highest mortality rate attributed to ambient and household air pollution is in Sierra Leone and equals 324.1 (per 100,000 population) and the lowest is in Canada were the mortality rate was estimated at 7. In Europe, the lowest mortality rate was noticed in Finland (7.2) and the highest in Bosnia and Herzegovina (79.8). For comparison, this factor was estimated in China at 112.7, in the USA at 13.3, India at 184.3, in Poland at 36.3, and in Germany at 16.0. As in developed countries, the main problem of air quality is high pollution concentrations in ambient air, while in developing countries it is still household emissions that have a significant impact on human health. It should be noted that still more than 40% of the population globally use polluting fuels such as wood, coal, charcoal, and crop waste for cooking. These fuels are burned in households and have a huge significant impact on indoor air quality. Globally, household air pollution was responsible for 3.8 million deaths in 2016. It was 7.7% of the global mortality (WHO, 2018). The data from the Institute for Health Metrics and Evaluation also show that approximately one third of deaths are linked to air pollution from indoor solid fuels and the rest from ambient air pollution (IHME, 2016). These estimation results show that the majority of deaths linked to ambient air pollution are caused by high concentrations of particular matter. New Delhi in India was the city most polluted by particulate matter in the world in 2016, with an annual mean concentration of PM2.5 of 173 µg/m^3 and PM10 of 319 µg/m^3 (WHO, 2018).

Residential emissions have also had a significant impact on ambient air quality, mainly at a local scale. Residential emissions released during the burning of polluting fuels for cooking or heating are the component of low-stack emissions. Low-stack emissions are defined as the emissions released into the air at below 40 m above ground level. Other sources classified as low-stack emissions are road transport, agriculture, dusty surfaces, industry, and heat plants where the stack height is lower than 40 m, etc. To assess the ambient air quality, standards for the protection for health or vegetation were developed for selected pollutants which are treated as indicators of air quality (EEA, 2018a; EU, 2018). The standards also have informative and educational significance for inhabitants and set goals for the authorities. The European Union (EU) sets limits or target values, which are generally less restrictive than air quality guidelines (AQG) and reference levels (RL) developed by the World Health Organisation (Table 5.1). In EU-28, around 8% of the urban population is exposed to PM2.5 concentrations above EU legal standards. For PM10, this factor equals 19%, ozone 30%, NO$_2$ 8%, and B(a)P 24%. Taking into account the air quality guidelines of the World Health Organisation, 85% of the EU-28 population is exposed to levels of PM2.5 above reference concentrations. For PM10, this factor equals 52%, ozone 98%, NO$_2$ 8%, and B(a)P 90%. In EU-28, where the total population is approximately half a billion, years of life lost (YOLL) attributed to PM2.5, NO$_2$, and O$_3$ are estimated at 4,415,000, 795,000, and 180,000, respectively (EEA, 2018a). The numbers presented above show that air quality is a crucial problem and significant efforts should be made to avoid emissions and improve air quality. In order to decrease the negative health impact of ambient air pollution, it is crucial to know its sources and the quantity of emissions.

5.3 THE IMPACT OF EMISSIONS FROM THE RESIDENTIAL FUEL BURNING SECTOR ON AIR QUALITY

The estimations presented by Karagulian et al. (2015) for 2015, show that globally, in cities, the main sources contributing to PM2.5 concentrations are

TABLE 5.1
Air Quality Guidelines (AQG) and Reference Levels (RL) Sets by The World Health Organisation (WHO) and Limit/Targets Values Set by the European Union (EU) for Concentrations of Selected Pollution in Ambient Air

	WHO		EU			
Pollutant	Averaging Time	AQG/RL	Averaging Time	Limit/Target Value	Permitted Excess Each Year	Unit
SO_2	10 minutes	AQG: 500	1 hour	limit value: 350	24 times	µg/m³
	24 hours	AQG: 20	24 hours	limit value: 125	3 times	µg/m³
NO_2	1 hour	AQG: 200	1 hour	limit value: 200	18 times	µg/m³
	year	AQG: 40	year	limit value: 40		µg/m³
CO	1 hour	AQG: 30				mg/m³
	8 hours	AQG: 10	8 hours	limit value: 10		mg/m³
C_6H_6	year	RL: 1.7	year	limit value: 5		µg/m³
PM10	24 hours	AQG: 50	24 hours	limit value: 50	35 times	µg/m³
	year	AQG: 20	year	limit value: 40		µg/m³
PM2.5	24 hours	AQG: 25				µg/m³
	year	AQG: 10	year	limit value: 25		µg/m³
Pb	year	AQG: 0.5	year	limit value: 0.5		µg/m³
As	year	RL: 6.6	year	target value: 6		ng/m³
Cd	year	AQG: 5	year	target value: 5		ng/m³
Ni	year	RL: 25	year	target value: 20		ng/m³
B(a)P	year	RL: 0.12	year	target value: 1		ng/m³
O_3	8 hours	AQG: 100	8 hours	target value: 120	25 days	µg/m³

traffic—25%, unspecified sources of human origin—22%, residential fuel burning—20%, natural dust and salt—18%, industrial activities—15%, and others 22%. The contribution of PM10 concentrations in emission from traffic equals 25%, unspecified sources of human origin—22%, natural dust and salt—22%, domestic fuel burning—15%, industrial activities—18%, and others 20%. However, the share of these sources in different regions of the world varies significantly (Table 5.2). In Central and Eastern Europe, which is the most polluted region in the whole of Europe, 45% of PM10 and 32% of PM2.5 occurs in emission that are released from the domestic sector. In comparison, in Western Europe, only 7% of PM10 and 15% of PM2.5 emissions are released from the domestic sector. In this sector, a significant amount of fossil fuels are burned to heat houses and flats. Often, fuels are burned in old, low-efficiency, high-emissions furnaces and boilers. For example, in Poland, one of the most polluting countries in Europe according the analyses made by the Institute of Environmental Protection, 90% cases of the emissions from the domestic sector were the main cause of exceeding standards of PM2.5 and PM10 concentrations (GIOŚ, 2018). A detailed analysis prepared for the Upper Silesian agglomeration

TABLE 5.2
The Share of Emissions from Domestic Fuel Burning to Overall Emissions Contributing to Urban Ambient PM2.5 and PM10 Concentrations

	Share of domestic fuel burning [%]	
Region	PM2.5	PM10
Africa	34	21
Central and Eastern Europe	32	45
Northern China	15	19
Southern China	21	7
India	16	11
Republic of Korea	5	3
Oceania	13	–
South Eastern Asia	19	–
USA	12	15
Northwestern Europe	22	24
Western Europe	15	7
Southwestern Europe	12	7
Low- and middle-income countries	22	18
High-income countries	17	15
World	20	15

Source: Adapted from Karagulian F. et al., 2015. 5(4), 475–483.

in Poland also shows that area emissions (mainly residential sector sources) have the highest contribution to local air quality, despite this region of Poland having a developed industry and road transport owing to a relatively large number of inhabitants (Figure 5.1). Only in the case of NO_x emission road transport in the main source.

It should be noted that emissions from the domestic (residential) sector is mainly released into the air during the season when people need to heat their homes and then the concentration of pollutants is very high and exceeds legal standards (Figures 5.2 and 5.3). Furthermore, as was mentioned above, the emissions from the domestic sector are released from low height chimneys (low-stack emissions), which means that pollutants are not transported over large distances and their highest concentrations are noticed close to the place of emission. In the case of the Upper Silesian agglomeration in Poland, the local sources (area, line, and point) contribute to local concentrations of 58%, 55%, 58%, and 81% of PM10, PM2.5, B(a)P, and NO_x, respectively (Figure 5.1). Research carried out in selected cities located in the Danube region showed that the percentage of PM2.5 attributable to the residential sector located in the city is lower than 10%; only in Prague did this share reach almost 15% (Thunis et al., 2017).

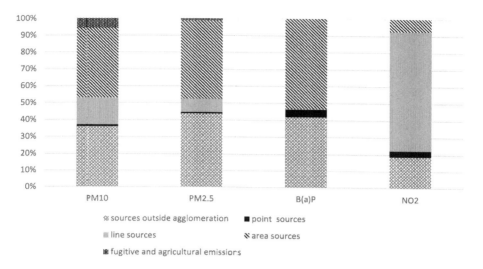

FIGURE 5.1 Causes of exceedances in Upper Silesian agglomeration. (Based on UMWS. 2017. The air protection program for the area of the Śląskie Voivodeship aimed at achieving the limit levels of substances in the air and the exposure concentration obligation.)

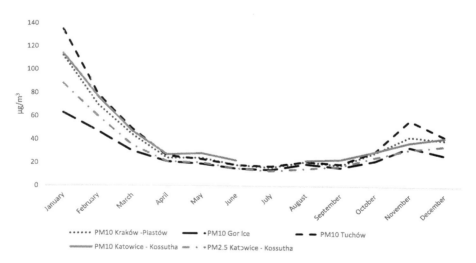

FIGURE 5.2 The monthly mean concentrations of PM10 and PM2.5 in 2017 at selected stations in Poland. (Data from WIOS-Krakow. 2018. Air quality monitoring system. Voivodeship Inspectorate for Environmental Protection in Kraków; WIOS-Katowice. 2018. Air quality monitoring system. Voivodeship Inspectorate for Environmental Protection in Katowice.) *At station Katowice–Kossutha data of PM10 concentration in July were not reported because of a technical break.

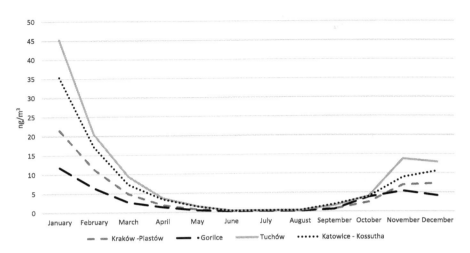

FIGURE 5.3 The monthly mean concentrations of B(a)P in 2017 at selected stations in Poland. (Data from WIOS-Krakow. 2018. Air quality monitoring system. Voivodeship Inspectorate for Environmental Protection in Kraków; WIOS-Katowice. 2018. Air quality monitoring system. Voivodeship Inspectorate for Environmental Protection in Katowice.)

5.4 EMISSIONS FROM THE RESIDENTIAL SECTOR

The residential (domestic) sector is one of the highest emission sources of pollutants, especially particulate matter, benzo(a)pyrene, and carbon monoxide. The emission inventories for various countries and regions show that the share of emissions from the residential sector compared to overall emissions can reach 65% for PM10, 81% for PM2.5, 67% for CO, 58% for nonmethane volatile organic compounds (NMVOC), and 80% for B(aP), respectively.

In the EU-28, the residential, commercial, and institutional sectors were the largest contributors of emissions of PM10, PM2.5, CO, black carbon (BC), and B(a)P in 2015. In the EU-28, from these sectors were emitted 17%, 14%, 39%, 56%, 17%, 48%, 45%, 5%, 21%, 16%, 12%, 12%, and 68% of total emissions of SO_x, NO_x, PM10, PM2.5, NMVOC, CO, BC, As, Cd, NI, Pb, Hg, and B(a)P, respectively (EEA, 2018b). This share varies within the EU-28 significantly. For example, in Poland, the contribution of the residential sector to overall national emissions is estimated at 61% of CO, 20% of NMVOC, 45% of PM10, 48% of PM2.5, and 80% of B(a)P (KOBIZE, 2018). In France, the share of the residential sector is estimated at 41% of CO, 21% of NMVOC, 38% of PM10, 42% of PM2.5, and 63% of B(a)P (EMEP, 2018). In Romania, the values for CO equals 67%, NMVOC—28%, PM10—65%, PM2.5—81%, and 71%—B(a)P. In contrast, only 8% of PM10, 14% of CO, 8% of NMVOC, and 15% of PM2.5 are emitted by the residential sector in the Netherlands. There, as in other countries, most of B(a)P comes from the residential sector—82%.

In China, 14% of SO_2, 42% of CO, 52% of PM10, 38% of PM2.5, 40% of BC, and 78% of OC are emitted by the residential sector (Hong et al., 2017; Li et al., 2017).

The contribution of residential and household emissions in total in India in 2011 is as follows: PM10—30%, PM2.5—40%, BC—53%, CO—67%, NO_x—10%, SO_2—5%, NMVOC—58%, and 66%—OC (Sharma et al., 2016; Venkataraman et al., 2018).

A marginal share of emissions from the residential sector was observed in New Delhi, which is one of the most polluted cities in the world; there is only 2% emission of PM10, 1% of SO_2, 4% of CO, and 1% of NMVOC from the residential sector (Gurjar and Nagpure, 2015). There, transport and roads are the main contributors to emissions. For example, 59% of PM10 came from road dust and 8% from transport. Road transport is responsible for approximately 70% of total emissions of NMVOC, CO, and NO_x.

In the USA, the emissions from fuel combustion in stationary sources beyond electric generation and industrial processes is also very low and equals 0.7% of PM10, 4% of CO, 0.5% of PM2.5, and 3% of SO_2 (EPA, 2018).

5.5 THE SHARE OF BIOMASS USE IN THE RESIDENTIAL SECTOR

The emissions from the residential sector depend on the structure of the technology and fuels that are used to heat homes and houses. Of course, the climate of a given country or region is also important. In areas where the ambient temperature is low and the cold season is longer, more heating is required. The type of construction and their thermal resistance influence energy and fuel use by particular buildings. In developed countries, houses are better insulated than in other regions. The contribution of residential emissions to total emissions is determined by the emissions from other sectors such as power plants, industry, transport, agriculture, etc.

In India in 2011, approximately 200 million tons of oil equivalent (Mtoe) of energy were used in both the household and residential sectors for cooking, lighting, and space heating. Almost 80% of this energy came from traditional biomass. More than 60% of rural households use wood fuel, 12% rely on crop residues, and 11% on cow dung cakes (Sharma et al., 2016). As in India, biomass remains the main source of emissions from the residential sector, more than 90% of pollutants such as PM10, PM2.5, OC, NMVOC, and CO are emitted by the residential sector as a result of biomass combustion.

In China, one third of households use biomass for cooking and 13% for heating (in rural areas this is 19% and 5% in urban areas). The share of solid fuels, including solid biomass such as wood, straw, and stalk, decreases in favor of gas and electricity for district heating. It is worth emphasizing that approximately 34.1% of Chinese families remain without heating during the cold season (Jingjing et al., 2001; Duan et al., 2014).

In the EU-28, the main source of heat is gas followed by electricity (24%) and solid biomass (15%) (Eurostat, 2018). Looking at the data presented in Figure 5.4, which shows the structure of energy consumption, it can be assumed that the main source of pollution released into the atmosphere from the residential sector is combustion of soild biomass. The development of a high resolution (7×7 km) anthropogenic aerosol emission inventory for Europe identified residential wood combustion as the largest source of organic aerosol in Europe (Denier van der Gon et al., 2015). Therefore, the correct estimation of emissions from biomass as well as the assessment of its impact on air quality is very important for the EU-28.

Monitoring and Modeling of Pollutant Emissions 135

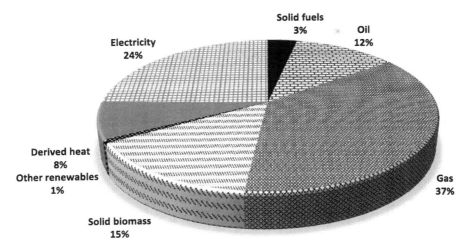

FIGURE 5.4 Final energy consumption in EU-28 in the residential sector split into energy carriers in 2016. (Based on Eurostat. 2018. *Energy Balance Sheets—2016 Data*, 2018 edition.)

The results of estimates from the EU show that almost 90% of total coal used in the residential sector in the EU-28 (9507 kiloton of oil equivalent (ktoe)) were burned in Poland (8554 ktoe) (Eurostat, 2018). There, in 2016, in the whole residential sector, 19,747 ktoe of energy were consumed, including 2662 ktoe (13.5%) of biomass. In comparison, two decades earlier (1996), in Poland 2412 ktoe of biomass were used for space heating, although the total energy consumed in the residential sector was estimated at 23,027 ktoe. Therefore, the amount of solid biomass increased, despite a reduction of total energy consumption in this sector (Eurostat, 2018).

5.6 ESTIMATION OF EMISSIONS FROM SMALL-SCALE BIOMASS COMBUSTION

The total emissions from biomass in the residential sector depends on the amount of burned biomass, its type, physical and chemical properties, devices used, and the conditions of the combustion process. Assessment and estimation of emissions data from the household/residential sector can be carried out with use of two approaches: top-down and bottom-up. Because of the specifics of this sector, where emissions are released from multiple small-scale sources, the emissions volume is estimated often with the use of a top-down approach. Emissions inventories at the national, regional, or city level are disaggregated to high resolution data based on population density or distribution of space for heating, etc. For a bottom-up approach, the emissions factors used are usually developed for various combustion installations and fuel types. The range of these factors can be very wide; however, it gives information about the emissivity of a given fuel. As is shown in Table 5.3, solid biomass can be a significant source of NO_x, CO, NMVOC, particulate matter, and B(a)P. For biomass, the range of emissions factors is the widest because of the diversity of the biomass itself (e.g., wood, pellets, straw) and

TABLE 5.3
The Range of Applied Emissions Factors for Various Technologies [g/GJ] of Utility Energy]

Pollutant	SO$_2$	NO$_x$	CO	NMVOC	PM10	B(a)P
Solid biomass	0–158	37.5–187	12.5–11764	25–3750	18–2338	0.005–1.8
Coal	0–1538	45–316	235–8750	312–1050	15–600	$7.9 * 10^{-05}$ −0.5
Gas	0.3–1.17	32–151	8.83–44.7	1.17–2.8	0.5–3.3	0
Oil	58–595	22–80	44–172	0.4–1.1	1.44–3.44	0–0.008
LPG	0.21–0.42	33–298	11–44	0.7–2.8	0.22–55	$0–5.9 * 10^{-07}$

Source: Based on Kubica, K. 2003a. Environment Pollutants from Thermal Processing of Fuels and Biomass. Thermochemical Processing of Coal and Biomass; Kubica, K. 2003b. Thermochemical Transformation of Coal and Biomass. Thermochemical Processing of Coal and Biomass; Kubica K. et al. 2003. Emission of pollutants from combustion of coal and biomass and its co-firing in small and medium size combustion installation. *Fourth Joint UNECE Task Force and EIONET Workshop on Emission Inventories and Projections in Warsaw*, 145–232; Hobson, M. 2005. Emission factors programme Task 7—Review of Residential and Small-Scale Commercial Combustion Sources; TUV. 2012. TÜV Rheinland Energy GmbH Am Grauen Stein; EMEP/EEA. 2017. EMEP/EEA air pollutant emission inventory guidebook 2016—July 2017.

the techniques of combustion used (e.g., stoves, boilers). The detailed emissions factors for various biomass fuels and technologies are presented in many scientific publications (e.g., Andreae and Merlet, 2001; Kubica, 2003a,b; Kubica et al., 2003; Hobson, 2005; TUV, 2012; Sharma et al., 2016; EMEP/EEA, 2017; Venkataraman et al., 2018). The use of emissions factors to estimate and monitor the pollutants from biomass, as well as from other fuels, can cause uncertainties. It is difficult to estimate the amount of fuel consumption, its quality, and parameters, as well as precisely determine its technology and combustion process parameters, especially if such estimation is done at the city or country level. However, owing to the specificity of the residential sector, it is not possible to measure the amount of emissions from individual chimneys, so using emissions factors seems to be the optimal method. The emissions inventory prepared using estimates and assumptions can be indirectly evaluated with the use of air quality modeling systems. The simulation of dispersion of atmospheric pollutants, which are as the inputs of the estimated emission data, gives the concentrations which are overestimated or underestimated compared to measurements of pollutant concentrations in ambient air. The obtained concentration discrepancy may result from overestimation or underestimation of estimated emissions. That study prepared for Europe showed the underestimation of emissions from organic aerosols, particularly during winter (Denier van der Gon et al., 2015).

5.7 MODELS OF SMALL-SCALE BIOMASS COMBUSTION

Biomass has a different composition, which mainly relates to its origins (e.g., wood, agriculture residues, and waste). Generally, biomass used for burning contains various volatile and nonvolatile compounds. Biomass also includes compounds that are not burned and remains in ash. Biomass consists of, among others, cellulose, lignin, lipids, proteins, starches, water, hemicelluloses, and simple sugars. The main chemical elements included in biomass are carbon, oxygen, and hydrogen. Additionally, a substantial amount of nitrogen, sulfur, and chlorine can be found in biomass (Magdziarz et al., 2011; Miljkovic et al., 2013).

The combustion of biomass can be split into three process: drying, pyrolysis, and combustion of gases and char in a bed (Williams et al., 2001).

Nowadays, many numerical methods exist to simulate the combustion process of biomass (Yongtie et al., 2017; Gómez et al., 2018; Rodriguez-Alejandro et al., 2018). Unfortunately, a relatively small amount of research is focused on analyses and prediction of emissions; however, a few computer programs to model the emissions from combustion processes of biomass exist and the results were published (Magdziarz et al., 2011; Zahirović et al., 2011; Kazakov and Frenklach, 2018). These programs based on various mechanisms and chemical models such as CHEMIKN, COMSOL, KINALC, DRM22, Kilpinen, and GRI. The mechanisms consist of different compounds and reactions. For example, in DRM22, 22 compounds and 104 elementary reactions are included. In the GRI mechanism, 49 species and 227 reactions were applied. The use of each mechanism can lead to very diverse results of gas concentrations in the combustion process (Mehrabian et al., 2014). The input data to model are provided: mass flow reactants, pressure, temperature profile, geometry of the test chamber, residence time, etc. (Zajemska et al., 2014).

The modeling results by Magdziarz et al. (2011) for wood biomass show that in the combustion chamber, the fuel oxidizes at temperatures above 600 K. Near this temperature, the CO concentrations are the highest and as the temperature rises, its concentration decreases. Above 1000 K, CO is rapidly reduced to CO_2, while at this temperature the concentration of NO rapidly increases. Similarly, the concentration of NO increases together with residence time and for CO the opposite dependence was observed. Additionally, it was shown that the content of SO_2 in the combustion chamber is at approximately the same level regardless of the combustion temperature. The model Chemkin-Pro was also used in the work of Radomiak et al. (2016). They analyzed wood pellets with 7% and 0% glycerin. In this case, the concentration of NO was the highest in the case of short residence time. The glycerin reduces the emissions of chlorine compounds.

The calculations of Yang et al. (2004) show that the primary air flow rate and moisture level in the fuel have an effect on the emission of the unburned gaseous fuel, for example, CO at the bed top. A lower air flow rate and a water content in fuel produce higher emissions of CO. Numerical studies with the use of PSR (perfectly stirred reactor) models were carried out to predict emissions of soot from coal, pine, and coal/pine fuels (Fitzpatrick et al., 2009). Owing to this study's results, the production of soot is more effective from pine that from coal. The PSR model was also used to determinate gaseous products from biomass pyrolysis (Lee et al., 2007). The C and H_2O mole fraction decreases with increasing of temperature. The reverse dependence is observed in the case of CO and H_2O. The maximum of concentrations of C_2H_2, C_6H_6, C_2H_4, and C_3H_3 are observed in the range of temperature 340–360°C.

An attempt to determine the concentration of pollutants (NO_x, HCI, and SO_2) in flue gases formed during the biomass (wood waste, straw, and hay) combustion and validation of them against the measurements was presented by Zajemska et al. (2014). The results show an acceptable agreement between numerical and experimental data. Likewise, compatible measurements and simulation results of CO_2 and CO concentrations from four tests were obtained by Gómez et al. (2016).

5.8 MEASUREMENTS OF POLLUTANTS FROM BIOMASS IN AMBIENT AIR

So far, models of biomass emissions and measurements have not pointed to any absolutely specific chemical tracers for the burning of biomass wood (AQEG, 2017). The main products of biomass burning as elemental and organic aerosols are emitted in large amounts from others sources.

As the biomass includes cellulose, one of the products of decomposition of cellulose is the organic compound levoglucosan. Potassium also can be used to identify the pollutants emitted from biomass combustion; however, a wide range of data on the relationship between potassium and biomass smoke concentration exist, because of the strong dependence of potassium emissions on the combustion process. Additionally, potassium appears in airborne sea spray and soil. The ratio of levoglucosan/K was reported at a ratio of 6.25–216 (Puxbaum et al., 2007; AQEG, 2017). The measurement of these compounds in the CARBOSOL sampling sites led to the following observations and conclusions: (1) for fires in small ovens the conversion factor (CF) for biomass smoke OC = CF * levoglucosan is approximately 5, while for fire places it is around 7–10, and for open wild fires it is above 10; (2) the highest monthly observation of wood smoke was 14 µg/m³; (3) biomass combustion contributes to organic matter 47%–68% at rural flat terrain sites during the winter; and (4) the share of biomass emissions in organic matter during the cold season is much higher than during rest of the year (Puxbaum et al., 2007). Harrison et al. (2012) proposed the relationship between biomass smoke mass and levoglucosan at 11.2. Based on this, the average concentrations of wood smoke during the summer and winter sampling periods were estimated at 0.23 µg m^{-3} in Birmingham and 0.33 µg m^{-3} in London.

The studies by Fine et al. (2001) and AQEG (2017) show that selected PAH retene also can be a good marker of biomass combustion. It was observed that the concentrations of retene are related to wood burning and there is no relationship between retene concentrations and emissions from traffic.

5.9 MODELING OF ATMOSPHERIC DISPERSION

Over the last three decades, owing to the development of computer systems and their capabilities, a modeling system, including systems dedicated for air quality modeling, have been developed. The systems for modeling air pollution dispersion allow for the following: (1) assessment of concentrations and deposition in places where the measurements are not conducted; (2) indication of emission sources; (3) assessment of the impact of sources (individual, whole sectors, or areas); (4) assessment of the quality of the emissions inventory; (5) assessment of the impact of efforts taken to improve air quality; and (6) prediction of air quality. Many numerical models of Eulerian (Polyphemus, ADOM, CAMx, PMCAMx, EMEP MSC-W, CYM-Hg, GEOS-Chem, ECHMERIT, MOZART, DEHM, GLEMOS, ADOM) and Lagrangian (HYSPLIT, RCTM) types have been developed to evaluate the atmospheric dispersion of various pollutants (gases, aerosols, heavy metals, and radionuclides) on regional, continental, and global scales (Zyśk et al., 2015). Among the Lagrangian models exist a group of Gaussian models (e.g., CALPUFF). A large part of the models consider pollution in the gaseous, aqueous, and particulate phases. In these models, both chemical reactions and physical transformations, as well as removal of pollutants through wet and dry deposition are included (Zyśk et al., 2015). As input, natural and anthropogenic emissions data (point, volume, area), meteorological, land use, boundary, and initial concentration data are used. Every model can be used for various works, including the assessment of pollution from biomass combustion.

The modeling concentrations of organic aerosols with wood combustion as the main source in Europe was carried out with the use of two models: PMCAMx and EMEP MSC-W. The simulation was based on a high resolution, bottom-up emissions inventory with, among others, the emissions of organic and elemental carbon. The results revealed the underestimations of organic aerosol emissions estimation in the cold season, especially for regions where wood combustion dominates residential heating (Denier van der Gon et al., 2015).

The impact studies of biomass burning from open fires with the use of the WRF-CHEM model showed that emissions from biomass have a strong effect on organic aerosols (primary and secondary) concentrations, but the impact on O_3 ambient concentrations was marginal (Lei et al., 2013). The opposite results for impact of biomass burning on O_3 concentration were obtained in work of Lin et al. (2014). In this work, concentrations of O_3 and also of CO and PM were 2–3 times higher during the biomass burning season than in other seasons. The results were obtained from a simulation run of the WRF-CHEM model and from satellite observation.

Work where models were used to evaluate the emission data were also carried out by Williams et al. (2012). The TM4 chemistry transport model was run to evaluate African biomass burning emissions estimates on global air quality. In this study, the concentrations of O_3, CO, and NO_x were considered.

The studies conducted for Tianjin, China with the use of a Gaussian plume model dedicated to atmospheric dispersion of PAHs was used to investigate the impact of various sources on air quality. In this model, 16 individual PAHs were taken into consideration. Biomass combustion contributed to 21.5% of overall concentrations of PAHs in ambient air (Tao et al., 2006).

Recently, modeling of atmospheric transport of pollutants emitted from biomass burning (fires) over Southeast Asia was also investigated with the use of both Eulerian (Models-3/Community Multiscale Air Quality—CMAQ) and Lagrangian (Numerical Atmospheric-dispersion Modeling Environment—NAME) dispersion models was done (Lin et al., 2013; Hertwig et al., 2015). It made it possible to assess the impact of these phenomena on air quality and the exposure to people.

In modeling work of Zhou et al. (2013), BC was treated as the tracer of natural biomass burning. The impact of biomass burning (mainly wildfires) to the composition of tropical tropopause layer (TTL) and lower stratosphere (LS) was investigated and CO was used as the "tape recorder" (Duncan et al., 2007).

Systems for modeling air pollution dispersion have been involved mainly in analyses of natural biomass burning. The episodes of wildfires in selected regions have had a significant impact on air quality. The research shows the impact mainly on O_3, CO, and organic aerosols concentrations. A few simulations run for anthropogenic biomass burning show that from residential sectors, the significant impact is on organic aerosols concentrations and uncertainty in emissions inventories.

5.10 CONCLUSIONS

Biomass combustion in small-scale units, mainly in the residential sector, has the most significant impact on the local air quality, especially in regions were biomass dominates soild fuels used in the residential sector. To these regions belong the well-developed Western and Scandinavian countries of the European Union, as well as the developing countries with relatively low incomes such as India. The reduction of emissions from biomass will contribute to improvements in air quality. Still, despite many efforts, local air quality in urban and rural areas is poor. In the well-developed EU-28, almost 90% of the population breathes air with excess concentrations of B(a)P. It should be noted that the biomass impact on the environment and air quality is still relatively poorly recognized. There are still problems with estimation of emissions. The models of biomass combustion in small-scale units are not widely developed. The air quality modeling systems are used mainly for natural biomass emissions. Very important from the global viewpoint is that biomass is the fuel with almost zero emissions of CO_2. Keeping in mind the emissions of greenhouse gases and their impact on the Earth, it is worth taking care of the of local air quality.

REFERENCES

Andreae, M.O., Merlet, P. 2001. Emission of trace gases and aerosols from biomass burning. *Global Biogeochemical Cycles*, 15(4), 955–966.

AQEG. 2017. *The Potential Air Quality Impacts from Biomass Combustion*. Report from the Air Quality Expert Group (AQEG) to the Department for Environment, Food and Rural Affairs. Scottish Government; Welsh Government; and Department of the Environment, Northern Ireland.

Denier van der Gon, H.A.C., Bergström, R., Fountoukis, C., Johansson, C., Pandis, N., Simpson, D., Visschedijk, A.J.H. 2015. Particulate emissions from residential wood combustion in Europe – revised estimates and an evaluation. *Atmospheric Chemistry and Physics*, 15, 6503–6519.

Duan, X., Jiang, Y., Wang, B., Zhao, X., Shen, G., Cao, S., Huang, N., Qian, Y., Chen, Y., Wang, L. 2014. Household fuel use for cooking and heating in China: Results from the first Chinese environmental exposure-related human activity patterns survey (CEERHAPS). *Applied Energy*, 136, 692–703.

Duncan, B.N., Strahan, S.E., Yoshida, Y. 2007. Model study of the cross-tropopause transport of biomass burning pollution. *Atmospheric Chemistry and Physics Discussions, European Geosciences Union*, 7, 2197–2248.

EEA. 2018a. *Air quality in Europe — 2018 report*. EEA Report No 12/2018.
EEA. 2018b. *European Union emission inventory report 1990–2016 under the UNECE Convention on Long-range Transboundary Air Pollution (LRTAP)*. EEA Report No 6/2018.
EMEP. 2018. The European Monitoring and Evaluation Programme [WWW Document]. URL http://www.emep.int/
EMEP/EEA. 2017. EMEP/EEA air pollutant emission inventory guidebook 2016—July 2017.
EPA. 2018. 2014 National Emissions Inventory Report.
EU. 2018. Directive 2008/50/EC of the European Parliament and of the Council of 21 May 2008 on ambient air quality and cleaner air for Europe.
Eurostat. 2018. *Energy Balance Sheets — 2016 Data*, 2018 edition.
Fine, P., Cass, G., Simoneit, B.R.T. 2001. Chemical characterization of fine particle emissions from fireplace combustion of woods grown in the northeastern United States. *Environmental Science & Technology*, 35, 2665–2675.
Fitzpatrick, E.M., Bartle, K.D., Kubacki, M.L., Jones, J.M., Pourkashanian, M., Ross, A.B., Williams, A., Kubica, K. 2009. The mechanism of the formation of soot and other pollutants during the co-firing of coal and pine wood in a fixed bed combustor. *Fuel*, 88.
GIOŚ. 2018. Assessment of air quality in zones in Poland for 2017. National report from the annual air quality assessment in the zones performed by the WIOŚ according to the rules specified in art. 89 of the Act—Environmental Protection Law.
Gómez, M.A., Porteiro, J., Chapela, S., Míguez, J.L. 2018. An Eulerian model for the simulation of the thermal conversion of a single large biomass particle. *Fuel*, 220, 671–681.
Gómez, M.A., Porteiro, J., de la Cuesta, D., Patiño, D., Míguez, J.L. 2016. Numerical simulation of the combustion process of a pellet-drop-feed boiler. *Fuel*, 184, 987–999.
Gurjar, B.R., Nagpure, A. 2015. Indian megacities as localities of environmental vulnerability from air quality perspective. *Journal of Smart Cities*, 1(1), 15–30.
Harrison, R.M., Beddows, L.H., Hu, L., Yin, J. 2012. Comparison of methods for evaluation of wood smoke and estimation of UK ambient concentrations. *Atmospheric Chemistry and Physics Discussions*, 12, 6805–6838.
Hertwig, D., Burgin, L., Gan, C., Hort, M., Jones, A., Shaw, F., Witham, C., Zhang, K. 2015. Development and demonstration of a Lagrangian dispersion modeling system for real-time prediction of smoke haze pollution from biomass burning in Southeast Asia. *Journal of Geophysical Research: Atmospheres*, 12605–12630.
Hobson, M. 2005. Emission factors programme Task 7 — Review of Residential and Small-Scale Commercial Combustion Sources.
Hong, C., Zhang, Q., He, K., Guan, D., LI, M., Liu, F., Zheng, B. 2017. Variations of China's emission estimates: Response to uncertainties in energy statistics. *Atmospheric Chemistry and Physics* 17, 1227–1239.
IHME. 2016. Global Burden of Disease Study 2015. Global Burden of Disease Study 2015.
Jingjing, L., Xing, Z., Beilijia, B., DeLaquil, P., Larson, E.D. 2001. Biomass energy in China and its potential. *Energy for Sustainable Development*, 5(4), 66–80.
Karagulian, F., Belis, C.A., Dora, C.F.C., Prüss-Ustün, A.M., Bonjour, S., Adair-Rohani, H., Amann, M. 2015. Contributions to cities' ambient particulate matter (PM): A systematic review of local source contributions at global level. *Atmospheric Environment*, 475–483.
Kazakov, A., Frenklach, M. 2018. Reduced Reaction Sets based on GRI-Mech 1.2.
KOBIZE. 2018. Poland's Informative Inventory Report 2018 Submission under the UN ECE Convention on Long-range Transboundary Air Pollution and the DIRECTIVE (EU) 2016/2284. The National Centre for Emissions Management.
Kubica, K. 2003a. Environment Pollutants from Thermal Processing of Fuels and Biomass. Thermochemical Processing of Coal and Biomass.
Kubica, K. 2003b. Thermochemical Transformation of Coal and Biomass. Thermochemical Processing of Coal and Biomass.

Kubica, K., Ranczak, J., Matuszek, K., Hrycko, P., Mosakowski, S., Kordas, T. Kubica 2003. Emission of pollutants from combustion of coal and biomass and its co-firing in small and medium size combustion installation. *Fourth Joint UNECE Task Force and EIONET Workshop on Emission Inventories and Projections in Warsaw*, 145–232.

Lee, D.H., Yang, H., Yan, R., Liang, D.T. 2007. Prediction of gaseous products from biomass pyrolysis through combined kinetic and thermodynamic simulations. *Fuel*, 86, 410–417.

Lei, W., Li, G., Molina, L.T. 2013. Modeling the impacts of biomass burning on air quality in and around Mexico city. *Atmospheric Chemistry and Physics*, 2299–2319.

Li, M., Liu, H., Hong, C., Liu, F., Song, Y., Tong, D., Zheng, B., Cui, H., Man, H., Zhang, Q. 2017. Anthropogenic emission inventories in China: A review. *National Science Review*, 4(6), 834–866.

Lin, C.-Y., Zhao, C., Liu, X., Lin, N.-H., Chen, W.-N. 2014. Modelling of long-range transport of Southeast Asia biomass-burning aerosols to Taiwan and their radiative forcings over East Asia. *Tellus B: Chemical and Physical Meteorology*, 66, 1.

Lin, N.-H., Tsay, S.-C., Maring, H.B., Yen, M.-C., Sheu, G.-R., Wang, S.-H., Chi, K.H. et al. 2013. An overview of regional experiments on biomass burning aerosols and related pollutants in Southeast Asia: From BASE-ASIA and the Dongsha Experiment to 7-SEAS. *Atmospheric Environment*, 78, 1–19.

Magdziarz, A., Wilk, M., Zajemska, M. 2011. Modelling of pollutants concentrations from the biomass combustion process. *Chemical in Process Engineering*, 32(4), 423–433

Mehrabian, R., Shiehnejadhesar, A., Scharler, R., Obernberger, I. 2014. Multi-physics modelling of packed bed biomass combustion. *Fuel*. 122, 164–178.

Miljkovic, B., Pešenjanski, I., Vicevic, M. 2013. Mathematical modelling of straw combustion in a moving bed combustor: A two dimensional approach. *Fuel*. 104, 351–364.

Puxbaum, H., Caseiro, A., Sanchez-Ochoa, A., Kasper-Giebl, A., Claeys, M., Gelencsér, A. 2007. Levoglucosan levels at background sites in Europe for assessing the impact of biomass combustion on the European aerosol background. *Journal of Geophysical Research*. 112.

Radomiak, H., Bala-Litwiniak, A., Zajemska, M., Musiał, D. 2016. Numerical prediction of the chemical composition of gas products at biomass combustion and co-combustion in a domestic boiler. *Energy and Fuels*, 14.

Rodriguez-Alejandro, D.A., Zaleta-Aguilar, A., Rangel-Hernández, V.H., Olivares-Arriaga, A. 2018. Numerical simulation of a pilot-scale reactor under different operating modes: Combustion, gasification and pyrolysis. *Biomass and Bioenergy* 116, 80–88.

Sharma, S., Kumar, A., Datta, A., Malik, J., Das, S., Mahtta, R., Lakshami, C.S., Pal, S., Mohan, I. 2016. *Air Pollutant Emissions Scenario for India*. The Energy and Resources Institute.

Tao, S., Li, X., Coveney JR., R.M., Lu, X., Chen, H., Shen, W. 2006. Dispersion modeling of polycyclic aromatic hydrocarbons from combustion of biomass and fossil fuels and production of coke in Tianjin, China. *Environmental Science and Technology* 40 4586–4591.

Thunis, P., Degraeuwe, B., Pisoni, E., Trombetti, M., Peduzzi, E., Belis, C., Wilson, J., Vignati, E. 2017. Urban PM2.5 Atlas: Air Quality in European cities.

TUV. 2012. TÜV Rheinland Energy GmbH Am Grauen Stein.

UMWS. 2017. The air protection program for the area of the Śląskie Voivodeship aimed at achieving the limit levels of substances in the air and the exposure concentration obligation.

Venkataraman, C., Brauer, M., Tibrewal, K., Sadavarte, P., Ma, Q., Cohen, A., Challyakunnel, S. et al. 2018. Source influence on emission pathways and ambient PM2.5 pollution over India (2015–2050). *Atmospheric Chemistry and Physics* 18, 8017–3039.

WHO. 2018. World Health Organisation. Global Health Observatory (GHO) data. Mortality from household air pollution.

Williams, A., Pourkashanian, M., Jones, J.M. 2001. Combustion of pulverised coal and biomass. *Progress in Energy and Combustion Science* 27, 587–610.

Williams, J.E., van Weele, M., van Velthoven, P.F.J., Scheele, M.P., Liousse, C., van der Werf, G.R. 2012. The impact of uncertainties in African biomass burning emission estimates on modeling global air quality, long range transport and tropospheric chemical lifetimes. *Atmosphere*, 3.

WIOS-Krakow. 2018. Air quality monitoring system. Voivodeship Inspectorate for Environmental Protection in Kraków.

WIOS-Katowice. 2018. Air quality monitoring system. Voivodeship Inspectorate for Environmental Protection in Katowice.

Yang, Y.B., Sharifi, V.N., Switchenban, J. 2004. Effect of air flow rate and fuel moisture on the burning behaviours of biomass and simulated municipal solid wastes in packed beds. *Fuel*, 1553–1562.

Yongtie, C., Wenming, Y., Zhimin, Z., Mingchen, X., Boon, S.K., Subbaiah, P. 2017. Modelling of ash deposition in biomass boilers: A review. *Energy Procedi*, 143, 623–628.

Zahirović, S., Scharler, R., Kilpinen, P., Obernberger, I. 2011. A kinetic study on the potential of a hybrid reaction mechanism for prediction of NOx formation in biomass grate furnaces. *Combustion Theory and Modelling*, 15.

Zajemska, M., Musiał, D., Radomiak, H., Poskart, A., Wyleciał, T., Urbaniak, D. 2014. Formation of pollutants in the process of co-combustion of different biomass grades. *Polish Journal of Environmental Studies*, 23, 1445–1448.

Zhou, D., Ding, K., Huang, X., Liu, L., Liu, Q., Xu, Z., Jiang, F., Fu, C., Ding, A. 2018. Transport, mixing, and feedback of dust, biomass burning and anthropogenic pollutants in eastern Asia: A case study. *Atmospheric Chemistry and Physics*, 18, 16345–16361.

Zyśk, J., Roustan, Y., Wyrwa, A. 2015. Modelling of the atmospheric dispersion of mercury emitted from the power sector in Poland. *Atmospheric Environment*, 112, 246–256.

6 Calculation Approach to Fulfill Emission and Energy Efficiency Limits of Individually Built Tiled Stoves (Kachelofen)

Thomas Schiffert

CONTENTS

6.1 Introduction 146
6.2 New Developments 146
6.3 Nominal Heat Output 147
6.4 Maximum Fuel Load 147
6.5 Part Load 148
6.6 Combustion Chamber Dimensions 148
6.7 Minimum Flue Length 148
6.8 Determination of Optimal Fuel Turnover (m_{BUopt}) 149
6.9 Determination of Optimum Excess Air 150
6.10 Design Air Volume Flow 150
6.11 Combustion Gas and Exhaust Gas Volume Flow 151
6.12 Air Density 151
6.13 Flue Gas Density (ρ_G) 151
6.14 Combustion Chamber Temperature 151
6.15 Combustion Gas Temperatures in the Flue Gas 152
6.16 Temperatures in the Connection Pipe and the Chimney 152
6.17 Calculation of Static Pressure 152
6.18 Determination of Flow Velocity 152
6.19 Calculation of Frictional Resistance 152
6.20 Calculation of Resistance Owing to Direction Change 153
6.21 Function Control 153
    6.21.1 Pressure Condition 153
    6.21.2 Dew Point Condition 153
6.22 Energy Efficiency 153
6.23 Conclusion 153
References 154

6.1 INTRODUCTION

Individually built tiled stoves are one of the oldest heaters used by people. They have a long tradition in many parts of Europe, especially in Central and Eastern Europe. In Austria, there are about 450,000 tiled stoves in a population with 3.8 million households. Examples can be found in many countries and many different living conditions. In some rural areas, these stoves were mortared and not tiled. In main towns, and especially in castles, many very beautiful and expensive stoves are found. The famous castle of Schönbrunn in Vienna has more than 100 tiled stoves, some of them partly decorated with gold. In the bathhouse of Rasputin inside the Kremlin, there are beautiful tiled stoves, too. In the center of Vienna, there is a staircase called "Hafnersteig," which means "staircase of the stove builders," and it dates back to the year 1274, when this craft was already in existence.

From a technical point of view, there is a major difference between individually built tiled stoves and all other kinds of stoves and boilers. Each tiled stove is a unique appliance, not just the outer shell of the stove but also all its technical parts. Combustion chambers vary in volume and configuration, and flue channels are built in endless different ways. Therefore, the typical approach of type testing does not fulfill the needs of a tiled stove.

The Austrian Tiled Stove Association has therefore created technical guidelines to calculate the draft conditions, the energy efficiency and the dew point of the flue gas individually built tiled stoves in a way that the fulfillment of emissions and efficiency requirements can be guaranteed (Baumgartner and Hofbauer, 1997). After many years of research and testing, the guidelines were published around 1993. Afterwards, around 20 stoves, assessed using the guidelines, were tested at the Technical University of Vienna. The stoves all had different heat outputs, sizes, and varying configuration, depending on the combustion chambers and different flue channels used, some with a few changes of direction between different sections of the flue channels, some with many. All of the tested stoves fulfilled the requirements of the Austrian *Agreement pursuant to Art. 15a BV-G concerning protective measures for small heaters* (legislation), which came into force also starting from 1999 and deals with emissions and energy efficiency thresholds. As a result of these positive tests, the Austrian government accepted the proof of the calculations for individually built tiled stoves.

6.2 NEW DEVELOPMENTS

In 2015 in Austria, the already strong thresholds for emissions and efficiency requirements got even tougher. Therefore, a new design for the combustion chambers of tiled stoves had to be developed, especially to fulfill the requirements for emissions of volatile organic carbon. This led to the innovation of the Eco+ firebox, where combustion air is supplied through air slots all around the combustion chamber instead of just through the front of the firebox. The calculation itself did not need to change. The tests for these stoves with the Eco+ firebox showed that they fulfilled the new, strict requirements for gaseous and particulate emissions—carbon monoxide <500 mg/m³, nitrogen oxides <120 mg/m³, volatile organic carbons <50 mg/m³, and particulate emissions <30 mg/m³. The calculation now also guarantees energy efficiency based on a net calorific value of more than 80%.

Individually built tiled stoves are not within the scope of the ecodesign regulation 2015/1185 *Ecodesign requirements for solid fuel local space heaters*, as the scope states that the regulation shall not apply to *solid fuel local space heaters that are not factory assembled, or are not provided as prefabricated components or parts by a single manufacturer which are to be assembled on site* (Article 1, "Subject matter and scop" 2. (d)). Therefore, the calculation method finally can be used to deal with national regulations as it is already common in some countries in Europe such as Austria, Slovakia, and Switzerland.

The following sections show essential parts of the tiled stove design method established by the Austrian Tiled Stove Association, which further led to the Austrian standard, ÖNORM B 8302 and since 2010, the European standard EN 15544 *One off Kachelgrundöfen/Putzgrundöfen (tiled/mortared stoves) – Dimensioning* (Baumgartner and Hofbauer, 1997).

6.3 NOMINAL HEAT OUTPUT

The nominal heat output (P_n, in kW) of the tiled stove is either based on the heat load of the heated area or defined by contract between the stove manufacturer and the customer.

6.4 MAXIMUM FUEL LOAD

The maximum fuel load, m_B is the amount of fuel necessary to get the nominal heat output, P_n. It depends on the nominal heat output, P_n and the nominal heating time, t_n of the net calorific value, H_u of the fuel and the efficiency, η. The maximum fuel load, m_B will be increased, the greater the required nominal heat output, P_n and the larger the nominal heating time, t_n) is and the smaller the higher the net calorific value, H_u and the energy efficiency, η are. This results in the following relationship:

$$m_B = P_n \times t_n / (H_u \times \eta)$$

m_B: max. fuel load in kg,
P_n: nominal heat output in kW,
t_n: nominal heating time in h,
H_u: net calorific value in MJ × kg⁻¹,
η: energy efficiency [–].

Since the net calorific values, H_u for air-dried firewood (about 15% water content) have no significant differences for the calculation, an average value of 4.16 kWh × kg⁻¹ (about 15 MJ × kg⁻¹) is used. The energy efficiency, η is fixed at 0.78 for further calculations.

With these definitions, the equation above simplifies the calculation of the maximum fuel load as follows:

$$m_B = P_n \times t_n / 3.25$$

The nominal heating time, t_n may be 8–24 hours.

6.5 PART LOAD

For part load (m_{Bmin}), the minimum fuel load is defined as 50% of the maximum fuel load.

6.6 COMBUSTION CHAMBER DIMENSIONS

In determining the dimensions of the combustion chamber of a tiled stove, two requirements must be observed. On the one hand, the combustion chamber must be able to absorb the maximum amount of fuel, and on the other hand, the basic requirement for optimal combustion has to be created. Therefore, the minimum combustion chamber is limited by the maximum fuel load. The maximum size is limited by requirements of the quality of the combustion. This is for maintaining a minimum temperature in the combustion chamber. The temperature in the combustion chamber is related to the size of the surface of the combustion chamber. With rising surface and under otherwise identical conditions, the combustion gas temperature decreases, as more heat from the combustion chamber over the larger surface is stored.

The size of the combustion chamber interior is calculated according to the following formula:

$$O_{BR} = 900 \times m_B$$

O_{BR}: inner surface of the combustion chamber in cm²,
m_B: maximum fuel load in kg.

The value for the combustion chamber interior surface may be between $800 \times m_B$ and $1000 \times m_B$. When determining the internal combustion chamber surface, all side walls, floor, and ceiling, including the surface of the heating door and the outlet section of the combustion gases from the combustion chamber, have to be considered. The temperature in the combustion chamber depends on the size of the combustion chamber interior surface. As mentioned earlier, it decreases with increasing combustion chamber interior surface under otherwise identical conditions, because more heat from the combustion chamber over the larger interior surface is stored. A specific value for the combustion chamber interior surface of 900 cm²/kg of fuel provides good conditions for optimal combustion.

6.7 MINIMUM FLUE LENGTH

The flue length is the imaginary central line of the flue channels from the combustion chamber to the entrance of the connection pipe. The minimum fuel

Emissions and Efficiency: Calculation Approach for Tiled Stoves

channel length is the length of the flue channel at which an energy efficiency of 78% is achieved. This corresponds to a combustion gas temperature at the inlet of the connecting pipe of approximately 240°C. This flue channel length must not be shorter.

Dozens of measurements, both on the test bench and on site, revealed the following empirically determined relationship:

$$L_{min} = 1.3 \times \sqrt{m_B}$$

Lz_{min}: minimum flue length in m,
m_B: maximum fuel load in kg.

For example:

$$L_Z = 1.3 \times \sqrt{12.6} = 4.61 \text{ m}$$

6.8 DETERMINATION OF OPTIMAL FUEL TURNOVER (M_{BUopt})

The aim of the calculation is to design a tiled stove so that it has minimal emissions. Each tiled stove has, like all others stoves too, an optimal operating range. This optimal operating range can best be determined by expressing the CO emissions over the fuel turnover (burning speed). The fuel turnover is averaged over the burning time (fuel turnover in kg × h^{-1}).

A variety of tiled stoves were investigated and the results can be summarized as follows:

- Every tiled stove has an area with optimal fuel turnover, which gives low CO emissions (about 1000 mg × MJ^{-1}) and therefore low organic carbon and particulate emissions.
- The extent of this range varies by ±25% of the optimum fuel turnover (minimum of the curve) and can be specified.

One the one hand, the range of optimum fuel turnover is based on the interaction between the tiled stove and the chimney, and on the other hand on the maximum fuel load. If one considers the connection between the optimum fuel turnover and the maximum amount of fuel, the following linear relationship is obtained:

$$m_{BUopt} = 0.78 \times m_B$$

m_{BUopt}: ideal fuel conversion (burning speed) in kg × h^{-1},
m_{Bmax}: fuel load in kg.

For example, a tiled stove with a maximum fuel load of 12.6 kg has an ideal fuel conversion of 12.6 × 0.78 = 9.8 kg × h^{-1}. The maximum fuel load therefore burns for about 77 minutes.

6.9 DETERMINATION OF OPTIMUM EXCESS AIR

The excess air is given as the air ratio λ. The air ratio indicates how much more air is supplied to combustion than theoretically needed to complete combustion. Burning in a tiled stove is a process that is not constant, but changes continuously. The burn-up can be divided into three phases:

Specification	Air Ratio	Temperature	Draft
Starting phase	2.5–8	Low	Low
Main combustion phase	1.5–2.5	High	High
Closing phase	2.5–5	Low	High

For good combustion, a sufficiently high temperature is needed in the combustion chamber. In order to achieve this high temperature, the average excess air should not increase significantly above $\lambda = 3$. However, the excess air must not be too low. When λ in the main combustion phase is less than 1.5, there are high CO emissions because of a lack of local oxygen.

6.10 DESIGN AIR VOLUME FLOW

A tiled stove must be designed so that an amount of air is supplied that is necessary for optimal combustion. The air volume flow (V_{air}) is calculated as follows:

$$V_{air} = (m_{BUopt} \times L_{min} \times \lambda)/3600$$

m_{BUopt}: ideal fuel conversion (burning speed) in kg $\times$ h^{-1},
L_{min}: theoretical air requirement per kg of fuel in Nm^3 $\times$ kg^{-1},
λ: air ratio –.

In order to ensure combustion with minimal emissions, there be must a value of 2.95 for λ. The theoretical air requirement per kg of fuel is a fuel-specific value and is about 4.0 Nm^3 $\times$ kg^{-1} for wood fuel (15% water content). When these values are put into the above formula, the following simplified relationship for the air volume flow in the standard state (0°C, 1013 mbar) is obtained:

$$V_{air} = 0.00256 \times m_B$$

m_B: maximum fuel load in kg.

For a precise calculation, the fact that the actual temperature of the air is often not 0°C must be taken into account. The air pressure is not 1013 mbar everywhere, but changes with altitude. The decrease of air pressure with increasing sea level causes a decrease in density. Therefore a larger volume of air has to be provided. This happens by introducing two multiplication factors:

$$V_{air} = 0.00256 \times m_B \times f_t \times f_s$$

Emissions and Efficiency: Calculation Approach for Tiled Stoves

m_B: maximum fuel load in kg,
f_t: temperature correction factor –,
f_s: altitude correction factor –.

with

$$f_t = (273 + t_L)/273$$

t_L: air temperature, °C

$$f_s = 1/e^{(-9.81 \times z)/78624}$$

f_s: altitude correction factor –,
z: altitude, m.

6.11 COMBUSTION GAS AND EXHAUST GAS VOLUME FLOW

Knowledge of the combustion gas volume flow (V_{gas}) is important for the dimensioning of the flue channels. This can be done with known formulas of combustion calculation and a theoretical combustion gas volume per kg of fuel (4.8 Nm³ × kg⁻¹) can be determined as follows:

$$V_{gas} = 0.00273 \times m_B \times f_t \times f_s$$

m_B: maximum fuel load in kg,
f_t: temperature correction factor –,
f_s: altitude correction factor –.

6.12 AIR DENSITY

Air density (ρ_L) in the standard state (0°C, 1013 mbar) is 1.293 kg × m⁻³. Since air is not in the standard state, a correction for air density has to be carried out. The density decreases with increasing temperature and is determined by the temperature correction factor. Sea level is taken into account by the sea level correction factor.

6.13 FLUE GAS DENSITY (ρ_G)

The density of the combustion or exhaust gas in the standard state is 1.282 kg × m⁻³, assuming an average composition of the combustion gas. Temperature and sea level correction factors are used again.

6.14 COMBUSTION CHAMBER TEMPERATURE

The combustion chamber temperature is needed for the determination of the draft in the combustion chamber. Because of the constant specific combustion chamber surface, the same temperature for all combustion chambers is taken. The calculation can be done with sufficient accuracy using an average temperature of the combustion gases in the combustion chamber of 700°C.

6.15 COMBUSTION GAS TEMPERATURES IN THE FLUE GAS

If one considers the average temperature distribution of the burning cycle, then the temperature decreases over the flue length because the temperature difference between flue gas and surface decreases. This can be described best by an exponential function. Since at the combustion chamber exit ($L_Z = 0$) the temperature is 550°C and at the minimum flue length ($L_Z = L_{Zmin}$) the temperature is 240°C, the following relationship arises:

$$T = 550 \times e^{(-0.81 \times L_z/L_{zmin})}$$

L_Z: flue length in m,
Lz_{min}: minimum flue length in m.

6.16 TEMPERATURES IN THE CONNECTION PIPE AND THE CHIMNEY

These temperatures are calculated according to EN 13384-1.

6.17 CALCULATION OF STATIC PRESSURE

The static pressure (p_h) results from the different densities of the exhaust gases or flue gases, and from the outside air. This static pressure is the driving force for the draft of the stove. The calculation is done according to the following formula:

$$p_h = g \times H \times (\rho_L - \rho_G)$$

p_h: static pressure, Pa,
g: acceleration of gravity, m × s^{-2},
ρ_L: air density, kg × m^{-3},
ρ_G: flue gas or exhaust gas density, kg × m^{-3}.

6.18 DETERMINATION OF FLOW VELOCITY

The flow rate is calculated from the air or gas volume flow divided by the cross-sectional area. The flow velocity in the flue channels, in the connector, and in the chimney must not be below 1.2 m × s^{-1} or not exceed a value of 6 m × s^{-1}.

6.19 CALCULATION OF FRICTIONAL RESISTANCE

The frictional resistance (p_r) in a flue channel increases with increasing dynamic pressure and with increasing channel length, but decreases with increasing channel

Emissions and Efficiency: Calculation Approach for Tiled Stoves

diameters (hydraulic diameter). The roughness of the channel and the flow state is considered by a coefficient of friction.

6.20 CALCULATION OF RESISTANCE OWING TO DIRECTION CHANGE

The resistance pressure owing to direction change (p_u) is calculated by multiplying dynamic pressure with the drag coefficient, ζ.

6.21 FUNCTION CONTROL

6.21.1 PRESSURE CONDITION

The calculation is based on the comparison of the sum of all static pressures with the sum of all resistance pressures. The following condition must be met:

$$\Sigma p_r + \Sigma p_u \leq \Sigma p_h \leq 1.05 \times (\Sigma p_r + \Sigma p_u)$$

p_r: frictional resistance in Pa,
p_u: resistance owing to direction change in Pa,
p_h: static pressure in Pa.

6.21.2 DEW POINT CONDITION

In addition, at part load, the inner wall temperature at the chimney outlet is compared with the dew point temperature of the exhaust gas. The following condition must be met:

$$t_{i,2} \geq 45$$

$t_{i,2}$: temperature at the inner wall at the top of the chimney in °C.

6.22 ENERGY EFFICIENCY

The energy efficiency of the tiled stove is calculated by the following equation:

$$\eta = 101.09 - 0.0941 \times t_F - 6.275 \times t_F^2 \times 10^{-6} - 3.173 \times t_F^3 \times 10^{-9}$$

η: energy efficiency –,
t_F: flue gas temperature at the entrance of the connection pipe in °C.

6.23 CONCLUSION

Tiled stoves differ in one aspect from all other stoves; every stove is unique, in design as well as in its technical aspects. Therefore, the common approach of type testing does not fit. A calculation method is a proper approach for designing a tiled stove. The Austrian Tiled Stove Association undertook extensive research activities to find a calculation method to design tiled stoves that guaranteed the fulfillment of state-of-the-art emissions and energy efficiency thresholds. This calculation method considers the load of fuel, size, and configuration of the combustion chamber, diameter and direction of the flue channels, the altitude, the existing chimney, and many more parameters. It is already published as European Standard EN 15544 and accepted by some countries in Europe for complying with emissions and energy efficiency limits of tiled stoves.

REFERENCES

Baumgartner G., Hofbauer H., *Bemessung von Kachelöfen, Schriftenreihe des Österreichischen Kachelofenverbandes*, Wien, AT, 02/1997. ISBN: 3-901680-07-1.

Commission regulation (EU) 2015/1185, of 24 April 2015, implementing Directive 2009/125/EC of the European Parliament and of the Council with regard to ecodesign requirements for solid fuel local space heaters.

EN 13384-1, Chimneys – Thermal and fluid dynamic calculation methods, Part one: Chimneys serving one heating appliance.

EN 15544, One off Kachelgrundöfen/Putzgrundöfen (tiled/mortared stoves) – Dimensioning, Edition 2009-09-01.

Vereinbarung gemäß Art 15a B-VG über das Inverkehrbringen von Kleinfeuerungen und die Überprüfung von Feuerungsanlagen und Blockheizkraftwerken.

Part IV

Heat and Power Generation Systems Based on Biomass Thermochemical Conversion

7 Novel and Hybrid Biomass-Based Polygeneration Systems

Rafał Figaj, Maria Di Palma and Laura Vanoli

CONTENTS

7.1 Introduction 157
7.2 Solar-Biomass Hybrid CHP Energy Systems 158
    7.2.1 Solar Photovoltaic Panels: Biomass Hybrid Systems 159
    7.2.2 Concentrated Solar Power: Biomass Hybrid System 161
        7.2.2.1 Parabolic Trough Collector Combined with Biomass 161
        7.2.2.2 Solar Tower Combined with Biomass 168
        7.2.2.3 Linear Fresnel Combined with Biomass 169
        7.2.2.4 Stirling Dish Solar Combined with Biomass 169
7.3 Hybrid Biomass-Wind Systems 171
7.4 Hybrid Biomass-Geothermal Systems 177
List of Abbreviations 182
References 183

7.1 INTRODUCTION

The increase in global energy consumption and its environmental impact, owing to population increase and standards of living improvement, and the development of renewable energy sources determine that the energy paradigm must be elastic in order to face new challenges of energy sector [1] and to meet the goals of sustainable development [2].

In this framework, the utilization of renewable energy is particularly important, since it represents a reliable alternative to conventional fossil fuels [3]. However, most renewable energy sources are usually unstable and intermittent, due to their inherent nature [4], whereas conventional energy sources are more programmable and manageable. In addition to this, it must to be considered that renewable energy sources are also characterized by a relatively low energy density with respect to the area [5], which is also regarded as a disadvantage for renewable energy. This limitation affects the possibility of meeting user demand in terms of thermal and electrical energy by only one renewable energy source without energy storage; thus, it is evident that a single source of renewable energy is not adequate to sustain continuous energy production. In order to mitigate the problems related to each renewable energy source, the combination of two or more sources of energy in hybrid configurations is performed [6]. In such layouts, the operational features of each technology are used in the system in order to achieve the best integration between the energy sources and to achieve an improvement of the performance with respect to a single source system. In general, hybridization implies an energy supply flexibility for the system and also it may positively affect its reliability and environmental and economic profitability. Hybrid renewable energy systems are adopted for large- and small-scale installations; however, the latter are particularly interesting from the point of view of distributed generation [7]. In this context, biomass-based energy systems are playing a significant role because they are more suitable for heating/cooling and electrical energy applications with respect to other renewable energy-based systems, owing to their high potential for heat generation and limited constraints regarding energy source availability [8]. However, operation, environmental, energy, and economic advantages may be achieved in the hybridization of biomass with other renewable energy sources [9].

In this chapter, examples of integration of solar, wind, and geothermal energy sources with biomass for small-scale system applications are discussed in terms layout, operation parameters, and energy/economic performance characteristics.

7.2 SOLAR-BIOMASS HYBRID CHP ENERGY SYSTEMS

Hybrid small-scale combined heat and power (CHP) systems combining solar and biomass have been widely discussed in the literature. Solar energy is an attractive renewable energy source that can replace fossil fuel and solve better the problem of global energy demand without damaging the environment, solar energy being the cleanest and safest of the renewable technologies.

The solar energy systems exist in different forms:

- Photovoltaic panels (PV)
- Solar thermal panels
- Concentrated solar power systems (CSP)

Photovoltaic panels include also photovoltaic-thermal systems (PVT) and concentrating photovoltaic-thermal (CPVT) collector systems. A photovoltaic system is able to transform global solar radiation into electrical energy, while PVT are solar devices, obtained by a conventional thermal collector whose absorber is covered by a suitable PV layer, which simultaneously provides electricity and heat. As regards CPVT collectors, they are simple PVT collectors placed in the focus of some reflectors.

Solar thermal collectors, which include nonconcentrating collectors, are:

- Flat plate solar collectors
- Vacuum collectors

Concentrated solar power systems (CSP) exist in different types based on the geometry and the position of the concentrator with respect to the receiver:

- Parabolic trough collector (PTC)
- Solar tower (heliostats)
- Linear Fresnel (LF)
- Parabolic dish collector

Novel and Hybrid Biomass-Based Polygeneration Systems

Concentrated solar thermal systems use beam solar radiation to heat a working fluid and to produce thermal energy. The integration of solar energy and biomass is relevant because during the day solar energy is available, while in the night, when the solar radiation is not present, biomass may be used to meet energy demands. Furthermore, solar-biomass technologies are advantageous and when used, allowing for a decrease in fossil fuel consumption and emissions of pollutants.

7.2.1 Solar Photovoltaic Panels: Biomass Hybrid Systems

PV panels capture solar energy and turn it into electricity. The elementary component of a PV panel is the PV cell in which the conversion of solar radiation into an electric current occurs. A PV cell consists of a thin slice of semiconductor material, generally silicon. When solar radiation strikes these semiconductors, they release electrons that free to move, which are channeled through the device producing the electric current. PV panels produce a direct current (DC), which must be converted through an inverter in alternating current (AC) for electrical appliances. In the context of PV systems, numerous research studies have been conducted on PV power plants for domestic and commercial uses [10].

A typical hybrid biomass-PV system for power generation is presented in Reference [11], which presents a case study in a rural unelectrified area of Bangladesh. This system, shown in Figure 7.1, is a hybrid mini-grid used to supply electrical energy. In particular, it consists of PV modules, which produce electricity when solar energy is available and a biomass fueled generator, supplied with rice husk. The biomass generator is used when solar radiation is absent, and it is coupled to the main AC bus to offer power balancing. As a consequence of the intermittency problem of PV panels, a diesel generator and battery storage are used as back-up.

For this system, three configurations based on the availability of the different combinations of solar energy, biomass generator, diesel generator and battery storage resources are evaluated. A technoeconomic analysis is carried out using Hybrid

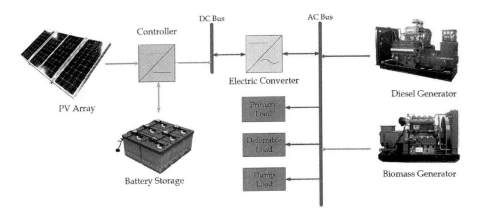

FIGURE 7.1 A typical hybrid PV-biomass power system. (From M.S. Islam et al. *Energy*. 145, 2018, 338–55.)

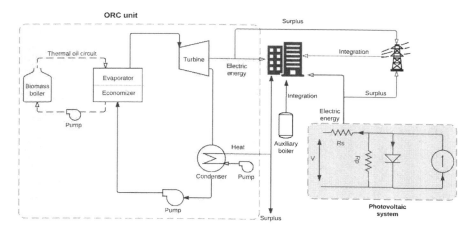

FIGURE 7.2 Scheme of the hybrid PV-biomass ORC integrated power system. (From J. Settino et al. *Energy Procedia*. 126, 2017, 597–604.)

Optimization Model for Electric Renewable (HOMER) software to identify the optimal configuration that satisfies the load demand. The configuration of hybrid mini-grid consists of a 103 kW PV field, 60 kW biomass generator, 40 kW diesel generator, 60 kW battery, and 77.2 kW converter. From the simulation results the annual electricity production to meet the load demand is 476.231 kWh/y, while the unit cost of energy is estimated as 0.188 $/kWh, which is higher than the residential consumers tariff in Bangladesh (0.065 $/kWh). These data outline that the grid parity is not reached, and this occurs even in the case of a full capital subsidy. However, for the proposed system the major advantage is found in the CO_2 emissions, since the analysis shows that it produces 75% less CO_2 with respect to existing technologies typically adopted to match the energy demands.

In Reference [12] Settino et al. presented a novel CHP system based on PV units and a biomass boiler integrated with an organic Rankine cycle (ORC), as shown in Figure 7.2. In this case, the solar energy is the primary energy source while the biomass fired ORC works when solar radiation is scarce or not available. Electrical energy can be integrated with the grid and an auxiliary boiler is used when the CHP thermal power is low.

The performance of the hybrid system is analyzed defining a thermodynamic model and considering the REFPROP database for thermodynamic properties of organic fluid. The working fluid for the ORC is toluene, the minimum temperature of the cycle is 150°C, while the maximum one is 300°C. Thermal power of the CHP system proposed is 70.5 kW_{th}, while the ORC system has nominal electrical power of 14.1 kW_{el} and PV nominal power is 37.3 kW_{el} at the maximum evaporation temperature (300°C).

Defining the ORC electrical and cogeneration efficiency as:

$$\eta_{el} = \frac{P_{el}}{Q_{th}}$$

Novel and Hybrid Biomass-Based Polygeneration Systems

$$\eta_{cog} = \frac{P_{el}}{Q_{th} - \dfrac{Q_{cog}}{\eta_{th,ref}}}$$

where, P_{el} is the ORC electrical power output (kW), Q_{th} is the thermal input of the biomass boiler (kW), Q_{cog} is the thermal power from the condensation process used for cogeneration (kW), $\eta_{th,ref}$ is the reference efficiency of a conventional boiler that is used to produce separately, and η_{cog} is the cogeneration efficiency. At the maximum evaporation temperature, the electrical and cogeneration efficiencies are 14.6% and 82.6%, respectively.

In the proposed simulation of the system, a comparison with a PV unit is carried out, and from the results it is found that the ORC overcomes the intermittency of the solar energy and increases the self-consumed electrical and thermal energy; but when the solar radiation is high, the ORC system can be switched off. On the other hand, this determines a minor convenience in the use of the ORC unit because the operation hours are constrained by the production of solar electrical energy.

7.2.2 Concentrated Solar Power: Biomass Hybrid System

Concentrated solar power (CSP) is one of the latest technologies for the exploitation of solar energy; they are used to produce electricity by supplying high temperature heat to a plant based on a thermodynamic cycle. CSP technologies generate electricity concentrating the incident solar radiation onto a receiver it where circulates a heat transfer fluid, which is then transferred to a power generation system. Hybrid CSP and biomass CHP plants are a promising solution for the efficient generation of thermal and electrical power. Various novel schemes have been proposed in the literature for CSP-biomass hybridization.

7.2.2.1 Parabolic Trough Collector Combined with Biomass

Parabolic trough collector (PTC) technology is the most mature among the ones used in CSP systems. PTCs are made of long parabolic shaped mirrors where a receiver tube is located on the focal point of the mirror. The receiver tube could contain synthetic thermal oil, molten salt, or pressurized water. The temperature reaches 400°C for thermal oil, 550°C for molten salt, and 500°C for pressurized water [13]. Usually produced heat is used for powering a boiler of Rankine cycle in order to produce electricity.

A typical solar and biomass hybrid power generation system using PTC units integrated with ORC is shown in Figure 7.3 [14].

In this study, the availability of solar and biomass resources for a 5 MW$_e$ hybrid power is investigated for a region of Delhi in India. In the plant shown Figure 7.3, the solar PTC field is used to heat the heat transfer fluid (oil) to a maximum temperature of 290°C when solar energy is available. Hot water from the feed water heater is heated in the heat exchangers by the oil, and it is sent to the biomass boiler to generate superheated steam at a pressure of 60 bar and a temperature of 500°C, which is sent to a turbine at mass flow rate of 5 kg/s in order to produce electricity. To save the biomass feedstock, the biomass boiler runs at low load when solar energy is available

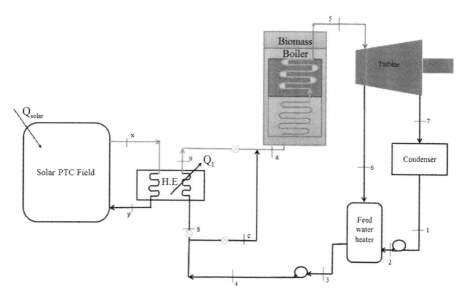

FIGURE 7.3 Scheme of the hybrid solar PTC-biomass Rankine cycle system. (From U. Sahoo et al. *Solar Energy*. 139, 2016, 47–57.)

through the PTC solar field; instead, when solar energy is absent, the biomass boiler runs at full load capacity.

A thermodynamic evaluation is carried out considering energy and exergy efficiency to assess the performance of a hybrid plant. From the results, it is seen that in the solar PTC and biomass integration systems, the solar energy is fully used and the advantage is the biomass feedstock saved. The integration of biomass and solar PTC technology for CHP applications is widely studied by researchers.

In Reference [15], Campo et al. 2015 described a thermodynamic analysis of a small-scale CHP system illustrated in Figure 7.4. This system consists of a biomass gasifier

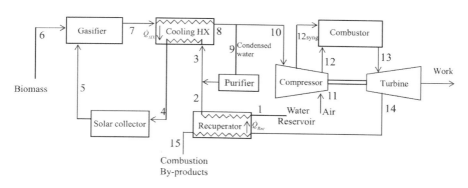

FIGURE 7.4 Biomass steam gasifier system diagram hybrid solar PTC-biomass system. (From P. Campo et al. *Energy Conversion and Management*. 93, 2015, 72–83.)

coupled with PTCs and a micro gas turbine (MGT). In detail, the system consists of two main parts: the first part "upstream" includes solar collectors that produce a steam at a high temperature (800–1200°C), a steam gasifier to produce syngas, and a cooling and water condensation system; while the second part "downstream" includes a micro gas turbine where syngas is used as fuel for the combustor.

In the upstream the following are present: the solar collector that converts liquid water to steam at a design temperature, the gasifier whose purpose is to produce syngas with high calorific value, and a heat recovery unit where water condensed by cooling is recycled. In this way, water consumption is reduced and the purifier removes present tar in the stream of condensed water. In the downstream, a gas turbine idealized by a Brayton cycle is present, where the syngas is compressed and then sent to a combustor to generate the exhaust gases used in the turbine to produce electricity. This combined heat and power system has an electrical power output of 20 kW_e. The mathematical model of the proposed system is developed in order to evaluate and optimize the performance of the plant in a parametric study, with the scope to minimize resource consumption, such as biomass feedstock and water, and to maximize the overall system efficiency achieving the optimal electric output.

The optimal condition of biomass rates is found in a range between 23 and 63 kg/h depending on different feedstock and different moisture content (wood, plastic, rubber, and municipal solid waste). Under optimal conditions, the system total efficiency is in the range between 30% and 43% for the selected feedstocks.

An emerging hybrid CHP system located in Salzburg in Austria is presented in Reference [16]. This system includes a biomass combustor, ORC technology unit, and PTCs situated in areas with low direct normal irradiance (DNI). The case study proposed is conducted in the framework of Project BIOconSOLAR funded by Austrian government. The nominal electric power output is 1.5 MW_e, while the thermal power output is 7.28 MW_{th}, which is sent to the district heating of the city of Salzburg. The configuration of the system is shown in Figure 7.5, in which it is noted that the biomass combustion occurs in a thermal oil boiler and exhaust gases pass through a heat recovery unit used to optimize the CHP system efficiency. The heat recovery unit consists of a thermal oil economizer, a combustor of air, and a water heat exchanger.

The solar thermal energy produced by the solar field is of particular importance, in fact for this system two feed-in points have been taken into consideration. Feed-in point 1 is located in the return of the thermal oil cycle, where the temperature of the heat transfer fluid (HTF) is about 260°C. In this case, the solar thermal energy is primarily used to produce steam for electrical power production in the ORC unit and to reduce biomass consumption. If the temperature of the produced solar heat is lower than 270°C, the solar thermal energy is fed into the water cycle of the district heating at a temperature level of about 100°C (feed-in point 2) in order to increase the heat production.

The simulation models of the solar field and biomass CHP system are verified with transient system simulation (TRNSYS) software, where a comparison between the climatic conditions with a low DNI in Salzburg and other two sites in Central Europe, Klagenfurt in south of Austria and Pisa in north of Italy, is performed. The analysis developed using a transient simulation model, estimates that the integration of a

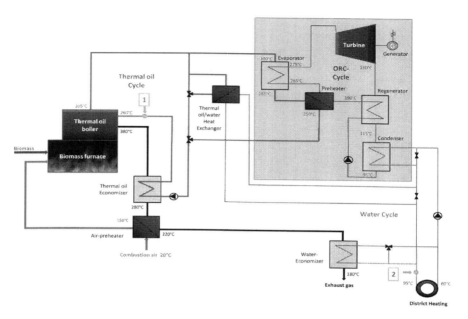

FIGURE 7.5 Flow sheet of the biomass CHP plant with ORC-power unit. (From R. Sterrer et al. *Energy Procedia*. 49, 2014, 1218–27.)

conventional biomass CHP plant with solar field is a promising approach to improve the economic performances of the system. However, the results of economic analysis show that under normal financial conditions the hybridization of the existing biomass CHP with solar system in Salzburg is hardly applicable, since the net present value (NPV) of the investment becomes negative. While for the same financial conditions but changing only DNI, for example, for climatic conditions in Pisa, the payback period decreases and after 25 years of service life, the resulting profit exceeds the total investment costs.

Another hybrid solar and biomass CHP system using a PTC field is proposed in Reference [17], consisting of a biomass boiler integrated with an externally fired gas turbine (EFGT) and an ORC unit, as shown in Figure 7.6. Solar thermal energy is produced by PTC technology using molten salts as the HTF provided at a maximum temperature of 550°C. In the plant, the ambient air is compressed to 3.5 bar, cooled to 30°C, next compressed to 12 bar at 180°C, and finally it is heated by molten salts of the CSP plant up to 500°C. Then air flow is heated to 800°C by exhaust gases of the biomass-furnace in the high-temperature heat exchanger (HTHE), and finally it is transferred to the gas turbine and ORC to produce electricity.

The hybrid solar-biomass combined heat and power investigated has an electric power of 2.1 MW$_e$ consisting of 1.4 MW$_e$ of EFGT and 0.7 MW$_e$ of ORC, and thermal power of 960 kW$_{th}$.

The thermodynamic analysis has been performed by Cycle-Tempo software considering the thermal input as the sum of biomass thermal (4523 kW$_{th}$) and thermal power produced by CSP plant (3673 kW$_{th}$).

Novel and Hybrid Biomass-Based Polygeneration Systems

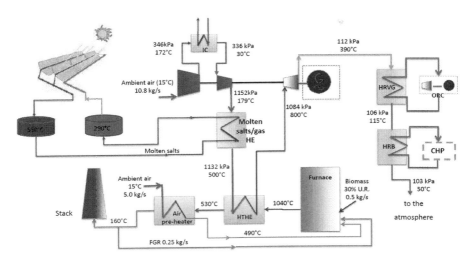

FIGURE 7.6 Layout of the hybrid solar PTC-biomass EFGT-ORC combined cycle power plant. (From A.M. Pantaleo et al. *Applied Energy*. 204, 2017, 994–1006.)

Desideri et al. [18] presented a solar-biomass trigeneration system located in Spain and developed in the framework of the BRICKER project. The plant is based on a parabolic trough solar field, a biomass combustor, and an ORC unit producing electricity. The system, shown in Figure 7.7, is composed of two main loops. In the first one, the solar field (SF) and biomass combustion boiler (BMB) produce thermal energy that is transferred by a HTF to an ORC in order to produce electric power (70 kW$_e$). The second loop is connected to the first one by means of the heat exchanger (HXI) and the condenser of the ORC unit. Moreover, in the second loop an adsorption chiller is included in order to match the cooling demand of the user during summer, and the connection to the thermal demand (TL) of the building. The

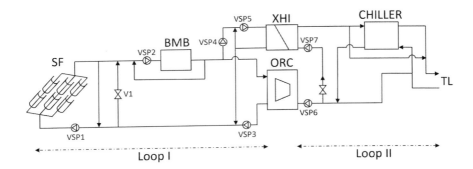

FIGURE 7.7 Schematic flow diagram of the PTC-biomass ORC integrated CHP system. (From A. Desideri, S. et al. Dynamic modeling and control strategies analysis of a novel small CSP biomass plant for cogeneration applications in building. *ISES Solar World Congress 2015, Conference Proceedings*, 2015. p. 12.)

HTF is a synthetic thermal oil (Therminoil ISP), that has good characteristics such as low operating pressure and high thermal stability up to 335°C.

A control strategy is implemented for the system, in which some technical restrictions are presented. For the solar field, high temperatures must be avoided in order to not damage the thermal oil. Moreover, the biomass boiler is equipped with a recirculation circuit and an internal control, which regulates the amount of biomass burnt at a constant mass flow of 9.5 kg/s to keep the temperature at the defined set-point. Controls of the adsorption chiller and HXI are required to respect the boundary limits.

A high-level control of the solar field is required in order to guarantee safe biomass boiler operation avoiding biomass shut-down. The solar field is equipped with a recirculation and a bypass stream. The ORC evaporator requires a constant thermal oil mass flow rate of 2.5 kg/s by running a variable speed pump (VSP1) at a fixed speed, avoiding high film temperature issues.

The control logic is developed by installing a PI (proportional integral) controller (PI1) on the bypass valve (V1) to regulate the biomass inlet temperature. In nominal conditions, the mass flow at the outlet of the ORC unit is preheated in the solar field and the bypass valve V1 is closed, but if the solar field outlet temperature increase, the bypass valve is opened to mitigate the temperature. When the solar field outlet temperature passes the set-point, the control system reduces the thermal power delivered by PTCs, while a sudden decrease of solar power is managed by the biomass internal control (PI2), which increases the biomass fed to the combustion chamber maintaining the operation temperature at its nominal value.

The heat exchange rate of HXI is controlled by a recirculation system with pump VSP4 and VSP5. The chiller works at a constant temperature to ensure a high coefficient of performance (COP) and a controller PI3 is implemented to maintain the chilled water temperature close to its set-point.

The control strategy is investigated using a dynamic model simulation, developed using the Modelica program based on the ThermoCycle library and adopted to examine the performance of CHP plant and to make the whole system safe and efficient. The focus of this model is to investigate the efficacy of the logic control of the oil loop, a critical point of the system. In particular, two solar conditions are investigated: the partial (PD) and total defocusing (TD) of the solar field. The main finding consist in the fact that the TD approach allows one to increase the solar fraction by 12% compared to the PD approach.

In Reference [19], Karellas and Braimakis proposed a thermodynamic modeling and economic analysis of a plant in Greece capable of operating either as a cogeneration or a trigeneration plant. As shown in Figure 7.8, the plant includes a PTC solar field and a biomass combustor, which provides heat to the ORC, while a vapor compression cycle (VCC) is used for the production of power and cooling, respectively. The ORC includes a series of heat exchangers where working fluid reaches the maximum temperature and then flows into an expander to produce electricity. ORC and VCC are connected and both using the same organic fluid. In this system, during the summer trigeneration mode, a part of the produced electric power is consumed by VCC to generate the cooling output of the system, while during the winter, the VCC can be disconnected because refrigeration is not necessary.

Novel and Hybrid Biomass-Based Polygeneration Systems

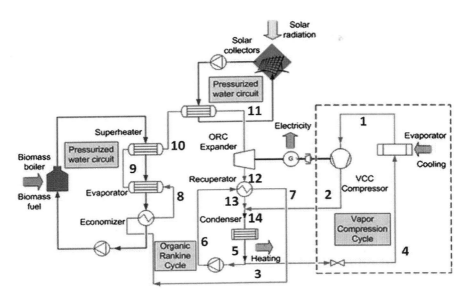

FIGURE 7.8 Scheme of the hybrid PTC-biomass boiler integrated with ORC–VCC cycle. (From S. Karellas, K. Braimakis. *Energy Conversion and Management*. 107, 2016, 103–13.)

The effect of the different parameters of this system is investigated using a thermodynamic analysis considering a net electricity output of 1.42 kW$_e$ with an electrical efficiency of 2.38% and a heating output of 53.5 kW$_{th}$. An economic analysis is also developed for a case study of an apartment, and a sensitivity analysis of a total capital investment (TCI) and biomass prices are carried out to consider their impact on the system cost. The results show that the profits generated from the reduction of fossil fuel and electricity consumption cause an important economy saving, leading to an internal rate of return (IRR) of 12% and a payback of 7 years.

Another solar–biomass trigeneration system is proposed in Reference [20] and shown in Figure 7.9. This system consists of an air/steam biomass gasifier, a steam generation with solar collectors (PTCs), an internal combustion engine, and a LiBr–H$_2$O absorption chiller. The biomass is preheated by air up to 200°C, reducing the biomass moisture to 10%. Then the dry biomass is fed into a fluidized bed gasifier using the preheated air and steam generated by the PTC field such as gasifier agents. Biogas produced is purified in a cyclone to remove ash and char and then it is used to preheat the air and produce domestic hot water. In this way, the purified and cooled biogas is fed into the internal combustion engine (ICE) to produce electricity. The exhaust gases from ICE firstly are sent to the LiBr–H$_2$O absorption chiller to provide cooling for users and afterwards to a heat exchanger to generate domestic hot water up to 80°C, and finally are released to the atmosphere at 120°C. The thermodynamic performance of this system is investigated using Aspen Plus and the results show that in the case of 5076 kW biomass energy input and 477 kW solar energy input, the electricity output power is 987 kW$_e$ and electrical efficiency is 17.8%. While the thermal heat power produced is 1988 kW$_{th}$, cooling generation is 843 kW$_{th}$, desiccant

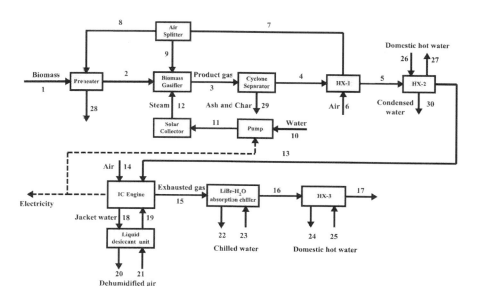

FIGURE 7.9 Scheme of the hybrid PTC-biomass gasifier integrated with ICE. (From X. Zhang et al. *Energy Conversion and Management.* 122, 2016, 74–84.)

power is 482 kW$_{th}$, and the thermal efficiency is 59.6%, with the overall efficiency system at 77.4%.

From simulations of the system, it appears that the introduction of the solar energy in this biomass trigeneration plant decreases the consumption of biomass and high PTC temperatures increase the energy efficiency of the system. In addition to this, the economic analysis of the system points out that the annual total cost saving ratio is 25.9%, the initial equipment capital cost and the operations costs are 51.5% and 33.9% of the annual total cost, respectively.

7.2.2.2 Solar Tower Combined with Biomass

A solar tower system is characterized by a field of flat reflecting panels, called heliostats, placed around a tower. The solar radiation reflected by the heliostats is concentrated on the receiver placed on the top of the tower, where the heat transfer fluid, circulating in the central receiver, absorbs the heat and converts solar radiation into thermal energy. The HTF used for tower solar system could be steam, molten salts, liquid sodium, air, or thermal oil.

Solar tower technology is able to reach temperatures from 400°C to above 1000°C and steam pressure up to 130 bar and it is highly efficient at producing electricity and heat.

It is worth noting that only some large-scale hybrid biomass-solar tower systems are investigated in the literature. For example, Khalid et al. [21] analyze a solar-biomass integrated plant for the generation of electricity, heat, and cooling. Energy and exergy efficiencies of the overall system are examined and are found to be 66.5% and 39.7%, respectively in order to provide performance information. The proposed multigeneration system contains two Rankine and two gas turbine cycles, while the thermal energy is collected through a heliostat tower and biomass combustion chamber. Compressed air comes into the biomass combustion chamber to produce biogas, which passes through the gas turbine to produce electricity. The exhaust gases obtained from Gas Turbine 1 are divided into two streams: a part of gases is sent to a Rankine cycle and the other part is sent into Gas Turbine 2 to produce additional electricity. Solar radiation on the heliostats transfer heat to a thermal oil (Dowtherm-A) to a high temperature used to heat water for the Rankine cycle. Furthermore, the thermal solar energy is used for the absorption cooling cycle. A similar system is investigated by Tanaka et al.[22], where a hybrid solar-biomass power system consisting of a fluidized bed gasifier (FBG), a gas turbine, a Rankine cycle, and a concentrated solar thermal process heliostat type. The heat produced by solar radiation is used to generate the steam in a heat recovery steam generator (HRSG) of the Rankine cycle.

7.2.2.3 Linear Fresnel Combined with Biomass

Linear Fresnel (LF) is considered a promising technology for CSP power production. LF collectors use an array of flat or nearly flat reflectors moved remotely starting from different positioning angles. The solar radiation is concentrated in a fixed receiver, consisting of one or more linear receiver tubes, and an optional secondary reflector is placed above the absorber tube to refocus the solar radiation. The HTF can be water or diathermic oil, which absorbs solar radiation. LF collectors are an economically convenient choice and appropriate for integration in rural areas.

The research presented in Reference [23] proposes the investigation of 17 different combinations of hybrid biomass and solar power. The solar power technologies investigated include a solar tower, PTC, and Fresnel, while the biomass systems considered are combustion and gasification. The results show that CSP solar tower systems combined with biomass gasification reach the highest peak of efficiency at 33.2%. The combination of LF and biomass combustion show that the increase of temperature and pressure on the system increase system efficiency. But in this study, only the technologies with an output electric power up to 5 MW$_{el}$ are considered.

Another hybrid system with LF for a large scale is reported in Reference [24] in which Lopez et al. propose the exergy analysis of two configurations of a hybrid LF solar-sugarcane cogeneration system localized in the tropical region of Brazil. The combination of solar energy and sugarcane is used to preheat the water for steam generation in order to reduce fuel consumption.

The sugarcane cogeneration system produces sugar and alcohol and in this plant steam burning sugarcane is generated using two steam generators; then, the steam is expanded into three turbines to produce electricity. Sugarcane is a seasonal crop and its availability is intermittent, therefore, the utilization of a solar system to generate steam was proposed, and the results indicated that the fuel saved was 10%.

7.2.2.4 Stirling Dish Solar Combined with Biomass

A solar system with parabolic dish collectors consists of a surface with a parabolic dish shape, which concentrates solar radiation and reflects to a receiver located on the focal point of the dish.

The HTF is a gas, in particular hydrogen or helium, which reach high temperatures above 1000°C. The heat collected is used directly for energy conversion, the generators typically used being Stirling engines or microturbines. The dish is moved during the day in order to collect all the DNI using two axis [25].

A hybrid system that uses two types of solar collectors is proposed in Reference [26]. The system proposed shown in Figure 7.10 is a solar-biomass generation system integrating a two-stage gasifier. Biomass gasification can be divided into two parts: pyrolysis and gasification. In this system, the solar energy is concentrated in the parabolic through solar collector and the steam is generated for the process of gasification, while the concentrating collector, such as heliostats and hyperboloid reflectors, are used to provide heat for tar cracking and char gasification at temperatures above 1000 K. Finally, the syngas produced is sent to a Bryton-Rankine combined cycle to generate power. In this system, the use of concentrated solar energy for biomass gasification allows one to achieve an overall energy efficiency of 27.9% and a solar to electricity efficiency of 19.9%.

In Reference [27], Burin et al. analyzed a CHP plant producing electricity and industrial process heat by adding a parabolic trough solar field to a biomass-based CHP plant located in Brazil in order to improve the overall capacity of the system. The cogeneration plant shown in Figure 7.11 presented a bagasse furnace, superheating system, and tubular air heaters to produce steam that flows into a condensing-extraction turbine (CEST) used in parallel with back-pressure turbine (BPST) to increase the power production. In Reference [28], the analysis of the system is further extended. In particular, the plant with the parabolic trough (PT) solar field is compared to plants with an LF solar field and a solar tower (ST). In the case of LF and ST, direct steam generation (DSG) was used, while in the case of the PT, thermal oil was implemented. The analysis for the considered system outlines that the solar tower is the most economically convenient solution among the three alternatives.

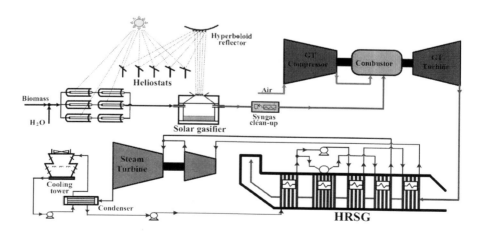

FIGURE 7.10 Schematic diagram of the novel solar-biomass power generation system. (From Z. Bai et al. *Applied Energy*. 194, 2017, 310–9.)

Novel and Hybrid Biomass-Based Polygeneration Systems

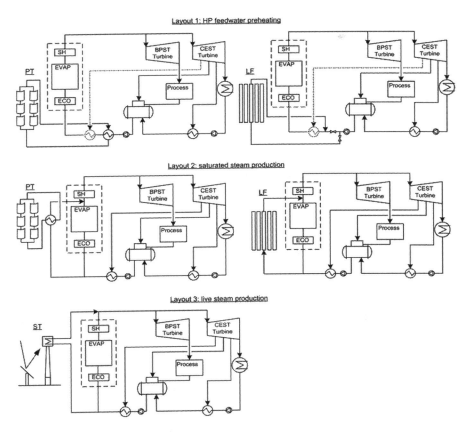

FIGURE 7.11 Process flow diagrams of the three studied integration layouts. (From E.K. Burin et al. *Energy*. 117, 2016, 416–28.)

7.3 HYBRID BIOMASS-WIND SYSTEMS

In the framework of hybrid small-scale biomass-based energy systems, the integration of wind energy is not as common as the one with a solar source. In fact, as outlined in the previous sections of this chapter, the literature studies and technical applications of hybrid biomass systems deal mainly with solar thermal collectors and PV panels. This is due to the fact that solar energy is more flexible than wind, since both thermal and electrical energy may be directly produced with the solar source by means of thermal collectors and PV panels, respectively; conversely, wind energy is used in general to produce electrical energy. Therefore, the systems integrating biomass and wind are dedicated to the production of electrical energy, or they are implemented in cogeneration or trigeneration layouts, where the waste thermal energy produced by biomass thermochemical conversion processes is used to match thermal and cooling demands.

The main advantage of the hybridization of biomass and wind energy installations consists in the possibility to reduce significantly the variability of electrical energy production occurring in wind turbine systems, owing to the intrinsic characteristics of the wind source. In particular, the biomass unit is operated in order to maintain a selected level of electrical power when the wind speed drops and it is not sufficient to run the wind turbines.

In Reference [29], the coupling of a biomass fired ORC with a wind turbine in an integrated system is presented. In particular, the two biomass and wind subsystems are set to operate in parallel in order to produce electrical and thermal energy for a domestic application. The presented system represents a possible installation, which allows matching the energy consumption requirements of a nearly zero energy house (NZEH) in the residential sector.

The layout of the hybrid system consisting of the biomass-fired ORC and wind turbine WT units is shown in Figure 7.12. In this layout, the ORC plant is operated in order to match the electrical energy demand of the user when the wind speed is not adequate, while the wind turbine is the primary source of electrical energy determining the operation point of the ORC unit. The proposed system is grid-connected since the electrical energy can be exchanged with the grid when needed; in addition to this, the thermal demand of the user is matched by an auxiliary boiler, which is used in case of low CHP thermal power.

In the simulated system, the ORC plant consists of an evaporator, a turbine, a condenser, and a pump system. In particular, the evaporator consists of two parts— the evaporator itself and the economizer used to increase the temperature of the working fluid to the saturated liquid one.

The evaporator of the ORC system is thermally powered by a biomass boiler operating with a thermal oil circuit, adopted in order to allow for proper operational conditions of the system, avoiding local overheating, and to prevent the organic fluid from becoming chemically unstable.

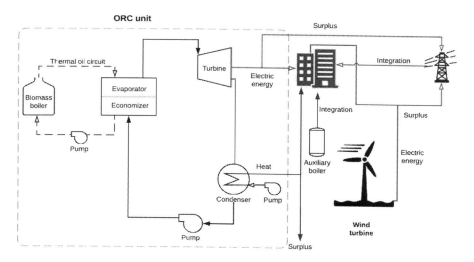

FIGURE 7.12 Scheme of the hybrid biomass fired ORC unit integrated with wind turbine. (From P. Morrone et al. *Energy Procedia*. 148, 2018, 986–93.)

Novel and Hybrid Biomass-Based Polygeneration Systems

The performance of the biomass ORC unit is investigated through a thermodynamic model developed on the basis of the literature and the REFPROP database, which are used to determine the thermodynamic properties of the organic fluid. The following assumptions are considered in order to perform the analysis: toluene as working fluid, steady state condition, negligible pressure drops and heat losses in the system components, and saturated vapor conditions at the turbine inlet. As regards the main operation conditions, condensation pressure and temperature are set to 0.39 bar and 80°C, while the nominal turbine pressure and inlet temperature are set to 32.76 bar and 300°C. Furthermore, in the analysis fixed efficiencies of the components are selected as well as the biomass lower heating value and humidity.

The simulation of the wind turbine is performed adopting real wind turbines' characteristics in five models with a nominal power in the range from 16.5 to 49.9 kW, and weather data from Palermo (southern Italy).

The selected user consists of a block of 40 dwellings, for which the energy demand is estimated taking into account the consumption of domestic lighting systems and appliances, including air conditioners during the hot season, whereas the thermal demand is based on typical space heating and hot water uses.

The results of the simulations show that the wind turbines with lower power (two units of 16.5 and 25.0 kW) are more suitable for this kind of system from the operating equivalent hours and a trade-off between surplus and self-consumed energy points of view. Both systems match between 69% and 79% of the yearly electrical energy demand. This yield is also achieved running only the biomass system; nonetheless, the hybridization with the wind turbines allows saving up to 50% of biomass and a reduction of the electrical energy surplus.

Biomass-wind systems are also developed for standalone applications, consisting of microgrid installations. In particular, an example of an integration of a biomass gasifier with a wind turbine system in a hybrid configuration is presented in Reference [30], where a case study of a village in India is considered. The system in this case is presented as a possible electrification solution for rural areas where the grid connection is not possible due to installation difficulties and economic unfeasibility. For this scope, the system is equipped with an electrical energy storage, which allows matching the electrical demand of the user in any operational condition. The system concept is very simple since the main energy flow of the installation is considered, as highlighted by the layout (Figure 7.13).

The simulation of the proposed system is performed by means of HOMER software and developing a case study based on measured data in an Indian village consisting of load demand and wind velocity. In order to model the system, simple models of the components are used to determine the electrical power of the wind turbine, power of the gasifier, and the operation of the battery bank. As outlined in Reference [30], an important aspect of this kind of system is the optimal sizing from an economic point of view. In detail, for this system, initial investment, operation, and maintenance costs as a function of the components' sizes and interest rates are considered in order to minimize the net present cost of the system. This optimization is performed taking into account the number of 25 kW wind turbines, the power of the biomass gasifier, and the number of batteries performing an hourly simulation for the energy balance of the system. For the selected case study, the optimal configuration consists

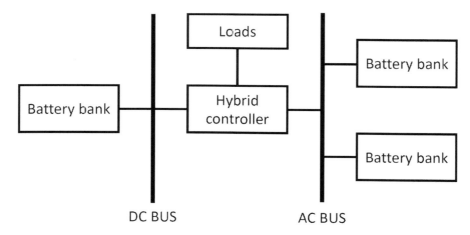

FIGURE 7.13 Scheme of the biomass gasifier-wind generator hybrid energy system. (From P. Balamurugan et al. *Energy Sources, Part A: Recovery, Utilization, and Environmental Effects.* 33, 2011, 823–32.)

of four wind turbines, 200 batteries, a 150 kW gasifier, and 100 kW converter, which allows one to achieve a cost for the electrical energy of 0.078 $/kWh. Moreover, the simulation preformed for this system configuration showed that the monthly mean power of both electrical production systems is comparable, with a slightly higher value for the gasifier unit. In addition, an important aspect of this system lies in the fact that it is more economically viable compared to a wind-diesel engine configuration.

Small-scale biomass systems integrated with wind energy are also coupled with a solar source. In particular, PV panels are adopted for the hybridization of biomass-wind systems in order to produce additional electrical energy. An example of this kind of system is presented in Reference [31], where a grid-connected hybrid renewable energy system including PV panels, a wind turbine, and a forest wood biomass gasifier are integrated. In particular, the gasifier unit fuels an internal combustion engine coupled with a generator group. The layout of the system is reported in Figure 7.14.

The key aspect of the proposed system in Reference [31] consists of the flexibility provided by the operation of the forest wood biomass engine and the grid connection, allowing a short response time in order to face the variability of renewables. Moreover, the presence of grid connection and the biomass unit with a supply storage system allows one to use a layout without electrical energy storage. In this way the system cost is not affected by the necessity of a battery bank installation.

For the PV and wind power subsystems, a 265 W module and wind turbine of a nominal power of 200 kW are selected. In this way, the proposed systems present a high grade of scalability since it is possible to change the size of both systems by varying the number of PV modules and/or wind turbines. Conversely, for the gasifier/internal combustion engine system, a fixed nominal power of 500 kW is assumed in order to partially match the average hourly electricity demand of the user, since the

Novel and Hybrid Biomass-Based Polygeneration Systems

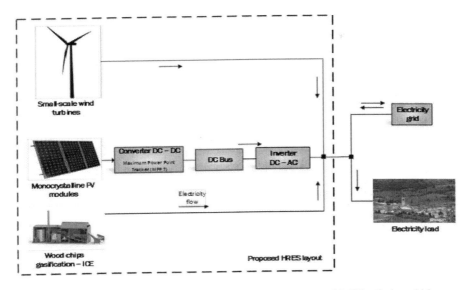

FIGURE 7.14 Scheme of the hybrid renewable energy system with PV, wind, and biomass source. (From A. Gonzalez et al. *Renewable Energy*. 126, 2018, 420–30.)

power of the unit cannot be varied arbitrarily. Concerning the biomass conversion technology, it consists of a forest woodchip downdraft gasifier with gas cleaning and a cooling section is adopted to feed the internal combustion engine.

In this type of highly hybridized system, the priority of energy source use is crucial along with the design condition of supply-demand matching. Taking into account the environmental aspects, the PV and wind turbine are used in the first place to match user demand, while the biomass subsystem is operated as back-up. However, when the renewable energy system power output is lower than the user demand, electrical energy is supplied by the public grid.

The hybrid nature of the system determines the necessity of an optimization of the system configuration based on economic and environmental criteria. In Reference [31], a genetic algorithm-based multiobjective optimization is performed, which determines a set of optimum solutions owing to the selection of different optimal condition criteria. In fact, high cost configurations allow one to achieve less environmental impact, while ones with lower costs present higher CO_2 emissions. Furthermore, for the proposed system reference configuration, it is possible to reduce the emissions by 50% by increasing the investment in the magnitude of 5%, whereas a further decrease in emissions is not possible without a significant increase in the investment cost.

A similar system consisting of a biomass combined heat and power-based microgrid is presented in Reference [32]. The other components included in this system consist of small-scale wind turbines, biomass gasifier, gas storage, PV modules, battery storage, thermal energy storage, and auxiliary boilers, as shown in Figure 7.15. It is worth noting that the proposed system integrates both electrical and thermal energy storage systems in order to manage optimally the energy flows supplied to the load.

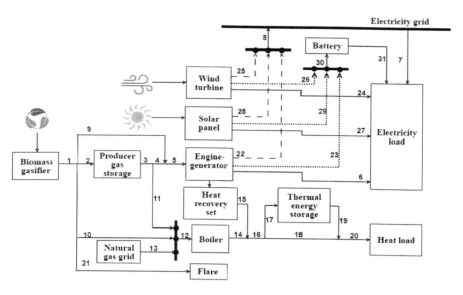

FIGURE 7.15 Layout of the microgrid biomass/wind/solar cogeneration system. (From Y. Zheng et al. *Renewable Energy.* 123, 2018, 204–17.)

In detail, the system operates in order to match the user electrical demand using the power of a biomass combined heat and power unit, wind turbine, PV panels, and battery discharge taking into account the limits and operation constraints of such equipment. All the power generation components are allowed to charge the battery system, when the control strategy is properly set. Moreover, the produced syngas is purified and cooled and may be used as (1) fuel for the internal combustion engine and the boiler; (2) stored in the gas storage tank; or (3) burnt by the flare if no economic demand exists and the gas tank is fully charged. In order to improve the system reliability, this one is charged in case of low user demand while it is discharged in case of high user demand to improve system reliability. Furthermore, the thermal energy produced by the cogeneration unit is used for direct heat uses, as well as the one produced by the auxiliary boiler.

As for the previous system, the optimization of the design (capacities) and operation (operation schedule) of the hybrid system is mandatory in order to achieve the best economic performance. For the case study reported in Reference [32], the nominal power or capacities for the components are: 100 kW for the biomass fueled engine; 10 kW for wind turbine; 10 kW for PVT field; 200 kWh for battery; 200 kWh for gas storage; and 150 kW for the boiler. In this optimization, the exclusion of some components may be considered, such as gas storage or an additional engine unit. Moreover, it is crucial to note that for the considered system, the design decisions are a function of the electrical energy and natural gas prices, and energy demand magnitude, and the convenience of this system compared to a conventional energy system is affected by utility pricing, its demand, and components' capacities.

7.4 HYBRID BIOMASS-GEOTHERMAL SYSTEMS

The hybridization of small-scale biomass installations is mainly performed with solar energy, and rarely with the geothermal one. This is due to the nature of the geothermal source [33], since zones with relatively high enthalpy geothermal energy are scarcely diffused. Moreover, typically hot reservoirs present a high availability of energy extraction and this determines the possibility to install large systems, instead of small-scale configurations. Conversely, in the case of a low enthalpy geothermal source, relatively small-scale systems may be applied by coupling a ground heat pump with biomass energy [34].

The combination of biomass and geothermal installation is performed in order to achieve an enhancement of the performance of the hybrid system with respect to separate installations [35]. In this way, synergistic interactions between geothermal and biomass sources are possible, which allows improving the thermal and electrical performance (power output and efficiency) as well as the economic profitability of the system.

In Reference [36], three schemes of geothermal power plants are presented. In particular, two of the schemes are based on the integration of a biomass energy source in order to the provide a part the thermal energy required to run the power cycles.

The three systems consist of an ORC power plant, a dual-fluid hybrid power plant and a single-fluid hybrid-fueled power plant. Apart from the first system where no biomass source is considered, the second system consists of a biomass fired steam cycle coupled with an organic fluid plant, thus in this installation two working fluids are adopted, as reported in Figure 7.16. In the upper cycle, water is used as a working fluid and the preheating and vaporization section is heated by the biomass boiler. After the expansion, water passes through a condenser which provides the thermal power to vaporize the working fluid of the bottom ORC cycle. In this way the water condensates and can be supplied to the pressurization pump, while the waste energy is used to power the evaporator of the second cycle. In addition, the ORC integrates a heat exchanger supplied by the geothermal fluid in order to preheat the organic fluid entering the vaporizer.

The third system presented in Reference [36], consists of a hybrid single power cycle plant where a geothermal source is used to preheat the working fluid while the biomass combustion in the boiler is adopted for the vaporization of the working medium, as shown in Figure 7.17. In this case, both water or organic working fluids may be used, with the only limit being the organic fluid critical temperature, which must be relatively high in order to avoid a high temperature difference between the output of the biomass boiler and the working fluid. The advantage of the dual-fluid system consists of the use of the waste heat of the steam cycle condenser to power in part the ORC unit. In this way, the condenser of the conventional cycle operates at a pressure higher than the atmospheric one owing to required driving temperature of the ORC, and this avoids sealing issues for the heat exchanger. In addition to this, the steam turbine operates always with dry stream, a condition which allows increasing the lifetime of the steam turbine.

The three systems are analyzed assuming different working fluids (water and organic fluids), and temperatures of the geothermal well. In this context, the

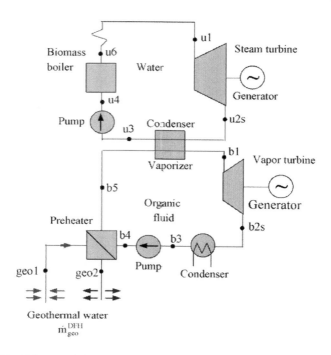

FIGURE 7.16 Scheme of the dual-fluid-hybrid biomass geothermal power plant. (From A. Borsukiewicz-Gozdur. *Geothermics*. 39, 2010, 170–6.)

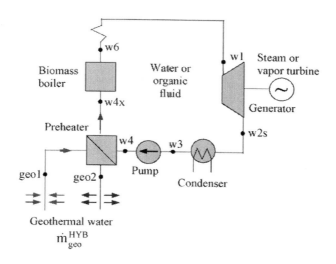

FIGURE 7.17 Scheme of the single-fluid biomass-geothermal hybrid power plant. (From A. Borsukiewicz-Gozdur. *Geothermics*. 39, 2010, 170–6.)

investigation regarding the above described systems found that the type of working fluid significantly affects the thermal power of the Rankine cycle, since the range is from several hundreds of kW to a few MW. In particular, for the systems integrating the biomass-geothermal energy source, the minimum thermal power of few hundreds of kW is achieved in the case of the water and hybrid system reported in Figure 7.17, while the same system achieves a power between 9 and 10 MW in the case of cyclohexane. The other system (Figure 7.16), working with water and an organic fluid achieves a power ranging from about 3–5 MW, depending on the organic fluid used among R236fa, R245fa, and R365mfc.

Another hybrid biomass-geothermal system is presented in Reference [37]. The considered plant shown in Figure 7.18 includes a hybrid cogeneration system based on an enhanced geothermal system (EGS) and biomass, used to meet some of a university campus heating and electrical demand. In this configuration, EGS is used to match the base load thermal demand of the user, while the biomass auxiliary heat source is used to match the user demand during peak demand hours. The geothermal system consists of a producer well and injection well, used to extract the geothermal fluid stored in the enhanced geothermal reservoir and to reinject the used fluid back in order to avoid reservoir depletion, respectively. The geothermal fluid is supplied to the district heating system of the campus during the winter heating season, whereas it is used during summer to thermally power an ORC. In this way, the geothermal source is used all year long allowing its optimum exploitation. Moreover, the geothermal fluid, after being used by the district system or by the ORC, is supplied to a dryer of wood biomass, and then is reinjected back to the geothermal reservoir. The biomass energy-based system consists of a gasifier producing syngas using dried wood and a combined heat and power unit fueled only by this syngas production.

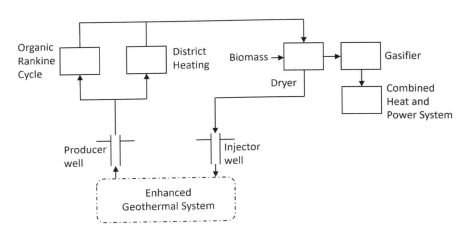

FIGURE 7.18 Hybrid geothermal-biomass co-generation system. (From K.F. Beckers et al. Hybrid Low-grade geothermal-biomass systems for direct-use and co-generation: From campus demonstration to nationwide energy player. *40th Workshop on Geothermal Reservoir Engineering Proceedings*, Stanford University, Stanford, CA (USA), 2015.)

An important aspect that must be considered for this kind of system lies in the availability of a geothermal and biomass energy source; as a consequence, an analysis of the site from a geological point view and of the local yield of wood biomass is mandatory in order to evaluate the system profitability. For the proposed system, the data regarding the geothermal reservoir are carried out by GEOPHIRES software, while the biomass source is evaluated on the basis of data carried out by project and literature data.

Furthermore, two sizes of the proposed hybrid system are analyzed in Reference [37]. The first one consists of a pilot plant integrating with only single production and reinjection wells for the geothermal source, and a small-scale biomass harvest area of 500 acres for the gasifier, which together supply about 20% of the campus heating load. The second case is based on a complete conversion system with six production and three reinjection wells coupled with 2500 acres for the biomass source. In particular, for the pilot plant a peak electrical generation capacity of 700 and 500 kW is achieved for the EGS and biomass-based ORC systems, respectively. As regards the larger scale system, the power of the EGS and ORC systems are 4.1 and 2.6 MW, respectively.

Along with the energy considerations, for the proposed system installation, economic considerations are mandatory. The increase of the cost between the two systems allows one to increase proportionally the thermal energy produced by the system and even more in the case of electrical energy, since for an investment of 47 M$, 76.2 and 2.4 GWh of thermal and electrical energy are produced, respectively, while for an investment of 217 M$ is possible to generate and use 379 GWh of heat and 15.2 GWh of electrical energy. An additional consideration that must be taken into consideration for this kind of system is the effect of uncertainty regarding technical parameters such as drilling costs, discount rates, and geothermal and biomass reservoir performance on the capital investment, levelized cost of energy, and heat and electricity output.

The previous systems allow one to produce thermal and electrical energy; however, the development of much more complex installations is possible when such renewable energy sources are coupled. An example of this kind of system is presented in Reference [38], consisting of a novel small-scale polygeneration system powered by biomass and geothermal energy. The layout of the system is relatively complex, since it integrates eight units: biogas steam reforming system, two Kalina cycles, an absorption-compression heat pump unit, an absorption refrigeration cycle, a liquefied natural gas system, a humidification-dehumidification desalination system, and a domestic water heater unit. Therefore, the polygeneration character of the system lies in the production of biogas, desalinated water, and hydrogen. As concerns the operation of this system, biogas is used as primary energy source, geothermal energy is used in second place, while liquefied natural gas is used as an energy sink. A simplified version of the system layout is shown in Figure 7.19, whereas a very detailed and complete scheme is reported in Reference [38]. The complete layout includes several components for each subsystem, including heat exchangers, reactors, pumps, and compressors, etc., and it is based on an operation strategy that takes into account all the mass and energy flows of the system.

In such a complex system, the thermodynamic analysis is crucial in order to assess the operation point for each component and its performance as well as the one for the

Novel and Hybrid Biomass-Based Polygeneration Systems

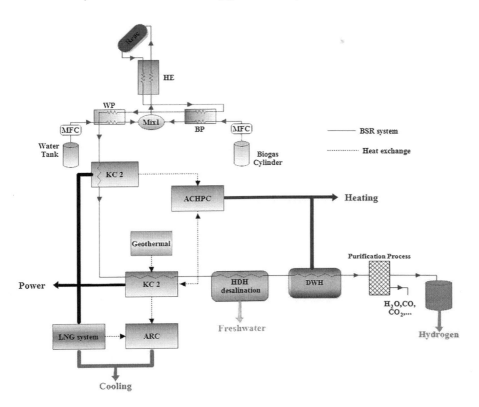

FIGURE 7.19 Scheme of the polygeneration hybrid biomass-geothermal system. (From H. Rostamzadeh et al. *Energy Conversion and Management.* 177, 2018, 535–62.)

whole system. Therefore, some assumptions must be considered for the analysis of the system, namely steady state conditions, no heat and pressure losses, constant air humidity, and desalinated water temperature.

The thermodynamic analysis carried out for the presented polygeneration system points out the main energy and mass flows within the plant, allowing for the evaluation of global performance. Under the conditions of Reference [38], the system achieves a cooling, heating, and electrical power of 1799, 538.1 and 443.4 kW, respectively, while the freshwater and hydrogen production is 367.9 L/h, and 292.9 kg/h, respectively. Moreover, for the same conditions the exergy and thermal efficiency results are 74.9% and 62.3%, respectively, determining the relatively high performance of the system. This performance may be also improved from the thermal point of view by changing some of the operation parameters of the system, for example, increasing the steam to carbon molar ratio, basic ammonia concentration of the absorption-compression heat pump, evaporator temperature of the absorption chiller, liquefied natural gas turbine expansion ratio, and the geothermal inlet temperature. The thermal efficiency increase may be also optimized as a function of reactor temperature, basic ammonia concentration of the first Kalina cycle, second Kalina cycle turbine expansion ratio, and desalination mass flow rate ratio. Furthermore, from the point of view of exergy efficiency, the maximum is achieved varying the reactor temperature, basic ammonia concentration of the first Kalina cycle and its turbine inlet pressure, and desalination mass flow rate ratio, as for the thermal efficiency. In summary, the optimization of the proposed biogas-geothermal polygeneration system must be performed taking into account the parameters of each subsystem and their effect on each component and the whole system.

LIST OF ABBREVIATIONS

AC	alternating current
BFB	fluidized bed gasifier
BMB	biomass boiler
BPST	back-pressure turbine
CEST	condensing-extraction turbine
COP	coefficient of performance
CPVT	concentrating photovoltaic thermal
CSP	concentrated solar power
DC	direct current
DH	district heating
DNI	direct normal irradiance
DSG	direct steam generation
EFGT	externally fired gas turbine
EGS	enhanced geothermal system
HTF	heat transfer fluid
HTHE	high-temperature heat exchanger
HX	heat exchanger
ICE	internal combustion engine
IRR	internal rate of return
LF	linear fresnel
MGT	micro gas turbine
NPV	net present value
NZEH	nearly zero energy house
ORC	organic rankine cycle
PD	partial defocusing
PI	proportional integral controller
PT	parabolic trough
PTC	parabolic trough collector
PV	photovoltaic
PVT	photovoltaic-thermal
SF	solar field
ST	solar tower
TCI	total capital investment
TD	total defocusing
TL	thermal demand
VCC	vapor compression cycle
VSP	variable speed pump
WT	wind turbine

REFERENCES

1. S. Jacobsson, A. Bergek. Transforming the energy sector: The evolution of technological systems in renewable energy technology. *Industrial and Corporate Change*. 13, 2004, 815–49.
2. H. Lund. Renewable energy strategies for sustainable development. *Energy*. 32, 2007, 912–9.
3. R. York. Do alternative energy sources displace fossil fuels? *Nature Climate Change*. 2, 2012, 441.
4. W. Su, J. Wang, J. Roh. Stochastic energy scheduling in microgrids with intermittent renewable energy resources. *IEEE Transactions on Smart Grid*. 5, 2014, 1876–83.
5. J. Twidell, T. Weir. *Renewable energy resources*. Routledge, 2015.
6. M. Deshmukh, S. Deshmukh. Modeling of hybrid renewable energy systems. *Renewable and Sustainable Energy Reviews*. 12, 2008, 235–49.
7. H.L. Willis. *Distributed power generation: planning and evaluation*. CRC Press, 2018.
8. M. Wegener, A. Malmquist, A. Isalgué, A. Martin. Biomass-fired combined cooling, heating and power for small scale applications–A review. *Renewable and Sustainable Energy Reviews*. 96, 2018, 392–410.
9. J. Wang, Y. Yang. Energy, exergy and environmental analysis of a hybrid combined cooling heating and power system utilizing biomass and solar energy. *Energy Conversion and Management*. 124, 2016, 566–77.
10. B. Norton, P.C. Eames, T.K. Mallick, M.J. Huang, S.J. McCormack, J.D. Mondol et al. Enhancing the performance of building integrated photovoltaics. *Solar Energy*. 85, 2011, 1629–64.
11. M.S. Islam, R. Akhter, M.A. Rahman. A thorough investigation on hybrid application of biomass gasifier and PV resources to meet energy needs for a northern rural off-grid region of Bangladesh: A potential solution to replicate in rural off-grid areas or not? *Energy*. 145, 2018, 338–55.
12. J. Settino, P. Morrone, A. Algieri, T. Sant, C. Micallef, M. Farrugia et al. Integration of an organic Rankine cycle and a photovoltaic unit for micro-scale CHP applications in the residential sector. *Energy Procedia*. 126, 2017, 597–604.
13. W.R.E. is Hot. Concentrating Solar Power Global Outlook 09.
14. U. Sahoo, R. Kumar, P. Pant, R. Chaudhary. Resource assessment for hybrid solar-biomass power plant and its thermodynamic evaluation in India. *Solar Energy*. 139, 2016, 47–57.
15. P. Campo, T. Benitez, U. Lee, J. Chung. Modeling of a biomass high temperature steam gasifier integrated with assisted solar energy and a micro gas turbine. *Energy Conversion and Management*. 93, 2015, 72–83.
16. R. Sterrer, S. Schidler, O. Schwandt, P. Franz, A. Hammerschmid. Theoretical analysis of the combination of CSP with a biomass CHP-plant using ORC-technology in central Europe. *Energy Procedia*. 49, 2014, 1218–27.
17. A.M. Pantaleo, S.M. Camporeale, A. Miliozzi, V. Russo, N. Shah, C.N. Markides. Novel hybrid CSP-biomass CHP for flexible generation: Thermo-economic analysis and profitability assessment. *Applied Energy*. 204, 2017, 994–1006.
18. A. Desideri, S. Amicabile, F. Alberti, S. Vitali-Nari, S. Quoilin, L. Crema et al. Dynamic modeling and control strategies analysis of a novel small CSP biomass plant for cogeneration applications in building. *ISES Solar World Congress 2015, Conference Proceedings*, 2015. p. 12.
19. S. Karellas, K. Braimakis. Energy–exergy analysis and economic investigation of a cogeneration and trigeneration ORC–VCC hybrid system utilizing biomass fuel and solar power. *Energy Conversion and Management*. 107, 2016, 103–13.

20. X. Zhang, H. Li, L. Liu, R. Zeng, G. Zhang. Analysis of a feasible trigeneration system taking solar energy and biomass as co-feeds. *Energy Conversion and Management.* 122, 2016, 74–84.
21. F. Khalid, I. Dincer, M.A. Rosen. Energy and exergy analyses of a solar-biomass integrated cycle for multigeneration. *Solar Energy.* 112, 2015, 290–9.
22. Y. Tanaka, S. Mesfun, K. Umeki, A. Toffolo, Y. Tamaura, K. Yoshikawa. Thermodynamic performance of a hybrid power generation system using biomass gasification and concentrated solar thermal processes. *Applied Energy.* 160, 2015, 664–72.
23. J.H. Peterseim, U. Hellwig, A. Tadros, S. White. Hybridisation optimization of concentrating solar thermal and biomass power generation facilities. *Solar Energy.* 99, 2014, 203–14.
24. J.C. López, Á. Restrepo, E. Bazzo. Exergy analysis of the annual operation of a sugarcane cogeneration power plant assisted by linear Fresnel solar collectors. *Journal of Solar Energy Engineering.* 140, 2018, 061004.
25. C.I. Hussain, B. Norton, A. Duffy. Technological assessment of different solar-biomass systems for hybrid power generation in Europe. *Renewable and Sustainable Energy Reviews.* 68, 2017, 1115–29.
26. Z. Bai, Q. Liu, J. Lei, H. Hong, H. Jin. New solar-biomass power generation system integrated a two-stage gasifier. *Applied Energy.* 194, 2017, 310–9.
27. E.K. Burin, L. Buranello, P.L. Giudice. T. Vogel, K. Görner, E. Bazzo. Boosting power output of a sugarcane bagasse cogeneration plant using parabolic trough collectors in a feedwater heating scheme. *Applied Energy.* 154, 2015, 232–41.
28. E.K. Burin, T. Vogel, S. Multhaupt, A. Thelen, G. Oeljeklaus, K. Görner et al. Thermodynamic and economic evaluation of a solar aided sugarcane bagasse cogeneration power plant. *Energy.* 117, 2016, 416–28.
29. P. Morrone, A. Algieri, T. Castiglione, D. Perrone, S. Bova. Investigation of integrated organic Rankine cycles and wind turbines for micro-scale applications. *Energy Procedia.* 148, 2018, 986–93.
30. P. Balamurugan, S. Ashok, T. Jose. An optimal hybrid wind-biomass gasifier system for rural areas. *Energy Sources, Part A: Recovery, Utilization, and Environmental Effects.* 33, 2011, 823–32.
31. A. Gonzalez, J.-R. Riba, B. Esteban, A. Rius. Environmental and cost optimal design of a biomass–wind–PV electricity generation system. *Renewable Energy.* 126, 2018, 420–30.
32. Y. Zheng, B.M. Jenkins, K. Kornbluth, C. Træholt. Optimization under uncertainty of a biomass-integrated renewable energy microgrid with energy storage. *Renewable Energy.* 123, 2018, 204–17.
33. M.H. Dickson, M. Fanelli. *Geothermal energy: Utilization and technology.* Routledge, 2013.
34. H. Li, X. Zhang, L. Liu, S. Wang, G. Zhang. Proposal and research on a combined heating and power system integrating biomass partial gasification with ground source heat pump. *Energy Conversion and Management.* 145, 2017, 158–68.
35. R. DiPippo. *Geothermal power generation: Developments and innovation.* Woodhead Publishing, 2016.
36. A. Borsukiewicz-Gozdur. Dual-fluid-hybrid power plant co-powered by low-temperature geothermal water. *Geothermics.* 39, 2010, 170–6.
37. K.F. Beckers, M.Z. Lukawski, G.A. Aguirre, S.D. Hillson, J.W. Tester. Hybrid low-grade geothermal-biomass systems for direct-use and co-generation: From campus demonstration to nationwide energy player. *40th Workshop on Geothermal Reservoir Engineering Proceedings*, Stanford University, Stanford, CA (USA), 2015.
38. H. Rostamzadeh, S.G. Gargari, A.S. Namin, H. Ghaebi. A novel multigeneration system driven by a hybrid biogas-geothermal heat source, Part I: Thermodynamic modeling. *Energy Conversion and Management.* 177, 2018, 535–62.

8 Application of Thermoelectric Power Generators in Small-Scale Heating Devices

Mariusz Filipowicz, Krzysztof Sornek, Mateusz Szubel and Maciej Żołądek

CONTENTS

8.1	Introduction	186
8.2	Fundamentals of Thermoelectricity and Thermoelectric Devices	186
8.3	Thermoelectricity Generation	187
	8.3.1 Materials and Efficiency of TEGs	189
	8.3.2 Electrical Properties of Electrical Circuits Applied with TEGs	190
8.4	General Overview of TEG Applications	193
8.5	Overview of Application of Biomass Devices	193
	8.5.1 Units with Direct Contact with the Combustion Chamber (DCCC)	195
	8.5.1.1 Stoves with Convective Cooling	195
	8.5.1.2 Stoves with Forced Air Cooling	195
	8.5.1.3 Stoves with Water Cooling	197
	8.5.2 Units Extracting Heat from Exhaust Gases	200
	8.5.3 Combined Heat Exchanger Systems	201
8.6	Case Studies Regarding Fireplaces/Stoves with Heat Accumulation and Integrated TEGs	202
	8.6.1 Case Study of a Stove with Heat Accumulation Modules	202
	8.6.2 Fireplaces with TEGs	206
	8.6.2.1 TEG No. 1: A DCCC	206
	8.6.2.2 TEG No. 2: A DCCC	207
	8.6.2.3 TEG No. 3: An FGHE	208
8.7	Summary	211
References		212

8.1 INTRODUCTION

Various types of small-scale biomass devices are still used in a considerable number of homes in rural and suburban areas in many countries around the word (such as northern USA, Canada, and northern Europe). Very simple or even primitive units are also widespread in Africa.

It may seem that the technological progress related to various household appliances has bypassed biomass technology (e.g., wood stoves). Power generation, especially thermoelectricity, may lead to reinventing this old biomass combustion technology for today's environmental regulations and requirements. A wider application of modern biomass units may significantly reduce pollution and could help millions of families reduce their dependence on gas and oil while also providing a small power surplus owing to the use of thermoelectric generators (TEGs).

The other problematic issue is that 1.6 billion people still have no access to electricity, especially in low income populations, living mostly in rural areas. As little as 10 W is often sufficient to cover basic needs such as a small light and a radio. To fill that gap, more than 16 GW power plants would be required. Individual TEGs coupled with biomass units (e.g., cooking stoves) are thus a potentially interesting solution to providing electricity [1].

In rural areas of developing countries, biomass energy accounts for approximately 90% of the total energy supply. Biomass combustion meets the basic energy needs for cooking and heating in rural households and for the heating process in traditional industries. In general, biomass is usually burnt in open fire stoves. These traditional stoves are characterized by low efficiency, which results in inefficient use of the scarce wood supplies [2].

The development of biomass-based TEGs supports distributed generation, biomass energy, and energy storage leading to increased energy efficiency. Of course, TEGs may be combined with residential solar photovoltaic (PV) systems, battery systems, and other power generation units, constituting a part of efficient local microgrids.

The variety and number of biomass units around the world is still very high. The peculiarities of their structures, the applied materials, and the purposes may vary widely [4]. This poses certain technical challenges for the proper design and efficient application of TEGs. Access to a certain amount of electricity is very important for biomass devices, as it allows for self-powering. Auxiliary devices such as electronic controllers, fans, pumps, etc., require constant power. Therefore, the lack of access to electricity may hinder the application of biomass units. The required power is not high—even as little as 30–70 W is sufficient, which provides a significant opportunity for the development of TEG technologies.

The introductory Section 8.1 shows that this goal is achievable, but not without certain obstacles.

8.2 FUNDAMENTALS OF THERMOELECTRICITY AND THERMOELECTRIC DEVICES

Generally, thermoelectric phenomena consist of a direct conversion of a heat stream into electricity if a temperature difference can be obtained during such a flow. Voltage

Application of Thermoelectric Power Generators

can be generated when a different temperature is observed at each side of the TEG. At the atomic scale, a temperature gradient causes charge carriers (electrons) in the material to diffuse from the hot side to the cold side. From the point of view of thermodynamics, TEGs are similar to heat engines, but their size and mass are smaller and the TEGs have no moving parts. On the other hand, TEGs are typically more expensive and less efficient.

In the following section, the idea of thermoelectric phenomena, the materials, and applications are briefly presented.

8.3 THERMOELECTRICITY GENERATION

Typical TE (thermoelectric) modules (also called Seebeck cells) are composed of a set of semiconductor components formed from two different materials (as in thermocouples). As shown in Figure 8.1, these components are connected thermally in parallel and electrically in series. Two ceramic plates are placed on each side for electrical insulation. When heat flows through a cell, the n-type components are charged negatively (excess of electrons) and p-type components are charged positively (deficiency of electrons), resulting in the formation of an electric flow [2].

The possibility of generating electricity using the thermoelectric effect depends on the Seebeck coefficient (S), the electrical conductivity (σ), and thermal conductivity (K) to maintain the temperature difference between the junctions. These properties are represented by the so-called Z figure-of-merit, where:

$$Z = \frac{S^2 \sigma}{k}. \qquad (8.1)$$

In a TEG, the temperature difference controls the external voltage. In Figure 8.2, electrical and thermal processes occurring in TEG are presented. The thermal processes are plotted in gray and the electrical processes in black [3].

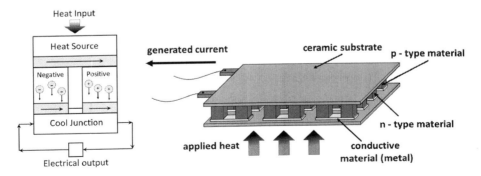

FIGURE 8.1 Principle of thermoelectricity (left) and TEG operation, connection of type p and type n semiconductors has been depicted (right).

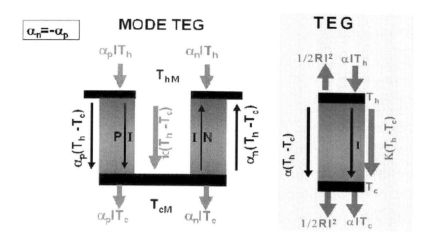

FIGURE 8.2 Thermal input-output and electricity generation in TEG with semiconductor *n-p* layers (left); conversion of electric current to heat during TEG operation (right). Thermal processes have been plotted in grey and the electrical processes in black. (Figure taken from C. Goupil et al., *Entropy*, vol. 13, pp. 1481–1517, 2011.)

The following processes have been illustrated:

- Generation of Seebeck voltage in the direction of heat transfer according to the formula:

$$V_0 = \alpha(T_h - T_c), \tag{8.2}$$

(the figure shows voltages generated on the *p* and *n* layers with opposite signs $\alpha_n = -\alpha_p$
- Conductive heat flow $K_{n,p}\Delta T$ (where K is the thermal coefficient)
- Heat generated at the internal resistance of TEG: $(1/2)RI^2$
- Peltier heat flow, $\alpha_{n,p}T_h I$.

A high-temperature heat stream $(\alpha_p IT_h)$ is provided to the hot side, and the heat stream $(\alpha_p IT_c)$ is removed from the cold side. Heat is also generated as a result of the flow of current I on internal resistance R and additional Peltier heat proportional to current I and temperature difference ΔT is generated. Therefore, the cooling system should be efficient, as all the heat is dissipated to keep the temperature difference.

The η energy conversion efficiency of thermoelectric devices is determined by the value of the figure of merit—*ZT* for thermoelectric materials according to Equation (2):

$$\eta = \left\{\frac{T_h - T_c}{T_h}\right\}\left[\frac{\sqrt{1+ZT_m}-1}{\sqrt{1+ZT_m}+\frac{T_c}{T_h}}\right], \tag{8.3}$$

where: T_m is the medium temperature of TEG: $1/2\,(T_h + T_c)$.

Application of Thermoelectric Power Generators

It is evident that the equation consists of two parts: the Carnot cycle efficiency and the efficiency of the TEG. The TEG efficiency depends mainly on the ZT parameter. For $ZT \to \infty$ the overall efficiency is equal to the Carnot cycle efficiency. Therefore, obtaining materials with a higher ZT parameter is essential for the applications discussed in the following sections.

8.3.1 Materials and Efficiency of TEGs

There are multiple materials that may be used for the construction of TEGs. Only several of those, however, are available on the market at a reasonable price. One such material, fit for ambient temperature applications, is bismuth telluride (Bi_2Te_3). New materials offering great potential for application, such as clathrates, skutterudites, Heusler alloys, Chevrel phases, and oxides, should soon leave laboratories [2].

Figure 8.3 presents the most important parameters of the selected materials—the ZT value (Equation 8.1) and the operating temperature.

It may be noted that devices using some of the materials presented in Figure 8.3 may operate at a wide range of temperatures, but the ZT parameter does not exceed 1.6. For Bi_2Te_3, which is the most common one, the value reaches only c. 0.9.

The efficiency of the TEGs (Equation 8.3) can be compared to other thermodynamic cycles—the results of such a comparison are presented in Figure 8.4.

In the case where the value of $ZT \approx 1$ (e.g., 0.9 for Bi_2Te_3), other competitive applications are more efficient. If a value of at least $ZT = 4$ is to be reached, TEGs could be competitive in relation to, for example, organic Rankine cycle (ORC) units and in the case of $ZT = 20$ the devices may reach efficiencies similar to Stirling

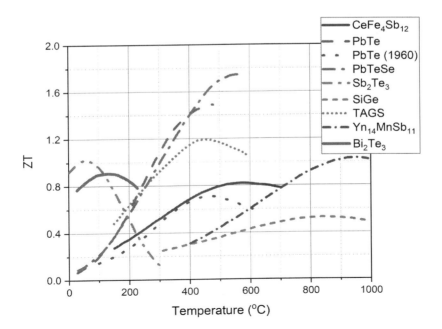

FIGURE 8.3 ZT values for selected materials against the temperature.

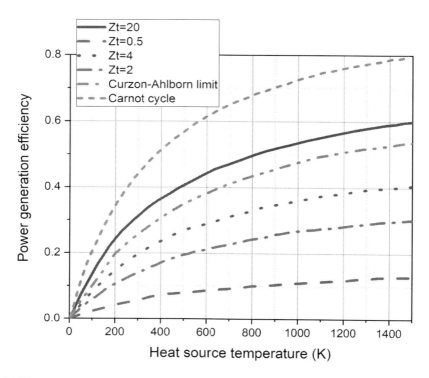

FIGURE 8.4 The efficiency of the thermoelectric devices compared to other cycles. The curves have been provided for $ZT = \{0.5, 2, 4, 20\}$.

engines on concentrated solar radiation or Rankine cycles in a coal power station. At this point, however, it seems rather far-fetched.

Taking all thermal and electrical processes occurring in the TEG into account, the general Sankey plot is presented in Figure 8.5.

Heat is generated in a combustion chamber as a result of fuel combustion—Q_f; heat dissipation before it reaches the thermoelectric generator has been marked as Q_c and Q_t—wastes in the combustion chamber and from the hot side of the TEG, respectively. Q_{teg} represents the heat delivered to the TEG, partially converted to electric power (P_{el}) and transferred to the cooling system—Q_c. In practice, part of the heat generated in the combustion chamber has an insufficient temperature to be used in the TEG—therefore it can bypass the TEG and be, for example, transferred to the boiler's water jacket (together with the heat Q_t-Q_j). The heat losses through flue gas have been marked as Q_e.

8.3.2 Electrical Properties of Electrical Circuits Applied with TEGs

The external wiring of TEGs is very important in terms of power generation applications [2,4]. The wiring diagrams are presented in Figure 8.6. The simplified diagram has been depicted in Figure 8.6 exhibits a TEG in practical application.

Application of Thermoelectric Power Generators

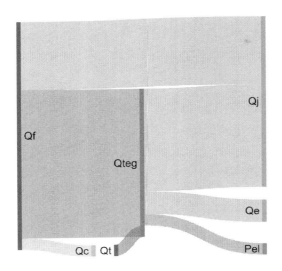

FIGURE 8.5 Sankey plot for the TEG generator working in a biomass unit.

The simplest wiring diagram is not efficient in terms of transferring the generated power to an external circuit and load. Based on the Ohm's law, conditions for an optimal circuit for a TEG generator [3] may be obtained.

Considering that the generated power P_{pro} is the difference between the delivered and the extracted power for the hot and cold sides—Q_{in} and Q_{out}, respectively—the following equation may be derived:

$$P_{pro} = Q_{in} - Q_{out} = \alpha I (T_h - T_e) - R_{in} I^2. \tag{8.4}$$

The designations are provided in Figure 8.6.

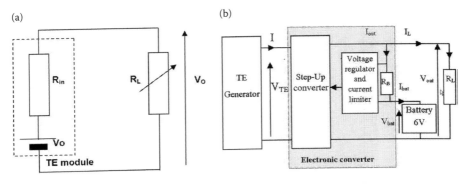

FIGURE 8.6 Electrical diagrams of a TEG generator (left), the simplest structure (right), TEG in practical application. (Merged diagrams from D. Champier et al. *Energy*, vol. 35, pp. 935–942, 2010, (left) and D. Champier et al., *Energy*, vol. 36, pp. 1518–1526, 2011 (right).)

The voltage measured at load depends on the internal resistance and load resistance:

$$V_{out} = V_0 \frac{R_L}{R_{in} + R_L} = V_0 \frac{M}{1+M}, \qquad (8.5)$$

(where M defines the load ratio $M = R_L/R_{in}$) and hence the load current I_{out}:

$$I_{out} = \frac{V_0}{R_{in} + R_l} = \frac{V_0}{R_{in}(1+M)}. \qquad (8.6)$$

Then power on the load can be presented as a multiplication of V_{out} and I_{out}, Equations 8.5 and 8.6, respectively:

$$P_{pro} = \frac{V_C^2}{R_{in}} \frac{M}{(M+1)^2}. \qquad (8.7)$$

The efficiency of the electric circuit can be defined as the power on the load and provided heat:

$$\eta = \frac{P_{Pro}}{Q_{in}} = \frac{\Delta T}{T_h} \frac{M}{M+1+(((M+1)^2)/ZT_h)-(1/2)(T/T_h)}. \qquad (8.8)$$

The efficiency can be expressed as a product of:

- Reversible Carnot efficiency, η_c, and

$$\eta_c = \frac{\Delta T}{T_h} = \frac{T_h - T_c}{T_h} \qquad (8.9)$$

- An irreversible factor

$$\frac{M}{(M+1+(((M+1)^2)/ZT_h)-(1/2)(\Delta T/T_h)}. \qquad (8.10)$$

Owing to the fact that $\partial \eta / \partial M = 0$, the maximum current is given by the expression:

$$\frac{\partial P_{pro}}{\partial I} = 0 = \alpha(T_h - T_c) - 2R_{in}I, \qquad (8.11)$$

$$I_P^{max} = \frac{\alpha(T_h - T_C)}{2R_{in}} = \frac{V_0}{2R_{in}}, \qquad (8.12)$$

and the maximal generated power is equal to:

$$P_{pro}^{max} = \frac{\alpha^2 \Delta T^2}{4 R_{in}}. \qquad (8.13)$$

Application of Thermoelectric Power Generators

Therefore, the maximum power is transferred to the electric load only for $R_{in} = R_L$. In practice, manual fulfillment of this condition is impossible. The diagram in Figure 8.6 (right), where the so-called maximum power point tracer (MPPT) regulator is applied, exhibits better properties. The idea behind this system is that the electronics system shown in the "electronic converter" box performs the function of a MPPT device. In such cases, the surplus of power can be stored in a battery, for example.

8.4 GENERAL OVERVIEW OF TEG APPLICATIONS

As presented in Figure 8.7, there is a wide range of possible TEG applications.

In practice, TEGs can be applied if a heat source is available and there is the possibility to transfer the heat to the surroundings, for example, by using a cooling medium. Generally, the main existing, implemented, and suggested applications include:

- Automotive industry: heat recovery from exhaust gases
- Combustion: for virtually all types of fuels and a variety of combustion units
- Waste heat recovery, for example, from industrial processes
- Small heat sources, for example, human body heat
- Space probes using radioisotope heat sources
- Domestic heating systems (e.g., boilers).

8.5 OVERVIEW OF APPLICATION OF BIOMASS DEVICES

Temperature and heat flux during biomass combustion near the combustion chamber are sufficient to allow for the use of TEGs. Also, exhaust gases usually create conditions that are sufficient for the application of TEGs. The generated power ranges are relatively small, but are possible to use in biomass combustion devices—for

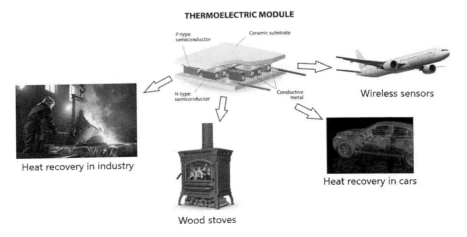

FIGURE 8.7 Some possible applications of TEGs.

example, to power the control electronics and some auxiliary devices (pumps, fans, etc.). The power surplus can be used for other applications.

The general idea behind biomass-based TEG applications is presented in Figure 8.8. MPPT relates to a special electronic unit applied to use the power generated by the TEG in the most efficient way.

As the thermoelectric units do not allow reaching high heat-to-electricity conversion rates (see Figure 8.4), it may not be feasible to build a domestic thermoelectric source based on combustible fuels to provide electric power for the entire household. TEGs may be feasible in situations where heat is transferred from high temperature levels to low temperature levels, and would otherwise be dissipated to the ambient air, or when the heat is available in large volumes without bearing additional costs [4]. The suitable points of heat transfer from biomass combustors are as follows:

- Inside the boilers, where combustion occurs at >500°C, but the heat is used for water heating to provide temperatures up to 80°C;
- Outside stoves, where efficiencies are rather low, typically less than 30%, and combustion of the fuel leads to stove surface temperatures exceeding 200°C; and
- In heat exchangers transferring high-temperature heat to low-temperature heat from heat carriers in biomass boilers (e.g., boilers with thermal oil jackets [5]).

A general classification of most TEGs applied with biomass devices is presented in Figure 8.9. The heat source, the cooling method, and the applied auxiliary devices are also depicted. This figure will serve as a guide for the remaining part of the chapter.

In the next part of the chapter, selected approaches are discussed and examples are provided. The following types of TEGs are presented:

- TEGs with direct contact with the combustion chamber (DCCC)
  - Convective cooling
  - Fan forced cooling
  - Water cooling

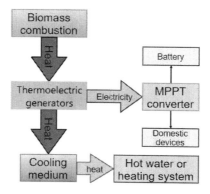

FIGURE 8.8 Idea of TEGs application in biomass combustion devices.

Application of Thermoelectric Power Generators

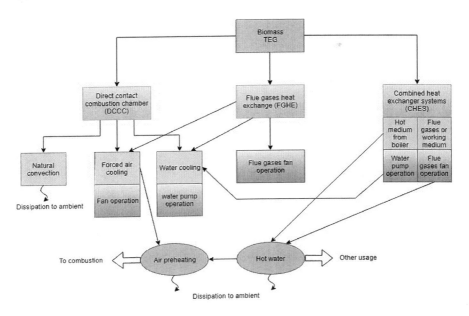

FIGURE 8.9 Diagram presenting the particular solutions of biomass-TEG units.

- Flue gases heat exchange (FGHE)
- Combined heat exchanger systems (CHES) for media heated by biomass combustion

8.5.1 Units with Direct Contact with the Combustion Chamber (DCCC)

8.5.1.1 Stoves with Convective Cooling

In this approach, the TEG generator is placed on the side wall of the stove and the cold side of the TEG is connected to a heat exchanger radiator with natural convection, for example. No fan for forced convection is applied. The example is presented in Figure 8.10.

The temperatures of the top surface are likely to rise well above 250°C, and similar temperatures are expected in case of side walls [4]. The structure of the TEG system (see Figure 8.10) includes a bolt fixed to the side of the stove, providing a good hot side thermal path. The cold side is cooled by natural convection using a rather massive finned aluminum heat sink.

In such a device, power of 4.2 W is generated with a single TEG module. The power is obtained when the hot side temperature reaches 275°C and the cold side temperature reaches 123°C—in such a case the temperature difference amounts to 152°C [4].

8.5.1.2 Stoves with Forced Air Cooling

The application of forced air cooling is a more complex approach, as it requires a fan consuming a part of the generated power; however, as a result, more power can

196   Biomass in Small-Scale Energy Applications

FIGURE 8.10 Cast iron stove with a TEG (on the left side wall) and a cross section of the TEG module with heat exchanger. (L. Kütt et al., *Renewable and Sustainable Energy Reviews*, vol. 96, pp. 519–544, 2018.)

be produced. Therefore, this approach is much more popular and several application examples may be provided.

As the first example, one should mention small-scale devices aimed at generating small amounts of power. Such devices may serve the charging of telecommunication units, for example. The devices can be used without access to any external electricity source. An example of a typical commercial solution is the Biolite camp stove or a similar device. Such devices have been described in Reference [6] and other literature sources. A typical unit is shown in Figure 8.11.

The Biolite unit (manufactured by Thermonamic, China), can be used for phone charging, as a USB connector has been supplied and the voltage and current are stabilized to maintain the parameters required by the USB standard. In this design, the heat is collected via a heat pipe within the chamber. The TEG module is cooled by fans and the air that was preheated is delivered to the combustion chamber of the stove. The

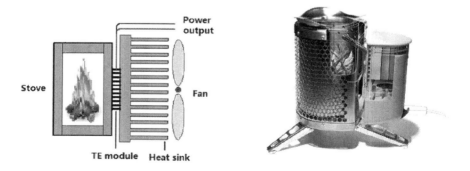

FIGURE 8.11 Biolite camp stove—the general idea and visualization. (Figure from H.B. Gao et al., *Applied Thermal Engineering*, vol. 96, pp. 297–310, 2016.)

system increases the amount of generated power and also improves the combustion of wood (or other biomass) in the small chamber. Of course, the device may be used for cooking by placing a pot or a container with water above the combustion chamber [6]. The battery provides electrical power to the fan during start-up and the TEG recharges the battery and powers the fan when the stove is hot. The Biolite camp stove was designed for outdoor adventures and emergency preparedness. Similar devices have been manufactured, for example, by Philips, Netherlands, where the stove is equipped with a stainless steel rod to receive the heat from the combustion chamber and transfer it to the TEG module to maintain the hot side of the TEG at a high temperature. The power output of the TEG is 4.5 W at a temperature difference of 240°C [6].

8.5.1.3 Stoves with Water Cooling

These devices may be placed on the back, rear, etc. side of a stove, but water has to be delivered for cooling. In general, the cooling efficiency is better, but it requires the presence of a water system (current water or a water container). In this case, it is also necessary to use water pumps providing sufficient flow. Therefore, some of the generated power has to be used for supplying power to the pump, mostly; however, the better cooling provides the power required by the pumps.

An example of such a unit has been presented in Figure 8.12.

Feeding a cooling water at 16°C, the pump used a power of approximately 5.7 W. The system operation was observed during various firewood feeding rates ranging from 1–9 kg/h. There was a total of 21 TEGs used all around the sides of the stove. During the 60 min preheating stage, the TEG system could not produce sufficient power to supply the water pump; however, at the lowest rate of wood supply, the output was measured to be above 16 W, a power sufficient for the pump. The highest fuel feed rate provided 155 W of electric output, but it has to be taken into account that 11 kW of heat has passed through to the cooling water. Thus, such power would only be available when a large enough hot water storage tank is provided or a solution with similar effects is applied.

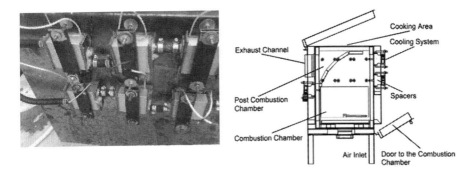

FIGURE 8.12 Water-cooled thermoelectric generator mounted on stove walls. Overview of the stove (right) and a view of an assembled chain of TEG generating modules (left). (Figure from L. Kütt et al., *Renewable and Sustainable Energy Reviews*, vol. 96, pp. 519–544, 2018.)

Another example [8] based on water containers is presented in Figure 8.13. Water from the tank is used for cooling the TEGs. This results in the increased temperature of the water. The hot water may then be used for various purposes.

In the system shown in Figure 8.13, the water pump consumes 8 W for a flow rate of 4.5 L/min. The total obtained power reaches a maximum of 42 W at a temperature difference of 250°C (hot side 300°C and cold side 50°C). The average power generation reached 27 W.

A very interesting example was provided by the Austrian Bioenergy Centre. Their system could provide a nominal power of 168 W with 16 modules arranged in two rings situated one over another, heated by the flame and hot gas from the inside and cooled with circulating water on the outside. The optimum flame temperature was found to be in the range of 1100–1150°C with excess air ratios of 1.3–1.6. The obtained electrical efficiency reached 1.5%, while the efficiency of the TEG was 3.5%. The test has exhibited that 45% of the heat generated by the stove flows through the TEG. The fuel heat input during the stove operation was in range of 6–13 kW [9–11].

Owing to their wide possibilities of combustion control (automated ignition, easy and clean fuel handling), and relatively low emissions, pellet stoves are an excellent biomass heat source for TEGs operation. Such stoves require power supply for their operation, which may be considered their disadvantage, especially when applied in remote areas.

To solve such technical problems, a special TEG unit was constructed and a computer fluid dynamics (CFD) calculation of heat transfer was carried out. The main goal was not to decrease the properties of the stove (efficiency, pollution emission,

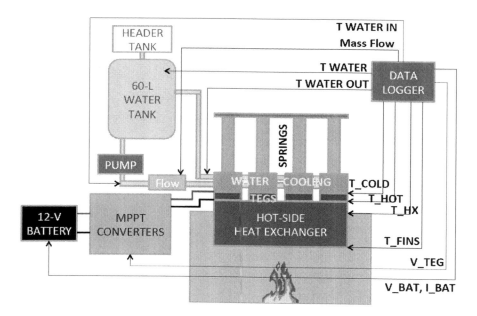

FIGURE 8.13 Diagram of a system with TEG generators and water tank used for cooling purposes. (Figure from A. Montecucco et al., *Applied Energy*, vol. 185, pp. 1336–1342, 2017.)

Application of Thermoelectric Power Generators

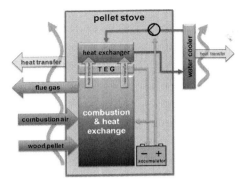

FIGURE 8.14 A general view of a pellet stove with a TEG unit. (Figure from I. Obernberger et al., *Biomass and Bioenergy*, vol. 116, pp. 198–204, 2018.)

etc.). Two of the most important concerns were to perform a complete burnout before the flue gases pass the TEG area and to maintain the uniformity of the temperature on the TEG's surface. The electrical output was in the range of 10–60 W and the value was correlated to the thermal power of the pellet stove. The maximum temperature difference was slightly higher than 200°C. The efficiency of the TEG was in the range between 1.4% and 2.2%, depending on the fuel load (2.2% for the nominal load). The electricity consumption of the TEG cooling system was in the range between 5 and 9 W, depending on the fuel load [10].

The system was also tested by conducting a load cycle simulation of the stove in actual operating conditions. The load cycle included four different load phases and three start-ups of the pellet stove and took 8 h as a whole. The system is presented in Figure 8.14.

The last example presents a very simple device used in developing regions of Africa. The TEG device is connected with a cooking pot [13]. Figure 8.15 demonstrates the general idea.

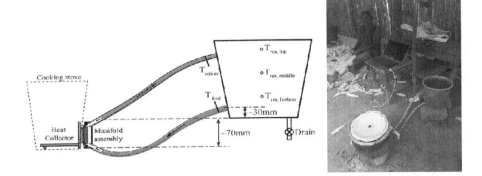

FIGURE 8.15 The general idea behind a TEG placed in the cooking stove (left) and a photograph presenting the operation of the device (right). (Figure from M.J. Deasy et al., *Energy for Sustainable Development*, vol. 43, pp. 62–172, 2018.)

The above TEG takes heat form the stove wall and is cooled by water placed in a tank. The water circulation is natural and therefore the positioning of the reservoir is very important. To establish a thermosyphon effect, the reservoir is placed higher than the TEG with the manifold. Moreover, the location of the connection of the return line to the reservoir should be below the surface level of the water in the reservoir to avoid air penetrating the return line.

Of course, during prolonged operation the average temperature of the reservoir will increase, and the user may choose to replace this warm water with colder water, which will increase the power output from the TEG system. Alternatively, after preparing a meal, the system may serve as a source of warm water, which may be used for other domestic purposes such as washing or cleaning. In this situation, the generator may be considered a very simple form of a cogeneration device. In outdoor conditions the generated power may reach a level of 3–4 W, sufficient for charging mobile electronic devices [13].

8.5.2 Units Extracting Heat from Exhaust Gases

This approach can be used if there is no possibility to install a TEG directly on the surface of a combustion chamber, for example, owing to the existence of an insulation layer or hindered access, and so on. Instead, it may be applied where an easy access to the exhaust channels is provided, for example, in a home system or when the exhaust channel is used as a part of a heat accumulation system.

The water container can be connected to a biomass stove—as presented in Figure 8.16. The TEGs are then placed inside the flue gas channel and fixed to the surface of the water container to achieve good contact with the heat source and the heat sink.

The obtained electrical power reached 7.6 W and provided electricity to power the fan for combustion improvement and basic electrical needs. The best location for the TEG is under the water tank, because the electricity can be produced without losing heat from the hot gas to the tank [7].

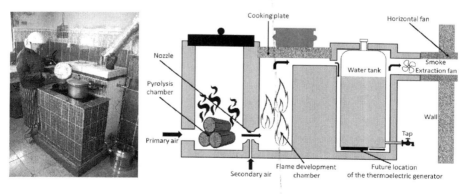

FIGURE 8.16 Stove with TEGs connected to a water tank. (Figure from L. Kütt et al., *Renewable and Sustainable Energy Reviews*, vol. 96, pp. 519–544, 2018.)

8.5.3 COMBINED HEAT EXCHANGER SYSTEMS

In this case, the thermoelectric modules operate inside a special dedicated heat exchanger between two streams of liquids: the heating liquid and the cooling liquid. Water and thermal oil may be used as a medium in practical applications, therefore, water-to-water or thermal oil-to-water thermoelectric heat exchangers can be considered, for example. The liquid may be heated in a variety of heating units including a biomass boiler. In the case where a water jacket is used, the maximum temperature may not exceed 95°C, but with an oil jacket, 200°C or higher temperatures are possible. The idea behind such units is presented in Figure 8.17.

Heat exchangers with TEGs allow the achievement of stable power generation, as the hot medium can be delivered at a controlled temperature and flow velocity. Stable temperatures can be achieved if, for example, a biomass unit is connected to a hot medium storage or in the case of a large volume of the jacket. After TEG cooling, the cold water may be used for domestic purposes.

This system may be considered an alternative to the standard heat exchangers allowing for power generation. However, thermoelectrics reduce the heat transfer and therefore the dimensions of such devices should be larger than standard heat exchangers. Despite this disadvantage, designs and prototypes of such heat exchangers are developed and their application is possible. To exemplify this, a commercial device that is ready for application should be mentioned— "Thermoelectric Assembly LL-210-24", which allows for the generation of 210 W of power [15].

The main difficulty related to this concept is the organization of flow of two separate liquid streams with different temperatures and the application of power-consuming pumps. Biomass installations, however, are often equipped with heat exchangers and even the application of oil as a thermal fluid and a working medium in a biomass installation does not pose a challenge. The concept of a system with a straw boiler is presented in Figure 8.18.

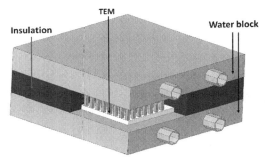

FIGURE 8.17 Heat exchanger with a TEG, two water blocks have been shown. (Figure from H. Lim et al., *Energies*, vol. 11, 2018.)

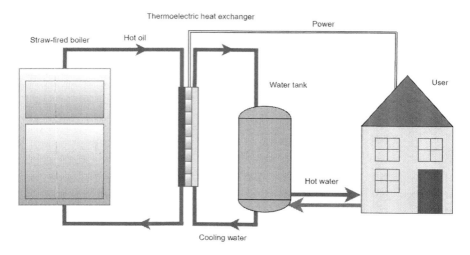

FIGURE 8.18 Diagram of a domestic installation providing hot water and electricity based on a thermoelectric heat exchanger.

8.6 CASE STUDIES REGARDING FIREPLACES/STOVES WITH HEAT ACCUMULATION AND INTEGRATED TEGS

This section presents two case studies related to the structures of fireplaces (stoves) with thermoelectric generation modules.

Case (a): A stove with ceramic modules as heat accumulation devices was used as a heat source; therefore, the only way to apply a TEG is to take the heat flux from the exhaust gases. Low-price thermoelectric modules and self-made heat exchangers were used. This example illustrates an FGHE system.

Case (b): A typical commercially-available fireplace for cheap and simple home application. The heat flux is taken from hot surfaces and the exhaust gas flow. Commercial TEGs not optimized for biomass units were applied. In this case, the DCCC and FGHE systems are illustrated.

8.6.1 CASE STUDY OF A STOVE WITH HEAT ACCUMULATION MODULES

In this case, stove-fireplaces with heat accumulation systems which are a combination of fireplaces and traditional accumulative stoves were used. Such structures are becoming increasingly popular. Within the system, the heat produced while burning wood is stored in an accumulative heat exchanger and released again for 12 h after the fire goes out. The resulting thermal efficiency of such devices reaches approximately 90% [16]. Such a high efficiency results also from the automation of the combustion process. The use of a specifically designed control system allows for control of the amount of air supplied to the furnace and the temperature of the flue gas in the exchanger (in order to avoid condensation resulting from insufficient temperature of the flue gas). The control of the air flow is achieved using one or more proportionally

Application of Thermoelectric Power Generators

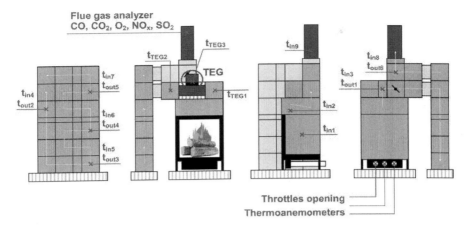

FIGURE 8.19 The diagram (and visualization) of the studied system. (K. Sornek et al. *Energy Conversion and Management*, vol. 125, pp. 185–193, 2016.)

opening valves operating in the range from 0% to 100%. The air flows in through different inlets providing good mixing of gaseous products from wood devolatilization and oxygen, thus allowing for clean combustion of the biomass.

The mass of the furnace ranges from 550 to 1050 kg in case of applying a meandering shape of the heat accumulation system. The average thermal power of the device is between 20–30 kW and the maximal value reaches 50–70 kW (during the burning process). A visualization of the test unit is presented in Figure 8.19.

In such a unit, only the heat from exhaust gases can be used. Therefore, a dedicated heat exchanger has to be applied. In practice, however, this heat exchanger may be considered a part of the system be manufactured along the unit. The final assembly of the system consists of special ceramic bricks and other components and thus, the TEG "box" may be included as a commercial element.

In the case under consideration, the TEG was installed in the exhaust channel—a special radiator was used instead of a ceramic block. Three particular radiator structures have been presented in Figure 8.20.

Three designs are described below:

A. The simplest design of the exchanger—a rectangular channel with the dimensions of 30 × 16 × 11.5 cm (length × width × height). In this case, the velocity of the gas remains constant. The large cross-section of the channel resulted in the fact that the flue gas traveled mainly near the channel axis, while the flue gas flow near the exchanger's walls was very small. The average velocity of the flue gas in the channel was about 1.1 ± 0.1 m/s. As a result, the surface temperature was relatively low (see Figure 8.20a).

B. The second design was a rectangular channel with a reducer. The reducer decreased the cross-section area of the channel by half (its height was decreased from 11 to 5.5 cm). In this case, the flue gas flowed at a higher

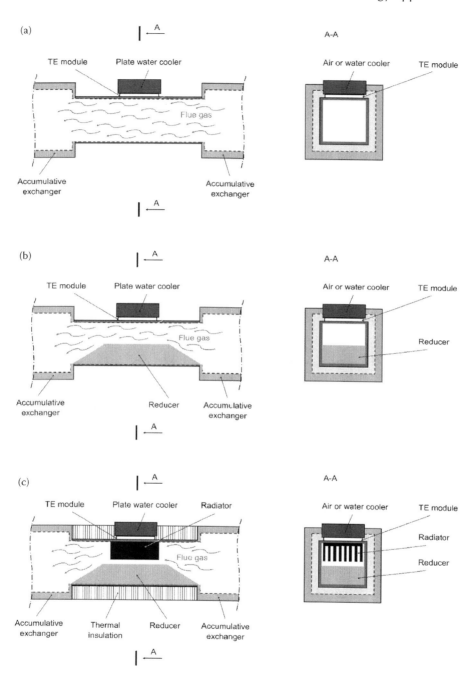

FIGURE 8.20 Structures of the heat exchangers designed for operation with a TEG—versions a, b, and c. (K. Sornek et al. *Energy Conversion and Management*, vol. 125, pp. 185–193, 2016.)

Application of Thermoelectric Power Generators

velocity compared to the first version, but the velocity did not exceed 2.0 m/s, and therefore the amount of heat transferred to the exchanger surface was also limited (see Figure 8.20b).

C. The third design included a radiator with fins that was used in addition to the reducer. The radiator consisted of elements (fins) arranged in parallel to the direction of the flow of the flue gas. This way, the surface for the heat transfer from the gases to the exchanger increased significantly. The radiator was mounted on the exchanger's wall using thermal paste with a high thermal conductivity coefficient. In addition, a thermal insulation consisting of a 5 cm thick layer of mineral wool was applied to eliminate the excessive cooling of the exchanger's surface caused by the ambient air (see Figure 8.20c).

A comparison of the obtained temperatures (for similar combustion conditions) are presented in Figure 8.21.

As may be noted, the highest temperature of the exchanger's surface was reached in case (c). This was caused not only by the introduced structural changes, but also owing to the application of the special insulation. Therefore, this structure was used in further tests where two thermoelectric generators based on Bi_2Te_3 were used:

- TEG no 1: with a maximum temperature difference of 330°C
- TEG no 2: with a maximum temperature difference of 150°C.

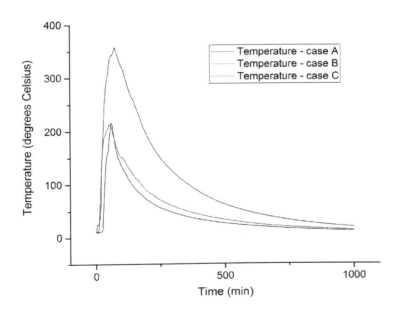

FIGURE 8.21 Variations of the surface temperatures of the exchanger during the combustion process for cases A, B, and C. (K. Sornek et al. *Energy Conversion and Management*, vol. 125, pp. 185–193, 2016.)

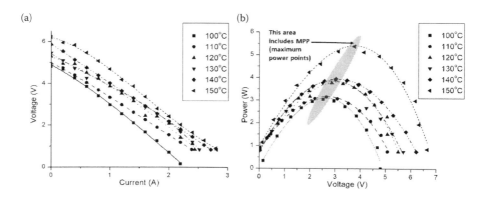

FIGURE 8.22 The voltage-current chart (a) and the power-voltage characteristics (b) of the TEG No. 2 for selected flue gas temperatures. (K. Sornek et al. *Energy Conversion and Management*, vol. 125, pp. 185–193, 2016.)

The best results were achieved for the TEG unit no. 2, as presented in Figure 8.22.

The presented curves correspond with the theory (see Equations 8.6 and 8.7), as the maximal current and power are clearly distinguishable. The position of the MPPT is clearly distinguishable and its variation depending on the temperature of the hot side may be noted. As shown, an electronic converter should be used, because a relatively small variation of load could significantly influence the power usage. In the example where the voltage on the load differs from the optimal $c.$ 1.0 V, the power decreases by 10%, while a variation of 2.0 V may significantly reduce the available power.

The maximal obtained power (approximately 6 W) corresponds to over a half of the value measured in laboratory conditions. It is a very good and promising result considering the simplicity of the developed solution.

The results of the conducted study exhibit the high potential for using TEGs to provide self-sufficient operation of stove-fireplaces with heat accumulation systems, even considering the limitations related to the exhaust gas temperature and flow [16].

8.6.2 Fireplaces with TEGs

In this approach, three specific solutions are considered [17]:

8.6.2.1 TEG No. 1: A DCCC

Device Is Mounted Directly on the Wall and Cooling Is Achieved by Means of Fans (Figure 8.23)

In this example, a maximal power of 18.8 W was obtained for the hot side temperature equal to 380°C and the estimated cold side temperature of 140°C (the temperature difference was approximately 240°C). The above refers to net power, excluding the power for the fans' operation. A commercial type of TEG provided by the Thermionic company was applied.

Application of Thermoelectric Power Generators

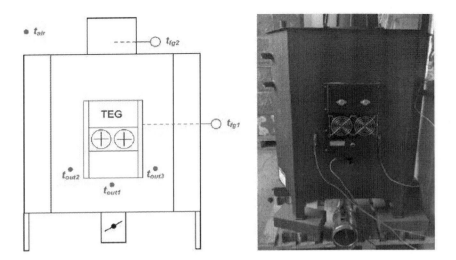

FIGURE 8.23 The rear wall of the fireplace with the TEG air-cooled by two fans. (K. Sornek et al. *Energy*, vol. 166, pp. 1303–1313, 2019.)

8.6.2.2 TEG No. 2: A DCCC

Device Is Mounted on the Wall with Water Cooling (Figure 8.24)

The maximal obtained power reached 31.2 W; however, a few watts (*c.* 5 W) should be subtracted for powering the water pump. The shape and dimensions of the TEG may pose a problem. In this type of commercial fireplace, only 70% of the active surface exhibits a high temperature. Therefore, a dedicated unit should be developed. The

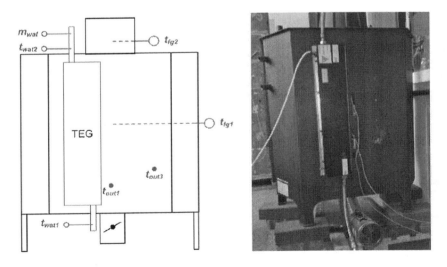

FIGURE 8.24 The rear wall of the fireplace with the TEG cooled by water flow. (K. Sornek et al. *Energy*, vol. 166, pp. 1303–1313, 2019.)

maximal power was obtained for the rear wall temperature of 380°C and estimated cold side temperature of approximately 70°C (a temperature difference of over 300°C).

8.6.2.3 TEG No. 3: An FGHE

Device Is Mounted on the Exhaust Gas Channel and the Modules Are Water-Cooled (Figure 8.25)

With the water cooling unit, a power output of 350 W was obtained. The flue gas should heat up the hot side of the module to 400°C or more. The number of single TEGs amounted to 36 and a serial-parallel connection was used. The matched power output may reach 350 W; however, in the case under consideration it was only 25 W (i.e., about 7% of the nominal power was obtained).

The fireplace used for the demonstrations is presented in Figures 8.23 through 8.25, and is a plate steel stove designed for burning seasonal hardwood (with a humidity up to 20%) as well as brown coal briquettes. The heating capacity of the unit is in the range from 8 to 16 kW. A typical structure of the stove contains fire clay (in form of ceramic plates), which increases the temperature in the combustion chamber due to low heat conducting properties to achieve more efficient combustion [17].

A comparison of the results for TEGs 1–3 is shown in Figures 8.26 and 8.27. The temperature of the hot side of the heating medium is indicated.

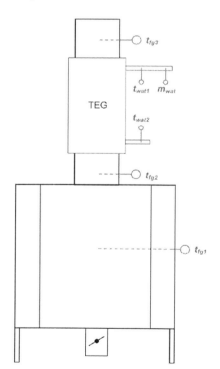

FIGURE 8.25 Water-cooled TEG device mounted to the exhaust gas channel. (K. Sornek et al. *Energy*, vol. 166, pp. 1303–1313, 2019.)

Application of Thermoelectric Power Generators

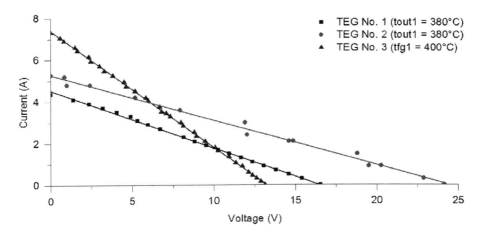

FIGURE 8.26 Current versus voltage characteristics of the demonstrated TEGs. (K. Sornek et al. *Energy*, vol. 166, pp. 1303–1313, 2019.)

Comparing the current-voltage characteristics of the tested TEGs, the highest short circuit current was observed in case of TEG no. 3, while the highest level of open circuit voltage was observed in case of TEG no. 2. In the case of TEG no. 3, the inclination angle of the characteristic is the highest, while the characteristics for TEG no. 1 and no. 2 behave in quite a similar manner. As a consequence, the power-voltage characteristics are also the most similar in the case of TEGs no. 1 and no. 2; however, the two examples differ in terms of the level of generated power.

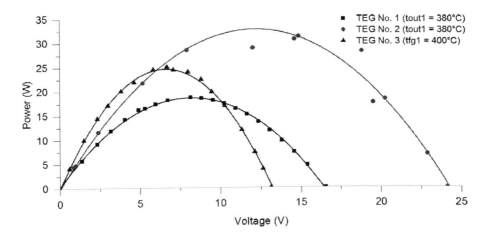

FIGURE 8.27 Power-voltage characteristics of the tested TEGs. (K. Sornek et al. *Energy*, vol. 166, pp. 1303–1313, 2019.)

TABLE 8.1
A Brief Summary of the Three Types of TEG Devices

Parameter	TEG no. 1	TEG no. 2	TEG no. 3
Heating element/medium	Hot plate	Hot plate	Flue gas
Cooling medium	Air	Water	Water
Cooling efficiency	Weak (140°C)	Good (70°C)	Good (80°C)
Heat transfer efficiency	Good	Good	Weak
Influence on stove operation	Minor	Minor	Significant
Required changes in control algorithm	Maximum temperature in combustion chamber	Maximum temperature in combustion chamber	Maximum temperature of flue gas, flue gas flow control
Modifications of the stove	Smooth wall, removal of fire clay from combustion chamber	Smooth wall, removal of fire clay from combustion chamber	Flue gas channel modification, installation of flue gas fan
Energy demand for operation	Low (fan operation)	Medium (water pump operation)	High (water pump and flue gas fan operation)
Cost of modification	Low	Low	Medium
Cost of investment in TEG	Low	Medium	High
Problems in exploitation	None	Nonuniform hot side temperature distribution	Cleaning flue gas channel
Possibility of application	High	Medium	Low

A summary of the three TEG units is presented in Table 8.1. Taking the actual conditions into account, TEG no. 1 is characterized by the widest range of possible applications. On the other hand, the unit still exhibits problems related to the low efficiency of the air cooling system. Despite the fact that the other tested TEGs have a more efficient water cooling system, certain exploitation problems render their use much less feasible.

To reach a wider range of applications of the TEG systems, it is necessary to undertake efforts aimed at adjusting the biomass devices as a heat source, adapting the control algorithm for more complex systems, and at applying TEG cooling systems with heat recovery (transfer to combustion chamber, production of hot water, etc.). Generally speaking, with reasonable efforts, the commercially available TEG devices can be efficiently used as typical fireplace units. The tests of the three TEGs presented in this chapter render the following parameters crucial: the size of the hot sides, the smoothness of the surface, the required heat flux, and sufficient temperatures.

The water cooling system is recommended to achieve higher power results. The system might be more efficient if the hot water could be used for other purposes. If hot air is useful in a given case (e.g., provided to the combustion chamber), then the air-cooled system may have an advantage by increasing the combustion efficiency. In such a case, the best solution may consist in a hybrid water-air-cooled system.

8.7 SUMMARY

There are several reasons to consider the application of TEGs in combination with a variety of biomass units such as wood stoves, fireplaces, pellet boilers, straw boilers, and so on:

- The price of thermoelectric modules, which are a component of TEGs, has dropped substantially because they are now mass-produced in China. The cost of a thermoelectric module has fallen below $2 per watt (uninstalled), compared to $3.50 per watt for solar panels (installed).
- The technical progress in cleaner biomass combustion and the regulations related to pollution emissions help to make new wood stoves cleaner and more efficient while allowing for integration of cordwood testing and automated features.
- A new generation of cleaner stoves is necessary, allowing for generation of electricity both for self-powering and a certain power surplus that could be used for other purposes; thermoelectric wood stoves can produce electricity up to 24 hours per day while eliminating load management concerns that are common in the case of solar and wind power.
- Lastly, the stoves are powered by local wood supplies, making the fuel low carbon and locally sourced.

Events such as the Fourth Wood Stove Competition focusing on the automation and electricity generation are important factors for the further development of biomass technology [18].

The problems related to these devices may be brought down to a simple statement: the power limits are a result of the units' efficiency. The thermodynamic limit is approximately 30%. This level of efficiency seems high, but the efficiency of thermoelectric modules reaches only $c.$ 20%. Finally, this results in a theoretical efficiency of $c.$ 6%. In practice, owing to several issues such as imperfect heat exchange, this efficiency is reduced to approximately 2%. In the presented case study, the problems consisted in the transfer of heat from flue gases to a solid body. To improve the efficiency of such heat transfer, large areas of heat exchangers are required, which, in turn, created structural problems. It is significant for the units not to obstruct the exhaust causing a pressure drop. Also, the deposition of dust on the elements of heat exchangers is a problematic factor—a layer of dust can significantly decrease the heat exchange. Therefore, the installation of TEGs in exhaust channels is not an efficient option. One of the key issues is that no industrial-scale and marketable solution has been introduced so far. It is easy to imagine that all industrial and domestic biomass devices could be equipped with TEGs, allowing the user to gain extra power. The demand for such devices is even accompanied by the market's interest. Besides the above, materials with a relatively high ZT are ready to be introduced in practice. Therefore, the efficiency of such generators will increase. If this is accompanied by price reductions and better applications, adapted to user demands for heat and electricity, the outlook is rather positive. It should be added that the emerging problems with energy supply, energy prices, and decrease in energy security may increase the need to store energy. In view of this chapter, biomass may be considered as a way to increase individual energy safety owing to its relatively easy storage conditions.

REFERENCES

1. D. Champier, J.P. Bédécarrats, T. Kousksou, M. Rivaletto, F. Strub, P. Pignolet, Study of a TE (thermoelectric) generator incorporated in a multifunction wood stove. *Energy*, vol. 36, pp. 1518–1526, 2011.
2. D. Champier, J.P. Bedecarrats, M. Rivaletto, F. Strub, Thermoelectric power generation from biomass cook stoves. *Energy*, vol. 35, pp. 935–942, 2010.
3. C. Goupil, W. Seifert, K. Zabrocki, E. Muller, G.J. Snyder, Thermodynamics of thermoelectric phenomena and applications. *Entropy*, vol. 13, pp. 1481–1517, 2011.
4. L. Kütt, J. Millar, A. Karttunen, M. Lehtonen, M. Karppinen, Thermoelectric applications for energy harvesting in domestic applications and micro-production units. Part I: Thermoelectric concepts, domestic boilers and biomass stoves, *Renewable and Sustainable Energy Reviews*, vol. 96, pp. 519–544, 2018.
5. K. Sornek, R.D. Figaj, M. Żołądek, M. Filipowicz, Experimental and numerical analysis of a micro scale cogeneration system with 100 KW straw-fired boiler. *CPOTE 2018: Proceedings of the 5 International Conference Contemporary Problems of Thermal Engineering: Energy Systems in the Near Future: Energy, Exergy, Ecology and Economics.* Gliwice, Poland, 18–21 September 2018, eds. W. Stanek, ISBN: 978-83-61506-46-1
6. D.L. Wilson, M. Monga, A. Saksena, A. Kumar, A. Gadgil, Effects of USB port access on advanced cookstove adoption. *Development Engineering*, vol. 3, pp. 209–217, 2018.
7. H.B. Gao, G.H. Huang, H.J. Li, Z.G. Qu, Y.J. Zhang, Development of stove-powered thermoelectric generators: A review. *Applied Thermal Engineering*, vol. 96, pp. 297–310, 2016.
8. A. Montecucco, J. Siviter, A.R. Knox, Combined heat and power system for stoves with thermoelectric Generators. *Applied Energy*, vol. 185, pp. 1336–1342, 2017.
9. W. Moser, G. Friedl, S. Aigenbauer, M.S. Heckmann, H. Hofbauer, A biomass-fuel based micro-scale CHP system with thermoelectric generators. *Central European Biomass Conference 2008*, Messe Center Graz, Austria, January 16–19, 2008.
10. W. Moser, G. Friedl, W. Haslinger, H. Hofbauer, Small-scale pellet boiler with thermoelectric generator. *Proceedings of 25th International Conference on Thermoelectrics*, Vienna, Austria, 2006.
11. W. Moser, S. Aigenbauer, M. Heckmann, G. Friedl, H. Hofbauer, Micro-scale CHP based on biomass-intelligent heat transfer with thermoelectric generators. *BIOENERGY*, Jyvaskyla, Finland, 2007.
12. I. Obernberger, G. Weiss, M. Kossl, Development of a micro CHP pellet stove technology. *Biomass and Bioenergy*, vol. 116, pp. 198–204, 2018.
13. M.J. Deasy, S.M. O'Shaughnessy, L. Archer, A.J. Robinson. Electricity generation from a biomass cookstove with MPPT power management and passive liquid cooling. *Energy for Sustainable Development*, vol. 43, pp. 62–172, 2018.
14. H. Lim, S.Y. Cheon, J.W. Jeong, Empirical analysis for the heat exchange effectiveness of a thermoelectric liquid cooling and heating unit. *Energies*, vol. 11, p. 580, 2018.
15. https://assets.lairdtech.com/home/brandworld/files/LL-210-24-00-00-00,%20rev05.pdf, last accessed 02.12.2018.

16. K. Sornek, M. Filipowicz, K. Rzepka, The development of a thermoelectric power generator dedicated to stove-fireplaces with heat accumulation systems. *Energy Conversion and Management*, vol. 125, pp. 185–193, 2016.
17. K. Sornek, M. Filipowicz, M. Żołądek, R. Kot, M. Mikrut, Comparative analysis of selected thermoelectric generators operating with wood-fired stove. *Energy*, vol. 166, pp. 1303–1313, 2019.
18. Fourth Wood Stove Competition. http://forgreenheat.org/2018-stovedesign/stovedesign.html. Accessed on July 16, 2019.

9 Straw-Fired Boilers as a Heat Source for Micro-Cogeneration Systems

Krzysztof Sornek, Mariusz Filipowicz and Karolina Papis

CONTENTS

9.1 Introduction 215
9.2 Straw as a Fuel 217
9.3 Straw Combustion Technologies 218
9.4 The First Approach: Implementation of an Oil Exchanger to an Existing Straw-Fired Boiler 220
9.5 The Second Approach: The Replacement of the Boiler's Water Jacket with an Oil Jacket 223
9.6 Power Generation in the Prototypical Micro-Cogeneration System with a Straw-Fired Batch Boiler 225
9.7 Conclusion 228
Acknowledgments 229
Nomenclature 229
References 229

9.1 INTRODUCTION

The presently available devices intended for biomass utilization in microscale applications are based on thermochemical processes of fuel conversion into heat: combustion, gasification, and pyrolysis.

Combustion of biomass is arguably the oldest known and most widely used controllable energy source on earth. It occurs with sufficient oxygen to completely oxidize the fuel. The first stage of combustion involves the evolution of combustible vapors from the biomass, which burn as flames. The residual material, in the form of charcoal, is burnt in a forced air supply to give more heat. The amount of heat that is produced is generally ~20 J of energy per dry kilogram of biomass and varies depending on species, for example. In order for combustion to be efficient and clean, the ingredients must be well mixed at the right temperatures for the right amount of time.

Biomass can be also converted into biofuels, both liquid and gaseous (e.g., biomethanol, bioethanol, biodimethyl ether, Fischer Tropsch fuels, biomethane, and biohydrogen). They can be an excellent alternative for traditional fuels used so far (e.g., oil, natural gas, and products of processing coal). Thermochemical processes of conversion of biomass are one of four main categories for preparing energy-efficient fuel. These include pyrolysis and gasification. Recently, great technological progress has been made in the production of biofuels and the issues relating to their environmental impact are much discussed, so that the technologies are becoming increasingly popular [1].

Gasification is the process of converting solid biomass into gas. Air is present during this process, but it is always substochiometric. The course of the process is affected by many parameters such as shape and size of the primary biomass, moisture, ash, and element content. Gasification process that produces pyrolytic oil and char can have a total thermal efficiency of about 70%. Gasification has three main stages. The first stage is autothermal heating; the required amount of heat is supplied by combustion of a portion of fuel. In the second phase, CO_2, H_2O, tar, and char are produced. The third stage, which occurs at high temperatures, is the reduction CO_2 and H_2O to CO, CH_4 and H_2. Tar is also mainly gasified. Char can be burned, lowering its concentration in the final product.

Pyrolysis is the thermal decomposition of biomass. It occurs in the inert atmosphere, which means the absence of oxygen. The main products of pyrolysis are charcoal, and liquid and gaseous products, which are generally CO, CO_2, hydrocarbons, and H_2. Their content the in final product depends mainly on the temperature to which the solid biomass is heated. The higher the temperature, the greater fraction of primary fuel is converted into useful products. The main advantage of pyrolysis biomass is the fact that the final products are burnt without any tar and they are cleaner during combustion. Moreover, their calorific value is much higher than solid biomass [2].

Among basic methods of biomass conversion to electricity, the most interesting options are gasification and combustion. Technologies based on biomass gasification are characterized by higher efficiency; however, they require higher financial investments (they are currently not widely available on the market). It has to be noticed that these drawbacks are related also to the energy systems based on the biomass combustion [3]. The main advantage of gasification systems is the possibility of using a wide range of biomass fuels [4]. Systems with Stirling engines, steam engines and steam turbines (operating in the Clausius–Rankine cycle—RC or organic Rankine cycle—ORC), as well as systems with thermoelectric generators have to be distinguished here [5,6]. Owing to the installed electrical power, the systems are divided into micro-cogeneration units (with a maximum power below 50 kW_e) and small-scale units (with installed capacity below 1 MW_e) [7].

Various types of boilers are used for different kinds of available biomass, and depending on the investor's expectations (concerning the operational parameters such as convenience of use), and fuel storage capacity. The most common solutions in case of residential buildings are multifuel boilers, pellet-fired boilers, and wood gasification boilers. However, straw-fired boilers are becoming increasingly popular on farms, housing states, schools, and other public buildings, as well as in industrial facilities, since straw is an inexpensive and widely available fuel.

The solutions available commercially include batch boilers for periodic and cyclic combustion of baled straw, devices based on ground straw continuous combustion using burning cigar method, and automatic devices for combustion of straw cut into 5–10 cm long pieces (also continuous operation). Owing to low prices and the possibility of using a wide range of biomass (e.g., wood, woodchips, energy willow, textile waste, sawdust, etc.), the first of the above-mentioned solutions has gained the greatest popularity in Poland.

The power output of typical straw-fired batch boilers varies between 40 and 700 kW and their application is primarily associated with the generation of low-temperature heat used in heating systems and domestic hot water systems. The heat also finds special applications in places such as greenhouses, drying rooms, distilleries, and so on. In the case of such applications, besides the typical heating boilers, it is also common to use oil-fired air heaters with thermal oil as the operating medium, heated to a temperature of 150–200°C. The possibility of generating heat with such high parameters justifies the attempts to use straw-fired devices (including batch boilers) as the source of heat supplied to small- and microscale cogeneration systems.

9.2 STRAW AS A FUEL

Straw is an agricultural crop by-product, made up of the dry stalks of cereals and legumes after removal of the grain and chaff. Typical straw crops are wheat, oats, barley, rice, rye, millet, and other varieties of grain. Straw used for energy purposes must meet certain technological requirements. Typically, straw grading is based on the heating value, moisture content, and the degree of wilting.

Low caloric value of freshly cut straw (yellow straw), high content of chlorine and alkali metals, as well as problems in storage, transport, and feeding to the combustion chambers mean that before burning straw has to be seasoned in order to eliminate harmful compounds by precipitation. A characteristic feature of straw being wilted is its gray color. A comparison of the calorific value of the basic types of straw depending on its moisture content is presented in Table 9.1.

TABLE 9.1
The Calorific Value of Different Types of Straw Depending on Its Moisture

Fuel	Moisture Content %	Caloric Value of Yellow Straw MJ/kg	Caloric Value of Gray Straw MJ/kg
Wheat straw	15–20	12.9–14.1	17.3
Barley straw	15–22	12.0–13.9	16.1
Rape straw	30–40	10.3–12.5	15.0
Corn straw	45–60	5.3–8.2	16.8

Source: I. Niedziółka and A. Zuchniarz, *Inżynieria Rolnicza*, 8, 2006, pp. 232–237.

The use of straw for energy purposes brings many benefits. The most important potential benefits of using straw as a fuel include:

- Reduction of carbon dioxide and sulfur oxide emissions to the atmosphere
- Significant reduction of energy production costs
- Reducing the consumption of conventional fuels
- Increasing the level of energy security on a local scale
- Avoiding costs of exporting and storing organic waste in landfill
- Economic activation of regions affected by unemployment (additional jobs)
- Limiting straw burning in the fields
- Improving the profitability of agricultural production.

Nevertheless, obstacles limiting the development of energy systems based on straw combustion and the specific properties of straw include:

- Large volume of straw in relation to its caloric value
- Heterogeneity of fuel
- High content of volatile parts
- Difficulties related to the use of straw and the need for specific boiler equipment [9].

9.3 STRAW COMBUSTION TECHNOLOGIES

The construction of straw-fired batch boilers is still under the evaluation. Owing to their simple construction and low maintenance costs, they are the perfect solution for a wide range of applications.

In the first-generation batch boilers, air was supplied to the combustion chamber through openings equipped with tilting flaps. Next, air flowed through the combustion chamber, supplying oxygen necessary to fuel combustion. Finally, flue gas from the combustion process flowed to the chimney. These types of straw boilers were characterized by low thermal efficiency (up to 35%–40%) and high emissions of carbon monoxide and dust into the atmosphere.

The second-generation batch boilers work on the basis of a counter flow combustion system. These boilers have two chambers: in the first chamber straw is gasified with insufficient oxygen, and in the second chamber, mixing and burning of carbon monoxide (CO) and organic compounds. This action is achieved by dividing the flowing air into two streams: the primary stream (directed into the combustion chamber) and the secondary one (which is mixed with the gasification products).

Initial solutions of batch boilers with a counter flow combustion system achieved an average thermal efficiency 70%–75%. The emission of carbon monoxide was in the range of 1000–4000 mg/m^3. Currently available batch boilers have a modified structure with a flame exchanger and improved secondary air supply system. Their efficiency may achieve a level of 87%, and the emission of carbon monoxide into the

Straw-Fired Boilers as a Heat Source for Micro-Cogeneration Systems

atmosphere is in the range of 250–1000 ppm. The scheme of a straw-fired batch boiler is shown in Figure 9.1.

Controlling the operation of the batch boilers is realized by changing the amount of air blown by the air fan to the combustion chamber and consequently, the instantaneous power of the device. The most common control signal to microcontroller is the temperature of the flue gas, the maximum value of which is defined by the user before the combustion process starts (as a rule, the exhaust gas temperature is set at 200–250°C). An additional protection, preventing overheating (boiling) of water is the temperature control in the water jacket.

The construction of batch boilers enables increasing their basic functionality with a power generation system. Among the technologies used in micro-cogeneration systems, technologies based on the use of heat of combustion (including steam engines, steam turbines, Stirling engines, Clausius–Rankine cycle, and organic Rankine cycle), are particularly interesting in the context of batch furnaces. The selection of a specific technology depends on numerous factors, including technical capabilities, investment and operating costs, and the required efficiency.

The construction of such types of boilers allows extending their functions to electrical energy production. Among the available technologies used in micro-cogeneration systems, particularly interesting are the ones that use heat of combustion (e.g., steam engines and turbines, Stirling engines, and installations with Clausius–Rankine cycle or organic Rankine cycle). The choice of a certain type of method depends on many factors, such as technical features, expected and exploitation costs, and the required efficiency. A comparison of selected operating parameters of the above-mentioned devices is presented in Table 9.2.

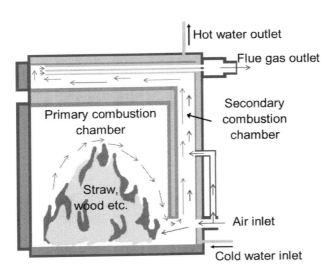

FIGURE 9.1 A scheme of a straw-fired batch boiler that works on the basis of a counter flow combustion system.

9.4 THE FIRST APPROACH: IMPLEMENTATION OF AN OIL EXCHANGER TO AN EXISTING STRAW-FIRED BOILER

The fundamental assumption that was made at the beginning of the investigation was that the basic functionality of the boiler as a heat source had to be maintained. Electricity generation was intended to be an additional feature which would not have a significant effect on the thermal efficiency of the device. An additional limitation was constituted by the low investment costs that should characterize the new system [11]. The new structure was created based on a modified Rankine cycle (RC); however, because it was very difficult to rescale the solutions applied in the average and large-scale systems, the steam turbine was replaced with a steam engine. The distribution of temperatures in the tested 180 kW_{th} batch boiler exhibits a quite high variability over time in different points of the primary combustion chamber. This is largely dependent on the batch size and the arrangement of straw bales. The flame temperature periodically exceeds 1000°C, thus being in the required range for optimum combustion of straw (850–1100°C). A far more even distribution of temperature occurs in the area between the post-combustion chamber and the fire-tube exchanger, where temperatures reach 500–700°C. So, in the first version of a high temperature oil system designed to transport heat from the boiler to the steam circuit, the use of heat from this part of the boiler was assumed. A special design of the oil exchanger was developed and adapted to the structural parameters of the batch boiler. It was placed in the space between the post-combustion chamber and the fire-tube exchanger. In order to maximize the use of available space, the assumed shape was a double meander, which allowed reaching the total length of approximately 20 m. The location oil of the exchanger between the post-combustion chamber and the fire-tube exchanger allowed for the reduction in the temperature of the flue gas, which leads to the minimization of the stack loss of the boiler (see Figure 9.2).

The hot oil heated in the oil exchanger was transported through a special circuit to the two shell-and-tube heat exchangers operating as an evaporator and superheater.

TABLE 9.2
Comparison of Selected Operating Parameters of Small-Scale Energy Conversion Technologies

Technology	Power Output, kWe	Electrical Efficiency, %	Total Efficiency, %	Investments Costs, EUR/kW_e	Possible Fuels
Rankine cycle	500–100,000	20–30	85–93	~1500	Most of solid, liquid and gaseous biofuels
Steam engines	20–5000	6–20	85–95	~1500	
Stirling engines	1–75	20–40	80–90	~3500	
Organic Rankine cycle	2–10,000	10–30	~85	~4500	

Source: M. Salamón et al. *Renewable and Sustainable Energy Reviews*, 15, 2011, pp. 4451–4465.

Straw-Fired Boilers as a Heat Source for Micro-Cogeneration Systems 221

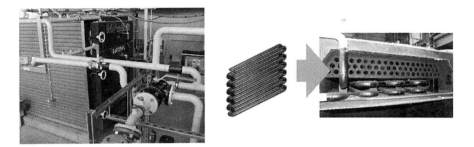

FIGURE 9.2 The location of the oil exchanger in the area between the post-combustion chamber and the fire-tube exchanger.

The maximum assumed operating parameters were 203°C in the case of temperature and 16 bars in the case of pressure. The special control and measurement system with a programmable logic controller (PLC), replacing the originally used microcontroller, was developed and introduced. The system was equipped with over 40 temperature sensors (located inside the boiler as well as in the oil circuit, steam-condensate circuit, and water circuit), an electromagnetic flow-meter, a flue gas analyzer, and inverters used for controlling the operation of the boiler's blowing fan and pumps. The dedicated visualization created using CoDeSys software allowed controlling the operation of the system and acquiring measurement data in any defined time range. The general scheme of the boiler and oil circuit, with marked position of the control and measurement elements, is shown in Figure 9.3.

From the perspective of the study, which was aimed at determining the possibilities of using a straw-fired batch boiler as a heat source for a microscale cogeneration system, it was crucial to identify the volume of the high-temperature heat that can be

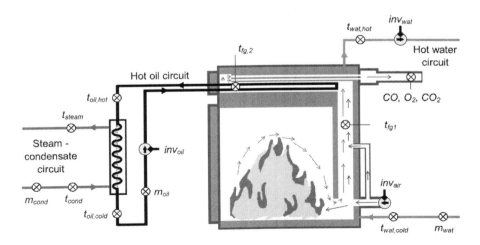

FIGURE 9.3 The general scheme of the boiler and oil circuit, with marked position of the control and measurement elements.

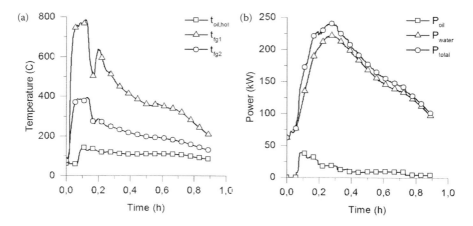

FIGURE 9.4 (a) The variations in the temperature of hot oil and flue gas, and (b) in power transferred to the oil and water circuits.

obtained from the unit. The following measurements were conducted: the temperature of flue gas in the post-combustion chamber of the boiler (t_{fg1}) and at the outlet from the fire-tube exchanger (t_{fg2}); the temperature in the water circuit ($t_{wat,hot}$, $t_{wat,cold}$); in the oil circuit ($t_{oil,hot}$, $t_{oil,cold}$); and in the steam-condensate circuit (t_{steam}, t_{cond}). In order to determine the instantaneous power obtained in the water circuit, the oil circuit, and the cooling water circuit, the flow of the operating medium was measured in each of the circuits (m_{oil}, m_{wat} and m_{cond}). Figure 9.4a shows the variations in the temperature of hot oil and flue gas and Figure 9.4b shows the variations in power transferred to the oil and water circuits.

The highest power of the oil circuit (nearly 40 kW) was reached in the initial phase of the combustion process. Next, the power of the oil circuit decreased to approximately 14–18 kW and maintained this level until the fuel feed was burned. The effect of the time shift between the maximum power of the oil and the water circuits resulted from different dynamics of heat transfer in case of water and oil:

- The water system contained *c.* 5 m³ of water compared to *c.* 0.1 m³ of the oil in the oil circuit.
- The volume of the water jacket of the boiler was larger than the volume of the oil heat exchanger (1.5 m³ vs. 0.04 m³).
- The surface of heat transfer was significantly higher in case of the water part comparing to the oil part.
- The specific heat of water was approximately two times higher as compared to oil.

The average ratio of the oil circuit power to the boiler's total power (defined as the sum of the water part power and the oil circuit power) was observed at a level of 6.6%. The maximum share of the oil circuit power in the boiler's total power output (38.2%) was reached at the time when the power of the oil circuit was at the peak value. Considering the time of the whole combustion process, an average of 61.0%

of the heat was transferred to the water tank, 4.2% to the oil circuit, and 34.8% of the heat was lost.

In the subsequent part of the process, the power of the oil circuit was definitely too low to be used efficiently for steam generation and consequently to power the steam engine. As a result of such a conclusion, the second approach of implementing a high temperature oil system to a straw-fired boiler was introduced.

9.5 THE SECOND APPROACH: THE REPLACEMENT OF THE BOILER'S WATER JACKET WITH AN OIL JACKET

Based on previously conducted studies, a biomass-powered micro-cogeneration system using modified Rankine cycle operation was developed. In this case, the priority for the boiler operation was high temperature heat generation and consequently, electricity generation. 100 kW$_{th}$ straw-fired batch-boiler was used as a high temperature heat source. This boiler was equipped with an oil jacket (replacing the typically used water jacket) and a dedicated fuel feeder (see Figure 9.5).

Thermal oil heated in the boiler was transferred to shell and tube heat exchangers, connected in series, and operated respectively as an evaporator and superheater (similarly to the first version of the system). An emergency oil-to-water plate heat exchanger was also used; it works only in time, when an oil temperature is too high. As before, the operation of the developed micro-cogeneration system was controlled by the control and measurement system based on a WAGO PFC200 modular PLC. The following measurements were conducted:

- Temperature of flue gas at the outlet from the boiler (t_{fg})
- Oil temperature ($t_{oil,1}$, $t_{oil,2}$, $t_{oil,2a}$, $t_{oil,3}$, $t_{oil,4}$)
- Oil flow (m_{oil})
- The temperature of the condensate and steam ($t_{st,1}$, $t_{st,1a}$, $t_{st,2}$)
- The pressure of the condensate and steam ($p_{st,1}$, $p_{st,1a}$, $p_{st,2}$)
- The concentrations of selected pollutants in the flue gas (CO, CO_2, and O_2).

FIGURE 9.5 Boiler (in the front) and evaporator and superheater (behind the fuel feeder).

The control of the system's operation was realized using inverters adjusting the flow of the inlet air, outlet flue gas (additional fan was installed at the outlet from the boiler), thermal oil, condensate, and cooling water [12]. All parameters were observed and recorded via CoDeSys software using a specially developed visualization. The general scheme of the oil circuit is shown in Figure 9.6.

During conducted tests, c. 100 kg of straw was burned. The moisture content of the straw was lower than 10%. The temperature of thermal oil was maintained between 190° and 210°C, whereas the flue gas temperature was maintained in the range of 320–340°C. The variations in the thermal oil temperature were low (which is an advantageous situation because it allows for the production of steam with relatively constant properties). However, the value of the flue gas temperature varied significantly, so continuous control was required (see Figure 9.7a). During the analyzed process, the maximum power generated in the boiler was c. 180 kW$_{th}$, while the typically power ranged from 75 to 100 kW$_{th}$ (in time, when the evaporator and superheater were working and the emergency heat exchanger was cut off). The variations in power transferred from the flue gas to the oil circuit, from the oil circuit to the steam circuit, and from the oil circuit to the emergency cooling water circuits are shown in Figure 9.7b.

Owing to the fact that oil is heated up directly in the boiler's jacket, a significantly higher amount of heat was transferred from the flue gas to the oil in comparison to the configuration described in Section 9.4. The only medium heated in the boiler was oil. The average level of heat transferred from the oil circuit to the steam-condensate circuit via the evaporator and superheater was 75% (during whole combustion process). However, in time, when the emergency heat exchanger was cut off, this value was a bit higher—80%. The steam was generated for 20–155 minutes, and effectively used to power the steam engine for 95 minutes. The use of a fuel feeder allowed for continuous straw combustion and maintained oil power and temperature at the required levels.

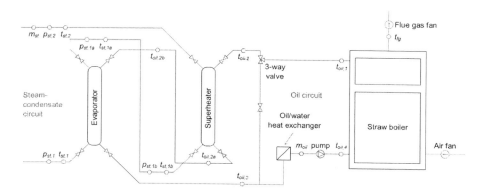

FIGURE 9.6 A general scheme of the oil circuit.

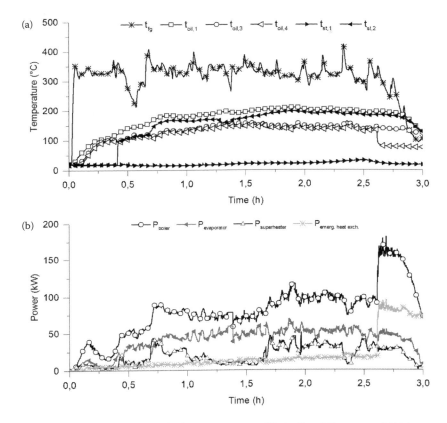

FIGURE 9.7 (a) The variations in the temperature of hot oil and flue gas and (b) in power transferred from the flue gas to the oil circuit, from the oil circuit to the steam circuit, and from the oil circuit to the emergency cooling water circuits.

9.6 POWER GENERATION IN THE PROTOTYPICAL MICRO-COGENERATION SYSTEM WITH A STRAW-FIRED BATCH BOILER

The most essential thing for power generation is to guarantee proper parameters of steam, especially high temperature, pressure, and flow of the steam. A reducing valve and a moisture separator were used to condition the superheated steam. After that process, The steam flows to the steam engine (20-horsepower, two-cylinder, double-acting). Cooled and expanded steam was condensed in shell and tube heat exchangers, and next pumped to a degasser. In the last step, liquid fluid was pumped to the evaporator to reuse in the next cycle.

The generator connected to the above-mentioned steam engine produced the electricity. The main components of the described micro-cogeneration system are shown in Figure 9.8.

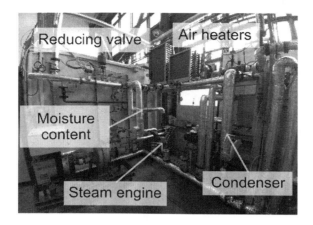

FIGURE 9.8 The main components of the steam-condensate circuit.

The simplified scheme of the steam-condensate circuit is shown in Figure 9.9. This scheme includes only the most important components, while the real configuration is much more complicated.

During the research carried out, the maximum temperature of the steam at the outlet from the superheater ($t_{st,2}$) was ~198°C and the inlet oil temperature ($t_{oil,2}$) was ~208°C. There were no significant fluctuations in the temperature of the condensate that was pumped to the evaporator ($t_{st,1}$); it varied only between ~18 and 22°C. The values of steam pressure were less stable; when the engine was working, pressure ($p_{st,2}$) oscillated from 2.5 to 5 bars, with a maximum value of ~5.8 bars before the steam engine was run (in general, the maximum value of pressure obtained in this installation was 10 bars). The main causes of such a situation was variation in oil temperature and changes in set of the regulating valve. The variations in the oil, steam, and condensate temperature and steam pressure are shown in Figure 9.10.

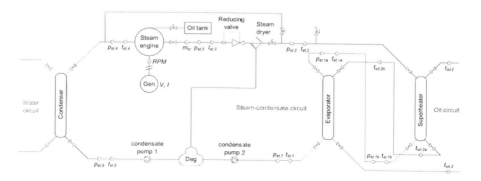

FIGURE 9.9 The simplified scheme of the steam-condensate circuit.

Straw-Fired Boilers as a Heat Source for Micro-Cogeneration Systems

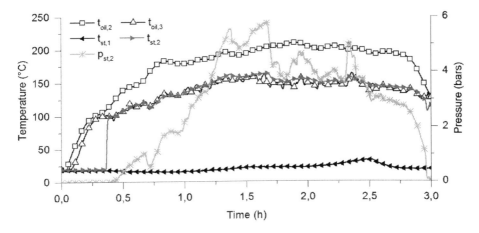

FIGURE 9.10 The variations in oil, steam, and condensate temperatures, as well as steam pressure during the analyzed process.

When the generator (connected with steam engine) was producing the biggest amount of power, steam mass flow (m_{st}) was about ~105 kg/h and the pressure ($p_{st,2}$) was equal to 4.3 bars. It allowed for the production of ~1.05 kW$_e$ of electric power. The variations in the steam pressure (before and after the steam engine) as well as in the engine's rotation speed are shown in Figure 9.11.

The electric load was equipped with 20 bulbs with a total power of 2 kW$_e$ and used to determine the current-voltage (I-V) and power-voltage (P-V) characteristics. The examples of such characteristics are shown in Figure 9.12. The maximum measured power (~1.05 kW$_e$) was generated for a current of ~3.65 A and a voltage of ~278 V.

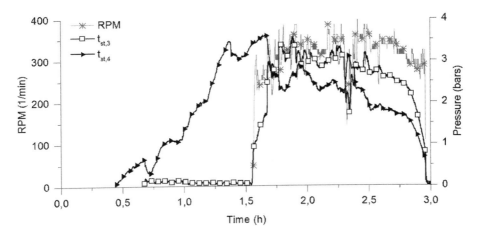

FIGURE 9.11 The variations in the steam pressure (before and after engine) and rotation speed of the steam engine.

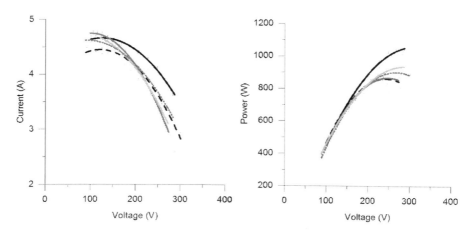

FIGURE 9.12 Current-voltage (I-V) and power-voltage (P-V) characteristics of the power generator.

The maximum efficiency of electrical system was only about ~0.95% (~1.05 kW$_e$ compared to thermal power of the boiler ~110 kW$_{th}$). This is a corollary of numerous limitation such as:

- Oil and steam temperature (e.g., resulting from the temperature limitations of fittings used in oil and steam-condensate circuits)
- Pressure and flow of the steam (e.g., resulting from achieved steam temperature)
- Too low heat exchange surface in the evaporator
- Too low power of the electricity generator
- Other construction and operating parameters.

Eliminating the above-mentioned limitations, the efficiency of the electrical system should be significantly higher; at the designing stage, the electrical efficiency was assumed at a level of 10%.

9.7 CONCLUSION

Research conducted so far shows that straw-fired boilers can be used as a heat source for combined heat and power generation in microscale systems (configuration shown in Section 9.5). However, further developments are required to achieve higher efficiency and more reliable operation of the system.

The configuration with a dedicated boiler with oil jacket is definitely a better option compared to a configuration with a typical water jacket and additional oil heat exchanger placed inside the boiler. In such a solution, heat generated in the boiler is transferred directly to the oil (while the average ratio of the oil circuit power to the boiler's total power in the first considered configuration was only ~6.6%). Then, ~75%–80% of heat transferred from the boiler to the oil circuit is next transferred to the steam-condensate circuit in an evaporator and superheater. As was concluded, the heat exchange surface of the evaporator should be higher to ensure higher stream of the generated steam. Moreover, owing to the fact that the actual configuration of the system is not designed to work with temperatures higher than 203°C, it is not possible to significantly increase the steam temperature (and consequently, its pressure). However, with temperatures ranging from 180° to 200°C, steam has a pressure ~8–10 bars, which is sufficient for proper steam engine operation. Another limitation in the considered micro-cogeneration system was connected to the maximum power generated in the existing electricity generator.

Development of a dedicated control and measurement system is crucial from the standpoint of operation of the micro-cogeneration system. Such a controller must include signals from all of existing circuits (oil, steam-condensate, and water) and provide efficient and safe operation of the whole system. Among other measured values, the most important will be flue gas temperature, hot oil temperature, temperature and pressure of the steam, hot water temperature, and condensate levels in the evaporator and the condenser. However, the operation of the following elements must be provided: oil, condensate and water pumps, air fan(s) integrated with the boiler, as well as electro valves and reducing valves in the steam circuit.

Another key functionality of the straw-fired batch boiler dedicated to the micro-cogeneration system is continuous fuel feeding. In the tested boiler the manual feeder was used, but in the case of the final product, an automation fuel feeding system should be developed.

ACKNOWLEDGMENTS

This work has been completed as part of the statutory activities of the Faculty of Energy and Fuels at the AGH University of Science and Technology in Krakow, "Studies concerning the conditions of sustainable energy development," using the data obtained during the fulfillment of the project entitled: "BioORC: Construction of cogeneration system with small to medium size biomass boilers."

NOMENCLATURE

CH_4	methane
CO	carbon monoxide
CO_2	carbon dioxide
H_2	hydrogen
H_2O	water
ORC	Organic Rankine Cycle
PLC	Programmable Logic Controller
RC	Rankine Cycle

REFERENCES

1. Y. Chhiti, M. Kemiha, Thermal conversion of biomass, pyrolysis and gasification: A review. *The International Journal of Engineering and Science (IJES)*, 2, 2013, pp. 75–85.
2. A. Siirala, *Assignment 8: Comparison of gasification, pyrolysis and combustion*, Aalto University School of Chemical Technology, 2013.

3. A. Rentizelas, S. Karellas, E. Kakaras, I. Tatsiopoulos, Comparative techno-economic analysis of ORC and gasification for bioenergy applications. *Energy Conversion and Management*, 50, 2009, pp. 674–681.
4. L. Dong, H. Liu, S. Riffat, Development of small-scale and micro-scale biomass-fuelled CHP systems – A literature review. *Applied Thermal Engineering*, 29, 2009, pp. 2119–2126.
5. A. Borsukiewicz-Gozdur, S. Wiśniewski, S. Mocarski, M. Bańkowski, ORC power plant for electricity production from forest and agriculture biomass. *Energy Conversion and Management*, 87, 2014, pp. 1180–1185.
6. K. Sornek, M. Filipowicz, M. Zoladek, R. Kot, M. Mikrut, Comparative analysis of selected thermoelectric generators operating with wood-fired stove. *Energy*, 166, 2019, pp. 1303–1313.
7. Directive 2004/8/EC of the European Parliament and of the Council of 11 February 2004 on the promotion of cogeneration based on a useful heat demand in the internal energy market and amending Directive 92/42/EEC.
8. I. Niedziółka, A. Zuchniarz, Analiza energetyczna wybranych rodzajów biomasy pochodzenia roślinnego. *Inżynieria Rolnicza*, 8, 2006, pp. 232–237.
9. A. Dyjakon, M. Penkala, Współspalanie biomasy z weglem w kotłach małej mocy (in Polish). *International Scientific and Technical Conference "Energetyka 2004"*, Wroclaw University of Science and Technology, 3–5 November 2004.
10. M. Salamón, T. Savola, A. Martin, C-J. Fogelholm, T. Fransson, Small-scale biomass CHP plants in Sweden and Finland. *Renewable and Sustainable Energy Reviews*, 15, 2011, pp. 4451–4465.
11. K. Sornek, M. Filipowicz, A study of the applicability of a straw-fired batch boiler as a heat source for a small-scale cogeneration unit. *Chemical and Process Engineering*, 37(4), 2016, pp. 503–515.
12. K. Sornek, M. Filipowicz, Study of the operation of straw-fired boiler dedicated to steam generation for micro-cogeneration system. *IOP Conference Series: Earth and Environmental Science*, 214(012108), 2019.

10 Straw Drying
The Way to More Energy Production in Straw—Fired Batch Boilers

Wojciech Goryl

CONTENTS

10.1 Introduction 231
10.2 Batch Dryers 232
    10.2.1 Three-Chamber Batch Dryer..232
    10.2.2 Duct-Grate Dryer..234
    10.2.3 Dryer with the Expansion Chamber 235
    10.2.4 Hybrid Dryer...235
    10.2.5 Dedicated Straw Batch Dryer 236
        10.2.5.1 Description of the Installation 237
        10.2.5.2 Measurement Methodology 237
    10.2.6 Results..241
        10.2.6.1 The Basic Drying Process..241
        10.2.6.2 The Modified Drying Process...242
10.3 Summary 243
References...245

10.1 INTRODUCTION

Most biomass dryers, as drying agents, use gases from the biomass combustion or process steam. Air, as a drying medium, is generally used for food material drying, because of the costs of the installation, efficiency, and size of the whole system. Biomass is characterized by low-bulk density and low calorific value. Moreover, it is quite a problematic fuel due to the heterogeneity of the material. The usefulness of biomass for energy purposes depends on the humidity, type, and composition. However, the basic problem of using biomass as a fuel is moisture content, which is primarily influenced by the period of harvest and the conditions of its storage. Dry biomass can reach a humidity of under 10%, but usually fresh biomass has more than 60% relative humidity, which is a very high value. This kind of fuel is almost impossible to combust [1–3]. For the straw combustion in biomass boilers, the humidity of the fuel has to be 10%–20%. This has a favorable effect on the combustion process owing to the catalytic effect of water vapor to the combustion of the excessive amount of volatiles which are in the biomass. The more compressed the straw (with higher density) and the weather conditions during its storage have a smaller impact on straw quality and the degree of moisture. However, too much compressed straw is very difficult to dry due to problems with the penetration of the drying agent through the straw layers.

Drying biomass for energy purposes could increase the energy efficiency of the combustion process [4–6], but this process is usually quite expensive. It is due to the need to use large amounts of energy for water evaporation [7]. The most commonly used biomass drying devices are drum dryers, pneumatic dryers, fluid bed dryers, and belt dryers [8,9]. Mostly, in these dryers, the exhaust gases from the fuel combustion [10] or process steam [9–11] are used as the drying agent. Moreover, the drying agent, which is air, is usually used for drying food materials [12–14]. Installations for the food industry are much more expensive, relating to investment and operating costs [15]. The above devices are used for drying powdered or crushed materials, so the drying process is faster and, at the same time, it is more energy-efficient. Sometimes, dry materials need to be dried in the form of bales, such as hay, straw, or willow, and the dryers must be specially designed for the type of drying material. The dryers presented in this chapter could be used for biomass drying in the form of bales.

10.2 BATCH DRYERS

If there is a need to dry biomass in the form of bales, for example, straw, hay, or energy willow, drum dryers, pneumatic dryers, fluid bed dryers, and belt dryers cannot be used. For this purpose, other installations should be used and strictly adapted to drying densified cuboidal or cylindrical bales. The literature presents a number of solutions for drying hay bales in the form of cuboidal, as well as cylindrical bales. For straw, there are practically no such dedicated solutions. Owing to the nature of the material, similar solutions may be used. Hay, as a feed, should have appropriate quality parameters to protect it against mold growth or rotting. Hay should have a moisture content that does not exceed 12%–15% if it is to be stored for a long period of time without any negative effects on its quality [16]. Straw for the combustion process should have a slightly higher humidity, of 15%–20%. One of the first solutions for drying hay in the form of bales, using forced air circulation, was described in 1947 [17]. A similar drying mechanism has also been described [18] where the air was used as a drying agent in the process of hay drying in small cuboidal bales. The hay was laid to a height of 4 m and dried for 4 months. After this time, it reached a humidity level of 15%; however, such a long drying time may cause molding in the outer layers of the hay [19]. The lower the material density, the easier it is to dry the material [16].

10.2.1 Three-Chamber Batch Dryer

In Reference [20] assumptions were made, and in Reference [16] the design and operation of a prototype hay dryer was described, in which it is possible to dry hay from 30% to 12% moisture content within up to 12 h. The dryer scheme is shown in Figure 10.1. It is a batch dryer housing one or two layers of 6 and 12 cuboid bales,

Straw Drying

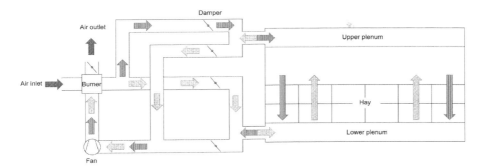

FIGURE 10.1 Schematic diagram of the three-chamber batch dryer, lighter—bottom-up operation, darker—top-down operation. (Based on Descouteaux S., Savoie P., *Applied Engineering in Agriculture*, 22(4), 2006, 481–489.)

respectively, of hay with dimensions of 80 × 90 × 245 cm (width × height × length). The dryer has three chambers. The largest, middle one is used as a drying chamber, and the other two—the lower and upper—serve to inject or outlet the drying agent. Between the chambers there is a perforated floor/ceiling (Flowforge type) to force the flow bottom-up or top-down of the drying agent. This dryer has the ability to dry the hay bales from bottom to top and top to bottom by changing the position of the dampers (see Figure 10.1, "Damper"). The drying process from bottom to top (during this operation there is an increased process of drying the bottom layer and the top layer is not dried) or from top to bottom (in this case there is an increased drying process of the upper bale layer) may be used. In addition, the device has the ability to recirculate the drying agent, which increases its efficiency.

The dryer has a propane gas burner with a capacity of 102 kW. It is used to get a hot drying agent. Moreover, the fan of 11.2 kW is used to press the drying agent (at a temperature of approximately 60°C) into the dryer. The flow of the drying agent depends on the number of bales in the dryer; with one layer it is approximately 2.2–3.3 m³/s, and with two layers 1.3–1.8 m³/s. For one layer it is possible to dry the hay within 5 hours from 21% to 12% humidity, 9 hours from 24% to 12% humidity, and about 14 hours from 34% to 12% humidity. With two layers, the drying time is extended twice at least. In addition, it has been found that drying bales with higher density takes longer than low density bales. For this reason, it is best to put bales of similar densities into the dryer to get comparable humidity in the entire volume of the material [21].

The dryer of the same type was used for drying very moist corn straw [22]. In this case, the bales were smaller, so it was possible to put nine bales in one layer and 18 bales in two layers. For one layer, drying from 56% to 19% took 52 hours with an average temperature of the drying agent at 45°C, and in the case of two layers—from humidity 56%–18%, drying time was 90 hours with an average temperature of 61°C.

It should be mentioned that the greatest influence on the drying process is the temperature of the drying agent and the change in the direction of its flow. The recirculation of the drying medium has a smaller influence.

10.2.2 Duct-Grate Dryer

In Reference [23], three methods of drying cylindrical hay bales with a diameter of 1.5 m and a length of 1.2 m have been described. The considered dryer could simultaneously dry nine bales of hay on three different air ducts. Figure 10.2 shows a schematic of three dryers connected to each duct. The ducts were located in a concrete floor and had a size of 0.6 × 0.6 m. The outlet of the drying agent took place through a 0.6 m diameter hole. Above the openings there were a metal grate on which the hay bales are laid. The connection between the grate and the inlet ducts is made of a flexible, leak-proof material. The air pressed into the bale's interior is heated by electric heaters up to 27°C. Each dryer is made in a different way.

The first one has a metal rim of 15 cm high and a diameter of 1.1 m, which is placed on plywood with a diameter of 1.8 m. In this plywood, a hole with a diameter of 0.9 m is made, through which hot air is pressed into the bale. The metal rim is designed to minimize the escape of the drying agent outside the bale.

The second one has an injected duct at the same diameter as the bale of hay. In addition, the side of the bale is tightly wrapped by foil to avoid the escape of the drying agent through the bale side.

The third one had a plywood with a hole of diameter 0.6 m, on which a hay bale is placed.

During the drying process of bales in two dryers (on first and third channels) a plate with a diameter of 1.2 m was placed—most probably in order to avoid the flow of the drying agent only along the bale's axis. The following measurements were made during drying:

- Temperature: Ambient, drying medium, bale interior in four different places
- Humidity: Ambient and the drying agent in the inlet channel
- Pressure in the inlet channel
- Weight of bales during the whole process

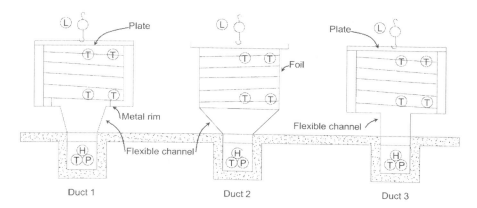

FIGURE 10.2 Schematic diagram of the duct-grate dryer: T—temperature, P—pressure, H—humidity, and L—mass measurement. (Based on Misener G.C. et al., *Canadian Biosystems Engineering*, 32, 1990, 263–268.)

Straw Drying

It was found that the bale wrapped by foil dried much faster than in the first and third cases. It could be caused by forcing the flow of the drying agent along the entire bale of hay. In the first and third cases, the drying agent most likely left the bale near the inlet, because there was much smaller flow resistance. In addition, a drop in overpressure in the inlet ducts was observed along with a humidity decrease in the bale.

10.2.3 Dryer with the Expansion Chamber

The article in Reference [24] presents the tests for the possibility of hay drying in cylindrical bales in three different configurations. The hay bales had a diameter of 1.5 m and a height of 1.2 m. The dryer is adapted to dry one bale at a time. The dimensions of the dryer are 1.6 × 1.6 × 0.4 m.

The drying agent is air heated by a 30 kW gas burner. The air reaches a temperature of 37–42°C. The heating of the air takes place before the inlet to the fan, which has a power of 1.1 kW and an output of 0.5–0.7 m^3/s. Then the drying agent goes to the expansion chamber. The outlet from the expansion chamber has a diameter of 100 cm and ends with a 10 cm high rim. The rim was installed to avoid losses resulting from the escape of the drying agent. The dried bale is placed on the rim.

Drying was carried out in the following configurations:

- The air was forced into the bale from the bottom through a hole with a diameter of 100 cm. The hole has a rim with a 10 cm high. The top of the bale has been covered so the air escapes only from the side of the bale.
- Similar to the above, with the difference that the bale from the bottom to the height of 50 cm was tightly wrapped in foil to force out the air from the bale at the highest possible height.
- In this case, the diameter of the outlet from the expansion chamber to 25 cm was reduced and a hole with a diameter of 15 cm was made along the bale of hay. The upper part of the bale was covered in order to radially spread the drying agent.

After the measurements, it was found that moisture filed in a bale of hay was not homogeneous, but it is possible to dry the bales of hay from 40% moisture content to 12%–15% in 7–12 hours.

10.2.4 Hybrid Dryer

It is also possible to combine two or more heat sources to achieve higher efficiency and lower costs of drying. In Reference [25] the possibility of drying hay in the form of cuboidal bales with a hybrid of a biomass boiler and solar energy is presented. The diagram of the installation is shown in Figure 10.3.

This installation has an air solar collector with an area of 458 m^2. There is an air inlet to the collector through the openings between the chamber wall and the roof. In the void under the roof, the flowing air heats up and reaches a speed of up to 4 m/s. In order to increase the heating power, a 146 kW boiler was installed, from which the

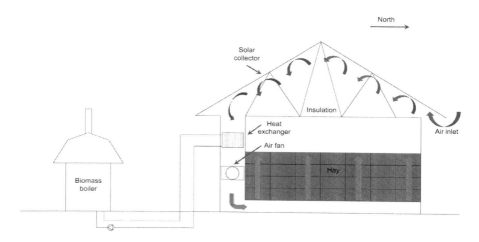

FIGURE 10.3 Schematic diagram of the hybrid dryer. (Based on Morissette R. et al., Drying baled hay with combined solar and biomass heat sources, The Canadian Society for Bioengineering, Paper No. CSBE11-203.)

heat is transferred to the water-air exchanger. Warm air preheated in the air collector flows through this exchanger, which increases the air temperature. For example, when the air temperature at the inlet to the solar collector reaches 30°C, at the exit of the collector is 35.5°C, and at the exit of the heat exchanger is 42°C. Then, using a 3.2 kW air fan with a flow rate up to 4.8 m³/s, the drying agent is forced under the floor to the drying material. This floor is perforated, which allows the free flow of the drying agent through the bed of moist material. Within 55 days, it is possible to dry almost 6000 small 22.5 kg hay bales from the initial humidity of 18.3%–11.1%, with a net energy consumption of 223.7 GJ, of which 192.4 GJ came from the sun and biomass, and 31.2 GJ is the electricity used to drive fans and circulation pump.

10.2.5 Dedicated Straw Batch Dryer

The above installations are used for hay drying but, with some modifications, those installations could be used for bale straw drying for energy purposes. The main problem in the above examples is the drying time, which is too long to be directly used for drying straw for a working boiler installation. For this purpose, several solutions with appropriate modifications should be considered to reduce the drying time. Additionally, in such a modified dryer, there is no need to use heated air as a drying agent, because the purity of the material is not important in this case. Heating the air with gas and liquid fuel burners, or using electric heaters, does not make economic sense. The costs of preparing such a drying agent are very high and could exceed the benefits achieved during the drying process. Owing to the fact that straw used for energy purposes does not have to meet high quality standards (as in the case of hay), it is possible to use direct hot fuel gasses, which come from combustion in a biomass boiler.

Straw Drying

Owing to the nature of fuel, straw boilers work effectively only with nominal power. Of course, it is possible to reduce power by reducing the airflow but, in practice, owing to the significant deterioration of energy efficiency, a power control system is not usually used. In commercial units, small batch straw boilers, dedicated controllers do not allow for such modulation of the power. For this reason, exhaust gases behind the heat exchanger reach temperatures of 250°C to even 300°C. Enthalpy of fumes coming out of the boiler is high and can be easily used for other purposes, such as for biomass drying.

Such utilized waste heat increases the efficiency of the entire system and helps to obtain much more energy fuel. In a further part of the chapter, a unique solution is presented to meet the assumed goals, that is, a straw batch dryer for drying fuel in the shortest possible time.

10.2.5.1 Description of the Installation

The main elements of the dryer used in the study include the drying chamber, the transport table, the discharge fan, the fire prevention system, the regulation assembly, the system for the control of operational parameters, the system of preextraction and separation of sparks, and the exhaust system. The heat used in the drying process comes from the biomass combustion in the 500 kW$_{th}$ batch straw boiler. In this type of boiler, fuel is manually fed into the combustion chamber. The boiler is a typical straw batch boiler with counter-current combustion. The boiler has two chambers. The first chamber is a primary combustion chamber where fuel undergoes gasification, and the second chamber is a secondary combustion chamber where the products from gasification are burned in the atmosphere of the secondary air. Next, fuel gasses flow through the heat exchanger and leave the boiler.

For safety reasons, the exhaust gases coming from the boiler are predusted in the sedimentation chamber (cyclone) due to the presence of large, sometimes incandescent, stalks of unburned straw. The fuel gasses are divided into two streams. The first stream of flue gasses is directed to the chimney. The second stream is used in the drying process. The gases are transported to a special mixer, where, if necessary, the process of mixing of the exhaust gas with ambient air takes place in order to obtain the desired temperature of the drying agent. It is necessary to control the drying agent temperature owing to the possibility of ignition of dried straw at a high temperature. After the mixing process, the fan presses the previously prepared mixture, using a specially designed nozzle, into the cylindrical bale of straw. A schematic diagram of the system is presented in Figure 10.4.

10.2.5.2 Measurement Methodology

The effectiveness of drying cylindrical straw bales by exhaust gasses was measured using a specially manufactured measuring system equipped with eight measuring probes. Each probe was equipped with temperature and humidity sensors. The system collected data every second. First, the probes were calibrated using samples of straw characterized by varying humidity levels. The measuring probes were made of metal, had a diameter of 0.5 cm each, and are of varied lengths: two probes each of 80 cm, 60 cm, 40 cm, and 20 cm. These lengths were matched to the bale diameter (it was 80 cm)

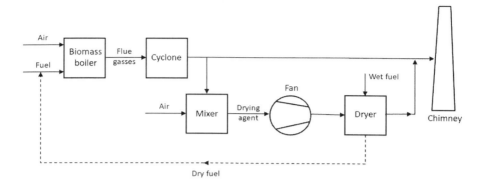

FIGURE 10.4 Schematic diagram of the dedicated straw batch dryer.

so that the longest probes were placed in the bale axis and the two 20 cm probes were placed closer to the surface of the bale. Figures 10.5 and 10.6 present the drying agent injection nozzle, two measuring planes, and probes which are used for temperature and humidity measurement.

The measurements were carried out in two planes. The measurement points were located at a distance ($x = 0$ – the place where the drying agent was pumped to the bale) of 40 cm (plane A, $x = 40$) and 100 cm (plane B, $x = 100$) from one end of the bale.

In addition, a dedicated measuring device was used to measure the temperature, humidity, and differential pressure of the drying medium. The measuring instruments were installed on a specially made drying agent monitoring channel. The channel is shown in Figure 10.7.

The monitoring channel consisted of two separated channels. In the longer channel at the top, humidity and temperature sensors had been installed. In the shorter channel, the venturi flowmeter was placed. The thermocouple was used for temperature measurement; however, the humidity of the drying agent was measured by the dew point method. The dew point is found by measuring the capacitance of the capacitor placed on the Peltier cell.

FIGURE 10.5 Visualization of the injection nozzle and measuring probes and planes.

Straw Drying

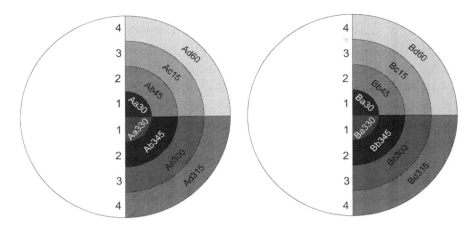

FIGURE 10.6 Visualization of placing probes in measuring layers 1, 2, 3, and 4.

The measurement elements were connected to a dedicated data logger, which had been synchronized with the temperature and humidity recorders in the straw bales. The measurement data (temperature, humidity, and flow of drying agent) were recorded on the SD card with an interval of 1 s.

For research purposes, two variants of the straw drying process were selected based on the number of holes along the bale axis and the use of a metal plate blocking the axial flow of the drying agent. These variants are defined in Table 10.1.

FIGURE 10.7 Location of the measuring channel: measurement of humidity, temperature, and differential pressure of the drying medium.

TABLE 10.1
Variants of Straw Drying

Variant of the Drying Process	Type of Modification	Additional Information
Basic	None	None
Modified-3—three holes	Three holes with a diameter of 60 mm	Metal plate with a diameter 1000 mm on the opposite side of the bale relative to the location of the drying medium inlet (Figure 10.8).

Variants of the drying process were selected on the basis of preliminary experiments and analysis of their results. The results led researchers to the decision to introduce the described modifications of the drying process.

Preliminary measurements in a basic drying process showed that the drying medium flows primarily along the bale axis, which caused the heating and drying of only a small part of the straw bale in its axis. This resulted in an under heating of the more external bale layers, wherein the temperature was low and the humidity remained at the same high level. Owing to this problem, modifications of the drying process were applied. Creating a hole along the bale axis helped to better distribute the drying agent in the bale. These modifications were introduced to intensify the drying process and to achieve homogenous temperature field and the smallest possible humidity field within the considered bale.

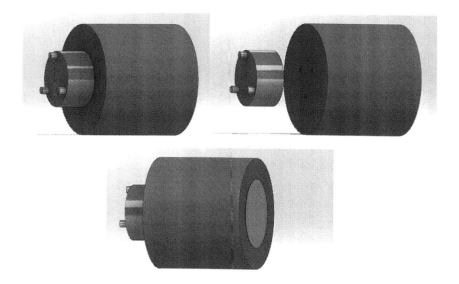

FIGURE 10.8 Visualization of the modified drying process—three holes along the bale axis and the metal plate opposite to the injection nozzle.

Straw Drying

10.2.6 Results

The results presented below include only results from measuring plane A. The measurements were carried out in two configurations (basic and modified) to examine the effect of the presence of the holes inside the bale of straw on the dynamic of cylindrical straw bales drying. In addition, in each drying process the temperature of the drying agent injected into the bale was recorded. The drying agent temperature is presented in charts in the form of a dash dot curve with the values given on the right axis. The basic and modified process of drying took 140 minutes.

10.2.6.1 The Basic Drying Process

Figure 10.9 presents the variation of humidity inside the bale of straw in the basic drying process (without the holes). The initial humidity of the bale was in the range of 23%–43%. The internal layers of the bale were drier, and the humidity increased along with the distance from the center of the bale. This was caused by the fact that the bale had not been covered while storing.

In the initial drying phase, one could note a sharp increase in the humidity at the inner layer of the bale represented by sensors Aa30H and Aa330H. This was due to the rapid temperature increase in this part of the bale, which contributed to the temporary increase in relative humidity. After the 30th minute of the measurement, a decrease in humidity in the first and second layer (the layers are presented in Figure 10.6) could be observed. This decrease continued until the end of the measurement. In the case of layers 3 and 4, a stable humidity oscillating in the range of 30%–35% was observed. The final humidity value after the measurement reached the value of 12%–35%.

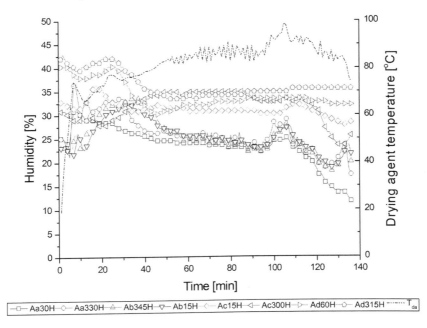

FIGURE 10.9 Basic drying process, humidity measurement.

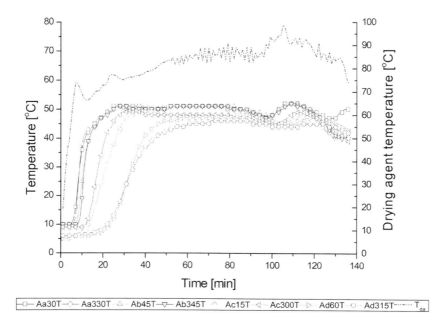

FIGURE 10.10 Basic drying process, temperature measurement.

Figure 10.10 presents the temperature changes inside the bale of straw in a period of 140 minutes of the drying process. Five minutes after the drying process was started, a rapid temperature increase was observed in layer 1. Subsequently, after another 2 minutes passed, rapid heating of layer 2 was observed. Layers 3 and 4, because of their distance from the center of the bale, heated up much more slowly, and the beginning of the temperature increase in these places could be observed from the 15th and 20th minute, respectively. The maximum temperature for layers 1 and 2 was achieved in the 25th minute of the measurement and its value was approximately 52°C. In the case of layer 3, the highest temperature of 50°C was reached in the 35th minute. The extreme plane was heated at the lowest rate and reached the lowest temperature, reaching a maximum of 46°C at approximately the 60th minute of the measurement. After reaching the temperature of 85°C, the drying agent temperature was maintained until the end of the combustion process.

10.2.6.2 The Modified Drying Process

Figure 10.11 presents the changes in humidity for the bale with the application of the three holes and the metal plate.

The initial humidity of the bale, which was subject to the measurement for the modified drying process, oscillated within 30%–45% and was comparable to the bale used in the basic drying process. The inner layers of the bale were drier and the further from the center of the bale the higher was the humidity.

In this case, problems occurred with igniting the straw in the biomass boiler, and thus the drying agent had a very low temperature. In the 18th minute of the

Straw Drying

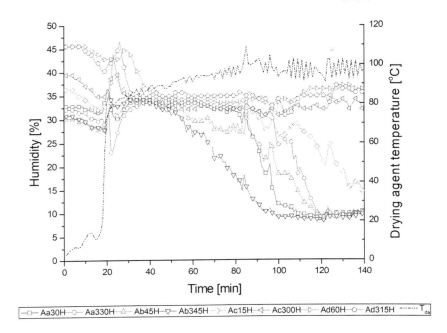

FIGURE 10.11 Modified drying process, humidity measurement.

measurement, it was possible to observe a sudden increase of the drying agent temperature, which proved that the fuel started burning in the combustion chamber. After a sudden increase in the drying mixture temperature, a sudden humidity increase was observed, and after approximately 5 minutes the humidity slowly declined. After 5 more minutes, the humidity in the whole bale was stabilized and reached 35%. Subsequently, until the 120th minute of the measurement, a sudden decrease of humidity was observed, and after another 20 minutes the humidity in layers 1 and 2 was stabilized at the level of 10%. Layers 3 and 4, just as in the previous measurements, were characterized by a similar humidity of approximately 35%.

Figure 10.12 presents the temperature changes inside the modified bale of straw. Until the ignition of the fuel in the combustion chamber, one may observe a similar temperature value throughout the entire bale of straw. In the 20th–25th minute of the drying process, a rapid rise in temperature in all layers was observed, until the temperature reached 50°C. Subsequently, a stable temperature value may be observed in all layers, until the 90th minute. After this period, the temperature in layers 1 and 2 started to increase and it reached a maximum of 90°C. The maximum temperature reached in layers 3 and 4 was 60°C. The drying agent temperature during the entire period of the drying process varied between 90 and 100°C.

10.3 SUMMARY

A summary of the drying results for two variants is presented in Table 10.2. The results of the experimental measurements clearly show that the modified drying

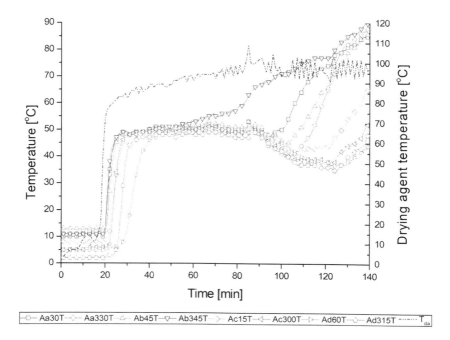

FIGURE 10.12 Modified drying process, temperature measurement.

process, with three holes made along the straw bale axis and with the metal plate on the opposite side of the bale from the drying agent injection site (variant modified-3), is the optimal solution. In this case, the drying time of the straw can be reduced by more than half compared to basic one.

In the modified case, with three holes, it was sufficient to dry the straw for about 280 minutes. It should be noted that in the existing system, only part of the exhaust is used to dry the straw. Considering that one cylindrical straw bale is burned in the boiler for about 120–150 minutes, it is able to dry another bale to a satisfactory humidity value using only fuel gases that come from the combustion of one bale. With the maximum possible use of the entire volume of exhaust gas for the fuel drying, it would be possible to shorten the drying time.

During measurements, an increase in humidity in the bales can be observed with a sudden increase in the temperature of the drying agent. The increase in temperature of the drying medium also caused a sudden increase in temperature in the individual bale layers. At a constant temperature of the drying medium, there was a slow process of straw drying only from the outer layers of straw stalk, which is visible in the graphs as a constant humidity value or a slow decreasing in the given measurement point. When the temperature rises rapidly, the water trapped inside the straw stalks begins to release rapidly from its interior after providing the appropriate amount of energy. This moisture went to the interstalks area causing a sudden increase in humidity in the area. Subsequently, the released moisture moved along with the flowing drying agent to subsequent bale layers, and in a given measurement area, the humidity stabilized and returned to the value from before the sudden increase in temperature.

TABLE 10.2
Results of Straw Drying in Different Variants

Variant of Drying Process	Drying Time [min]	Mean Weighted Humidity Before Drying	Mean Weighted Humidity After Drying	Mass of Evaporated Water [kg]	Drying Rate [kg/h]
Basic	140	33.8%	28.1%	8.3	3.56
Modified-3	140	39.6%	25.5%	24.8	10.62

Analyzing Table 10.2, it can be clearly determined that the optimal solution is a modified drying process with three holes, which allows for an increase in straw drying efficiency by approximately 70% compared to the basic process.

Comparing the obtained results with the literature for hay drying presented in previous sections, it can be concluded that the proposed straw drying configurations are very effective and allow the wet material to dry in a much shorter time. The temperature of the drying agent should be as high as possible to accelerate the drying process, but at the same time not too high to avoid temperature degradation of the dried material. The optimal temperature is around 100°C. In the studied cases, the flow of the drying agent is at least 10 times lower than in the solutions presented in the literature, and the drying time is much shorter, therefore it can be concluded that the drying efficiency in the dedicated batch straw dryer is much higher.

A dedicated straw dryer installation provides the user enormous financial benefits. Taking into account the results of drying straw in modified variant, it is possible to dry the straw from 40% to 25% humidity, which translates into a change in the heating value from 9 to 12 MJ/kg [26], that is, producing 30% more heat from a given mass of fuel. Therefore, using a straw dryer for a 250 kg bale (typical bale weight), it is possible to obtain 750 MJ, and considering the efficiency of heat generation in a biomass boiler, 615 MJ more heat.

In addition, taking into account 3 kWh electricity consumption by a fan used for forcing a drying agent during the drying process, the net energy gain will be about 600 MJ. The value of this heat is economically justified and, in an operational context, results in lower fuel humidity and less problems related to the combustion process in the boiler. The occurrence of such problems (including difficult fuel ignition and low exhaust gas temperature) was observed during the tests. Further, waste heat is used in this system, which increases the overall efficiency of the system, and there is no need to install a dryer powered by gas or fuel oil, which would entail high operating costs to obtain fuel with adequate humidity.

REFERENCES

1. Goryl W., Filipowicz M., The possibility of using flue gases as a medium for straw drying, *E3S Web of Conferences 10, 00136*, 2016.
2. Goryl W., Szubel M., Filipowicz M., Processes of heat and mass transfer in straw bales using flue gasses as a drying medium, *EPJ Web of Conferences 114, 02033*, 2016.

3. Goryl W., Filipowicz M., Experimental and numerical analysis of cylindrical straw drying, *EPJ Web of Conferences 143, 02031*, 2017.
4. Tesfaldet G. et al., Biomass drying for an integrated power plant: Effective utilization of waste heat, *Computer Aided Chemical Engineering*, 33, 2014, 1555–1560.
5. Luk H.T. et al., Drying of biomass for power generation: A case study on power generation from empty fruit bunch, *Energy*, 63, 2013, 205–215.
6. Verma M. et al., Drying of biomass for utilising in co-firing with coal and its impact on environment – A review, *Renewable and Sustainable Energy Reviews*, 71, 2017, 732–741.
7. Gebreegziabher T., Oyedun A.O., Hui Ch.W., Optimum biomass drying for combustion – A modeling approach, *Energy*, 53, 2013, 67–73.
8. Li H. et al., Evaluation of a biomass drying process using waste heat from process industries: A case study, *Applied Thermal Engineering*, 35, 2012, 71–80.
9. Haque N., Somerville M., Techno-economic and environmental evaluation of biomass dryer, *Procedia Engineering*, 56, 2013, 650–655.
10. Song H. et al., Influence of drying process on the biomass-based polygeneration system of bioethanol, power and heat, *Applied Energy*, 90(1), 2012, 32–37.
11. Bruce D.M., Sinclair M.S., Thermal drying of wet fuels: Opportunities and technologies, *EPRI Report (TR-107109 4269-01)*, 1996.
12. Ratti C., Hot air and freeze-drying of high-value foods: A review, *Journal of Food Engineering*, 49(4), 2001, 311–319.
13. Doymaz I., Convective air drying characteristics of thin layer carrots, *Journal of Food Engineering*, 61(3), 2004, 359–364.
14. Lewicki P.P., Design of hot air drying for better foods, *Trends in Food Science & Technology*, 17(4), 2006, 153–163.
15. Jchua K., Chou S.K., Low-cost drying methods for developing countries, *Trends in Food Science & Technology*, 14(12), 2003, 519–528.
16. Descouteaux S., Savoie P., Bi-directional dryer for mid-size rectangular hay bales, *Applied Engineering in Agriculture*, 22(4), 2006, 481–489.
17. Weaver J.W., Grinnells C.D., Lovvorn R.L., Drying baled hay with forced air, *Agricultural Engineering*, 28(7), 1947, 301–304.
18. House H.K., Stone R.P., Barn hay drying, *Ontario Ministry of Agriculture and Food*, 88–110, 1988.
19. Parker B.F. et al., Forced-air dring of baled alfalfa hay, *American Society of Agricultural Engineers*, 35(2), 1992, 607–615.
20. Descoteaux S., Savoie P., *Proceedings of the International Conference on Crop Harvesting and Processing*, Luisville, Kentucky, USA, 2003.
21. Savoie P., Joannis H., Bidirectional drying of baled hay with recirculation and cooling, *Canadian Biosystems Engineering*, 48, 2006, 3.53–3.59.
22. Savoie P., Descoteaux S., Artificial drying of corn stover in mid-size bales, *Canadian Biosystems Engineering*, 46, 2004, 2.25–2.34.
23. Misener G.C. et al., Drying of large round hay bales, *Canadian Biosystems Engineering*, 32, 1990, 263–268.
24. Roman F.D., Hensel O., Numerical simulations and experimental measurements on the distribution of air and drying of round hay bales, *Biosystems Engineering*, 122, 2014, 1–15.
25. Morissette R., Savoie P., Lizotte P.L., Drying baled hay with combined solar and biomass heat sources, The Canadian Society for Bioengineering, Paper No. CSBE11-203.
26. Boundy B. et al., Biomass energy data book, Appendix A – The effect of moisture on heating values, *Oak Ridge National Laboratory, Tennessee*, 2011.

11 Assessing the Feasibility of Renewable Energy Sources for Treatment of Biomass from Wastewater Treatment

Simona Di Fraia, Adriano Macaluso, Nicola Massarotti and Laura Vanoli

CONTENTS

11.1 Introduction 247
11.2 Methodology 250
    11.2.1 Thermal Drying Modeling 251
    11.2.2 Feasibility Study 252
11.3 Case Study 254
11.4 Results and Discussion 258
11.5 Conclusions 264
List of Abbreviations 265
References 267

11.1 INTRODUCTION

Sewage sludge is a biomass resulting from wastewater treatment. Its management is a challenge in wastewater treatment plants (WWTPs) due to stringent regulations and high operating costs related to sewage sludge treatment, reuse, and disposal [1]. Currently, the most common methods for sewage sludge management are reuse in agriculture, incineration, and landfilling [2]. Considering the recent Eurostat data, reported in Figure 11.1 [3], it can be observed that landfilling is decreasing, whereas the trends of both reuse in agriculture and incineration are increasing. Landfilling reduction can be mainly related to the implementation of Landfill Directive (EC Directive, 1999) [4], which requests a reduction in the amounts of biodegradable waste to be landfilled. Similarly, reuse in agriculture may be increased due to the European Sewage Sludge Directive 86/278/EEC, which promotes this practice; however, agricultural reuse is expected to decrease because of the increasing opposition owing to the harmful substances present in sewage sludge [5]. This could increase the diffusion of sludge

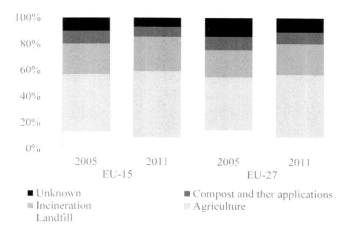

FIGURE 11.1 Sewage sludge disposal routes in EU-15 and EU-27 in 2005 and 2011.

incineration, as observed in Germany, Austria, Portugal, Slovakia, Hungary, and Belgium [6]. Beyond incineration, other thermal treatments are emerging, such as wet oxidation, pyrolysis, gasification, and co-combustion with other materials [7]. These treatments, aimed at energy recovery, are recognized as promising methods due to the conversion of sewage sludge into energy and fuel, making them more convenient than biochemical processes [8]. In terms of greenhouse gas emissions global warming balance, thermal oxidation processes show a best performance compared to agricultural spreading [9]. In the case of energy recovery, sludge should be dewatered to 20–40 wt% solids to carry out thermochemical processes with no additional fuel consumption [10]. In order to reach such water content, thermal drying is needed. By reducing the water content, thermal drying increases the heat value of sludge [11], reduces the mass to be handled, transported, and disposed [12], and sanitizes and stabilizes [13]. Despite these advantages, thermal drying is common only in few European countries [6], due its high thermal demand and economic costs [14]. A way to improve energy efficiency and to reduce the carbon footprint of sludge drying, and in general of wastewater treatment, is the use of renewable energy sources [15,16]. Their use can also be beneficial from an economic point of view, considering the increasing price and reduced availability of fossil fuels. As an example, natural gas prices have doubled over the last 10 years, and it is expected to double again by 2025 [17].

The most common renewable energy source used in wastewater treatment plants is the biogas produced through the anaerobic digestion of sludge [18]. It is estimated that between 39% and 76% of the total energy consumed in wastewater treatment can be supplied by biogas [19]. Biogas production strongly depends on organic matter concentrations in the initial wastewater. It can be used in a boiler to produce heat, to generate electricity through gas turbines, or in combined heat and power units [7]. The energy production can be used within the plant, sold to the national grid, or dispatched to proximal users [20]. In addition, biogas can be upgraded/refined to be converted to fuel transport [21]. In the case of on-site use, biogas from digestion is suggested for use as a fuel for drying, in order to make sludge drying cost-effective [22], even if it is not sufficient to cover the entire energy demand of the process [23]. Moreover, biogas production is variable both on a seasonal and an hourly basis, therefore, to mitigate the effect of fluctuations, a biogas storage system or integration with natural gas from the grid needs to be considered [24]. As an example, in a large Italian WWTP, internal biogas is integrated with natural gas purchased from the grid to fulfill thermal demand of the plant, including anaerobic digestion and thermal drying [25].

Another solution can be improving biogas production through codigestion of sewage sludge with other substrates, such as organic waste, fats, oils, grease, or highly concentrated "blackwater" (waste from toilets or urinals) coming from separate collection or algal biomass [26–28]. Codigestion not only increases energy production, but also reduces economic costs by sharing facilities and operation [29]. At the Strass WWTP in Austria, codigestion with kitchen waste increased electricity production to around 180% of the plant demand [28]. However, some problems with codigestion need to be considered. Variability of substrates properties can cause instability, inhibition, or overloading of the process [30]; an example is the presence of organic waste, which, by increasing nitrogen load in the water exiting the digester, may require an additional treatment step for nitrogen removal [28]. Other problems that limit the large-scale deployment of anaerobic digestion and biogas use are large investment costs, operations and maintenance costs, availability of incentive programs, and national regulations [31].

Thermal energy for sludge drying can be also supplied by using external sources. Müller and Stüben analyzed two case studies where the biogas produced in farms from agricultural wastes, manure, and other substrates was used to provide electricity and heat for sludge drying in a WWTP, finding significant ecological and economic benefits [32]. Bianchini et al. [33] developed an integrated facility where sewage sludge was dried by the flue gas of a waste-to-energy (WTE) power plant without the use of an intermediate heat exchanger. Employing this concept of plant integration was beneficial for both the WWTP, which decreased the sludge to be disposed of, and the WTE plant, which increased its energy efficiency owing the exploitation of waste heat. Another integrated system based on wastewater sludge and municipal solid waste (MSW) was suggested by Murashko et al. [34], who proposed coincineration of these two streams in order to improve the economic feasibility of a decentralized WWTP. The advantage is that MSW presents a higher heating value and availability than sewage sludge. The incinerator was equipped with a heat recovery system, which produced steam used to generate electricity through a turbine and heat through a condenser, in order to cover the entire demand of the WWTP. An integrated system for sludge treatment was proposed also by Li et al., who coupled a bubbling fluidized bed dryer and a circulating fluidized bed incinerator [35]. The advantages of such an integrated solution were the high efficiency, low pollutant emissions, and low investment and operating costs of the overall system. Moreover, combining drying and incineration, a high solids content suitable for self-sustaining combustion was obtained. Indeed, when energy recovery is considered for sewage sludge, an appropriate integration of drying and thermal treatment has been suggested to increase the efficiency of the overall system [15,36–38]. Despite this, it has to be noticed that currently in WWTPs thermal energy recovery appears to receive less attention than electrical optimization and self-sufficiency [39].

Owing to the contemporaneity of thermal and electrical energy demands, energy efficiency of WWTPs can be improved by using combined heat and power (CHP) units. Beyond conventional technologies, several innovative CHP systems have been proposed in literature. The combination of solid oxide fuel cells (SOFCs) and a microgas turbine both fed by biogas to supply thermal demand of the anaerobic digestion and a part of the electrical power demand in WWTP was demonstrated to be a cost-effective energy saving [40]. SOFCs for biogas applications have been found to be competitive with other cogeneration technologies also by Trendewicz and Braun [24], even if the effect of biogas quality, utility pricing, and incentives on the economic viability should be taken into account. Molten carbonate fuel cells fed by biogas were also analyzed; in this case the technology was found to be more efficient than conventional ones, but unfeasible from an economic point of view, owing to the very high capital and maintenance costs [41]. A life cycle assessment (LCA) was carried out to assess a WWTP with microalgae-based secondary treatment, thermal hydrolysis with steam explosion of microalgae, anaerobic codigestion of pretreated microalgal biomass and primary sludge, digestate composting, and biogas cogeneration. The coproduction of energy products was found to be fundamental to improve the life cycle performance of the system [42].

Beyond bioenergy, other renewables such as solar, wind, geothermal, and hydroelectric energy may also be used as an on-site energy source, depending on their availability [18,27]. However, only solar energy is commonly used for sludge drying. The first technical and economic feasibility analysis on the use of solar energy in WWTPs was carried out in 1978 [43]; the practice was demonstrated to be cost-effective in producing heat for sludge anaerobic digestion, therefore it was proposed also for sludge drying [44]. In the beginning, open solar drying was used, which was then replaced by greenhouse dryers, which allowed for a reduction in the drying area owing to sludge mixing and ventilation [45]. This significantly increases the evaporation rate per square meter up to three times more than conventional sludge beds [46], both in summer and winter. Greenhouse solar dryers can be improved by increasing floor temperatures by means of hot water produced through plate solar collectors [47] or a solar water heater [48]. Solar energy can be also integrated with other energy sources. A CHP system fueled by biogas and natural gas, coupled with a parabolic trough collector field, to supply electric and thermal energy to a real WWTP, can reach a primary energy saving of about 15% [49].

The use of other renewable energy sources for drying applications is limited to agricultural products [50,51]. Geothermal energy was used to dry fruit [52,53], pyrethrum, and maize [54].

In the following section, a methodology to assess the use of renewable energy sources for sewage sludge and wastewater treatment is described, along with the model to analyze thermal drying and estimate its energy demand. Then, a case study to apply the developed methodology is proposed. Finally the main conclusions are drawn.

11.2 METHODOLOGY

Depending on availability, the use of renewable energy sources can help WWTPs, which are high energy demand systems, to reduce their economic costs and carbon

Assessing the Feasibility of Renewable Energy Sources for Treatment 251

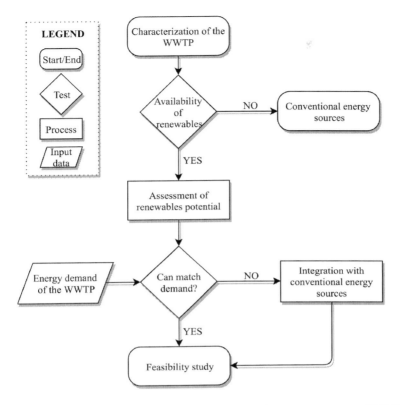

FIGURE 11.2 Flow-chart of integration between renewable energy sources and WWTP.

footprint. The design of most WWTPs operating today has not accounted for energy demand, and only recently, energy saving measures have been implemented in wastewater treatment management. In Figure 11.2, a flowsheet of the methodology to follow for including the use of renewables in WWTP is illustrated. Once the WWTP to be considered is identified, the availability of renewable energy sources, such as solar, geothermal, or bioenergy, should be considered, employing conventional energy sources only if necessary. In the case of available renewables, their potential use with different technologies should be assessed and compared to the energy demand of the considered WWTP. If the renewable energy potential and the WWTP do not match, conventional energy sources must be integrated. Finally, a study on the economic feasibility of the system has to be performed.

11.2.1 Thermal Drying Modeling

Considering the thermal energy demand, a significant consumption is related to sludge thermal drying. In order to estimate to compare its demand with the available renewable energy sources, a model to analyze thermal drying from an energy point of view is proposed.

The numerical model is based on the following mass and heat balances [55]:

$$(\dot{m}_{DF}\ Y)_{in} + (\dot{m}_{Sl}\ X)_{in} = (\dot{m}_{DF}\ Y)_{out} + (\dot{m}_{Sl}\ X)_{out} \tag{11.1}$$

$$(\dot{m}_{DF}\ h_{DF})_{in} + (\dot{m}_{Sl}\ h_{Sl})_{in} = (\dot{m}_{DF}\ h_{DF})_{out} + (\dot{m}_{Sl}\ h_{Sl})_{out} + q_L \tag{11.2}$$

where ṁ is the mass flow rate and Y is the humidity, both on dry basis of the desiccant flow (DF), X is the moisture content of sludge (Sl), h is the enthalpy, and q_L represents the rate of heat loss from the system. The desiccant flow is a vapor-gas mixture, whose enthalpy, h_{DF}, can be assumed as the sum of the enthalpies of the gas and the vapor, as [56]:

$$h_{DF} = c_{p,G}\ T + [c_{p,LW}\ T_S + \Delta H_v + c_{p,v}\ (T-T_S)]\ Y \tag{11.3}$$

where $c_{p,G}$, $c_{p,LW}$, and $c_{p,v}$ are the average specific heats of dry gas, liquid water, and vapor, respectively, T is the temperature, ΔH_v is the latent heat term for vaporization, and subscript S represents the saturation condition. A combined enthalpy is considered for the sludge, which is composed of the enthalpy of unit mass of the dry sludge and that of its associated moisture:

$$h_{Sl} = (c_{p,DS}\ T_S + c_{p,LW}\ X)\ T - X\ \Delta H_w \tag{11.4}$$

where $c_{p,DS}$ is the average specific heat of dry sludge and ΔH_w is the mean enthalpy of sorption to the material.

11.2.2 FEASIBILITY STUDY

The feasibility of the considered system based on renewable energy sources can be assessed through an economic and environmental analysis. The economic model proposed here is based on several indexes, dependent on investment costs, operation, and maintenance costs, as well as revenues. Investment cost, J_{tot}, is the sum of the costs related to the single components of the proposed system. Operation and maintenance costs, $J_{O\&M}$, include costs associated with operating and maintaining all the components of the system. The revenues in such systems are related to:

- Avoided cost of sludge disposal, R_{Sl}, since reducing the water content through thermal treatment decreases the mass of sludge to be disposed;
- Avoided electrical energy, $R_{el,purch}$, if the proposed renewable energy source and technology allow producing electrical energy, $\dot{P}$; when $\dot{P}$ is higher than the electricity needed by the WWTP, $\dot{P}_{WWTP}$, there is a surplus, $\dot{P}_{surplus}$, which can be sold to the national grid with a consequent additional revenue, $R_{el,sell}$.

Considering the specific cost of sludge disposal, c_{Sl}, and the avoided mass of sludge to be disposed, $(\dot{m}_{Sl,in} - \dot{m}_{Sl,out})$, R_{Sl} can be calculated as:

$$R_{Sl} = c_{Sl} \cdot (\dot{m}_{Sl,in} - \dot{m}_{Sl,out}) \tag{11.5}$$

In case of electricity production, the following scheme can be used:

$$\text{if } \dot{P}_{surplus} = 0 \text{ then } \quad R_{el,purch} = c_{el,purch} \cdot \dot{P}_{WWTP} \quad R_{el,sell} = 0 \tag{11.6}$$

$$\text{if } \dot{P}_{surplus} > 0 \text{ then } \quad R_{el,purch} = c_{el,purch} \cdot \dot{P}_{WWTP} \quad R_{el,sell} = c_{el,sell} \cdot \dot{P}_{surplus} \tag{11.7}$$

$$\text{if } \dot{P}_{surplus} < 0 \text{ then } \quad R_{el,purch} = c_{el,purch} \cdot \dot{P} \quad R_{el,sell} = 0 \tag{11.8}$$

In the last case, the cost for electricity to be integrated by the national grid, $J_{el,purch}$, has to be considered:

$$J_{el,purch} = c_{el,purch} \cdot (\dot{P}_{WWTP} - \dot{P}) \tag{11.9}$$

The economic indexes used for the analysis, based on the above mentioned costs and revenues, are simple payback (SPB) and net present value (NPV):

$$\text{SPB} = \frac{J_{tot}}{\text{AES}} \tag{11.10}$$

where AES is the annual economic saving, which is the sum of the revenues and the operation and maintenance costs of the system;

$$\text{NPV} = \text{AES} \cdot \text{AF} - J_{tot} \tag{11.11}$$

where the annual factor (AF), is determined as:

$$\text{AF} = \frac{1}{a}\left[1 - \frac{1}{(1+a)^N}\right] \tag{11.12}$$

with a discounting rate and N service life of the system.

Considering the thermal, E_{th}, and electrical energy, E_{el}, produced by renewable energy sources the avoided primary energy for electricity and heat production, and the consequent avoided CO_2 emissions can be calculated as:

$$PE_{el} = \frac{E_{el}}{\eta_{el,ref}} \tag{11.13}$$

$$PE_{th} = \frac{E_{th}}{\eta_{th,re}} \tag{11.14}$$

$$CO_{2el} = PE_{el} \cdot EF_{el} \tag{11.15}$$

$$CO_{2th} = PE_{th} \cdot EF_{th} \tag{11.16}$$

where η_{ref} is the average efficiency of the reference technology used for its production and EF is the emission factor related to the fuel used for energy production.

11.3 CASE STUDY

The proposed methodology is applied to a case study, based on the authors' previous work [57]. A WWTP designed to serve a district in a small island of Southern Italy, named Ischia, is considered. In fact, in small islands the lack of space and the environmental restrictions significantly limit the use of conventional technologies for electricity production and waste disposal, making the use of renewable energy sources very interesting [58,59]. Several studies have demonstrated that Ischia island presents an interesting geothermal energy potential [60]. In case of geothermal source exploitation, sustainability and profitability are highly affected by depth and temperature at which the geothermal fluid is available. Therefore, the data on the geothermal source in the area of the considered WWTP have been collected. It has been found that the geothermal fluid flow rate in the considered area is estimated to be 5.00 l/s [61], with a temperature of 140°C at a depth of 180 m from ground level [60]. Thermal energy for drying can be supplied by using a heat exchanger, whereas for electrical energy, an organic Rankine cycle (ORC) system is suggested as one of the most effective technologies to convert thermal energy from moderate-low temperature sources [62,63]. Geothermal energy exploitation can be maximized by using a cascade configuration at different thermal levels in order to obtain different products [64], and, depending on the temperature of geothermal source and the temperature required by the process, different arrangements can be taken into account [50]. For the analyzed case study, a parallel cascade configuration is considered. The system layout developed is sketched in Figure 11.3. The geothermal fluid extracted from the ground is split in two: a fraction is used to produce the desiccant flow for

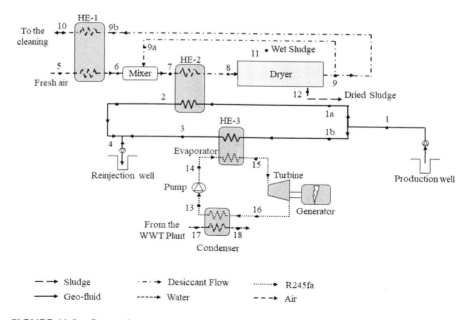

FIGURE 11.3 System layout.

the dryer, the rest feeds the ORC system. A part of desiccant flow exiting the dryer is recycled to increase the energy efficiency of drying process, and to reduce the exhausts sent to the cleaning system. The recycled desiccant flow is mixed with fresh air preheated by the dryer exhausts. The mixed flow is then heated by the geothermal fluid. The flow rate and temperature of the air heated by the geothermal fluid and the fraction of recycling are set to guarantee that the wet sludge is dried to a final solids content of at least 90%.

In order to guarantee sludge drying in case of temporary unavailability of the geothermal system, the installation of an auxiliary boiler is considered.

In parallel, the ORC system produces electric energy: a fraction is used to supply WWTP, including auxiliaries for the dryer, and pumping of the geothermal fluid, the surplus electric energy production is sold to the national grid. The ORC system operates with R245fa, which is a suitable working fluid for temperatures of the heat source up to 170°C [65]. The condenser operates with purified water coming from the wastewater treatment, to save fresh water, increasing the sustainability of the system.

The geothermal fluid is reinjected in the ground since it has been demonstrated that binary cycles with complete reinjection of geothermal fluid improve the sustainability of geothermal exploitation [66]. The developed system is modeled through the software ASPEN Plus (Advanced System for Process Engineering) [68].

In the case study analyzed, a convective belt dryer is proposed for sludge thermal treatment, due to the operation temperature which is compatible with that of the geothermal fluid, as well as its easy handling and high flexibility [67]. Thermodynamic properties of air are determined through Raoult's and Henry's laws, implemented in ASPEN Plus [68]. Sewage sludge is modeled as a carbonaceous fuel [69] by using two main algorithms implemented in ASPEN Plus, that calculate specific heat, density and enthalpy of coal and coal-derived substances, through statistical correlations based on the biomass ultimate, proximate, and sulfur analyses [70]. In the convective dryer, the Peng Robinson–Boston Mathias modified method is used since the system deals with multiple phases [69]. The convective dryer is considered to operate in cross-flow mode for the gas and plug flow for the solids. The kinetic of the process is implemented through the drying curve that expresses the evaporation rate depending on the water content [71]. The drying process aims at reducing the water content to a value lower than 10%.

The ORC module is calibrated in order to maximize the net power output. Regarding the thermodynamic cycle, the evaporation and condensation pressures are set by considering design specifications based on the pinch point difference temperature. The working fluid at the pump inlet is supposed to be saturated liquid, whereas at the expander inlet (microturbine) it is supposed to be saturated vapor. Since the system is supposed to operate under steady state conditions, both pump and turbine work at fixed isentropic efficiency, for which the Mollier approach is adopted.

$$\eta_{is,pump} = \frac{h_{14,is} - h_{13}}{h_{14} - h_{13}} \quad (11.17)$$

$$\eta_{is,turb} = \frac{h_{15} - h_{16}}{h_{15} - h_{16,is}} \quad (11.18)$$

All heat exchangers are calculated by using the logarithmic mean temperature difference (LMTD) approach, whose governing equations are [72]:

$$\frac{1}{U} = \frac{1}{h_{shell}} + \frac{D_{out}}{D_{in}} \frac{1}{h_{tube}} + \frac{D_{out} \ln(D_{out}/D_{in})}{2\lambda} \qquad (11.19)$$

$$\dot{Q} = UA \, \Delta T_{LMTD} F \qquad (11.20)$$

$$(\dot{m}h)_{hot,in} + (\dot{m}h)_{cold,in} = (\dot{m}h)_{hot,out} + (\dot{m}h)_{hot,out} \qquad (11.21)$$

$$\dot{Q} = (\dot{m}h)_{hot,in} - (\dot{m}h)_{hot,out} = (\dot{m}h)_{cold,out} - (\dot{m}h)_{cold,in} \qquad (11.22)$$

where U represents the overall heat transfer coefficient, h_{shell} and h_{tube} are the heat transfer coefficients used for shell and tube side respectively, A is the heat transfer area, D is the diameter of the tube, subscripts "i" and "o" refer to inlet and outlet conditions, and F is the correction factor of the specific heat exchanger.

The inlet temperature of condensation process is fixed at 20°C, while the maximum outlet temperature at 26°C (as recommended by the Italian legislation for water discharge at sea). The mass flow rate of R245fa is consequently determined.

The input parameters for modeling thermal drying and electricity production through the ORC are summarized in Table 11.1, where the numbering of Figure 11.3 is used. The sludge characteristics, in terms of proximate and ultimate analyses, are adapted from Reference [73]. As mentioned before, the geothermal fluid mass flow rate is split to produce heat for sludge drying and to power the ORC for electricity generation. The minimum mass flow rate of geothermal fluid needed for sludge drying is firstly determined. The remaining fraction is used as an input parameter of the ORC unit model.

The energy demand of the WWTP has been estimated, by taking into account the seasonal fluctuations due to tourism in the area during the summer months. The WWTP is designed for an equivalent population equal to 10,000 inhabitants, with an inlet concentration of biochemical oxygen demand (BOD) of 60 g/ab/d. The specific electricity consumption considering a membrane bioreactor (MBR) systems for organic matter removal with nutrients removal, followed by tertiary treatment is equal to 2.48 kWh/kg BOD [74].

For the economic analysis, the total investment cost is the sum of the costs of drilling, J_{well}, dryer, J_{Dryer}, heat exchangers, J_{AirHE}, and ORC, J_{ORC}. The cost for well drilling is assumed to be 500 €/m, whereas for the dryer, data from a market survey are considered. The cost of heat exchangers and ORC depends on their size:

$$J_{HE} = 150 \left(\frac{A_{HE}}{0.093} \right)^{0.78} \qquad (11.23)$$

$$J_{ORC} = 3.00 \cdot 10^3 \cdot \dot{P}_{net} \qquad (11.24)$$

TABLE 11.1
Input Parameters of Thermodynamic Analysis

Stream/Component	Parameter	SI Unit	Value
	Volatile matter, VM	%DM	72
	Carbon, C	%VM	51.0
	Hydrogen, H	%VM	7.40
	Oxygen, O	%VM	33.0
	Nitrogen, N	%VM	7.10
	Sulfur, S	%VM	1.50
	Sludge calorific value, LHV_{sl}	MJ/t DM	16,560
	$\dot{m}_{11}$ (wet basis)	kg/s	0.021
	T_{11}	°C	16.2
	X_{11}	%	75.0
Ambient air	p_5	bar	1.01
	T_5	°C	16.2
Desiccant flow	p_7	bar	1.01
	T_7	°C	130
	Recycled fraction	%	65.0
Geofluid	$\dot{m}_1$	kg/s	5.00
	$\dot{m}_{1a}$	kg/s	0.180
	$\dot{m}_{1b}$	kg/s	4.82
	T_1	°C	140
	p_1	bar	7.00
ORC	$\eta_{is,pump}$	%	70.0
	$\eta_{is,turb}$	%	80.0
	T_3	°C	70.0
	$\Delta T_{pinch,eva}$	°C	7.00
	$\Delta T_{pinch,cond}$	°C	5.00
	x_{13}	%	0.00
	x_{15}	%	100
	T_{17}	°C	20.0
	p_{17}	bar	2.00
	T_{18}	°C	26.0

An auxiliary boiler is also supposed to be installed to guarantee sludge drying in case of temporary unavailability of the geothermal system. Its cost is determined as:

$$J_{boil} = (175 \cdot \dot{Q}_{boil}^{-0.13} 10^3) \cdot \dot{Q}_{boil} \qquad (11.25)$$

The operation and maintenance costs are supposed to be 5% of the total investment cost. The other parameters for economic and environmental analyses are reported in Tables 11.2 and 11.3. The environmental analysis is carried out considering as reference technologies a boiler fueled by natural gas and electricity from national grid.

TABLE 11.2
Input Parameters of Economic Analysis

Parameter		Unit	Value
Specific cost of sludge disposal	c_{sl}	€/ton	160
Specific cost of electrical energy	$c_{el,purch}$	€/MWh	200
Feed-in tariff	$c_{el,sell}$	€/MWh	164
Discounting rate	a	–	5.00%
Service life	N	y	20

TABLE 11.3
Input Parameters of Environmental Analysis

Parameter		Unit	Value
Reference efficiency of the boiler	$\eta_{ref,boil}$	%	85.0
Reference efficiency of the national grid [75,76]	$\eta_{ref,nat.grid}$	%	48.8
Lower Heating Value of natural gas	$LHV_{nat.gas}$	kWh/Sm^3	9.59
Emission factor of natural gas [76]	$EF_{nat.gas}$	kg CO_2/MWh	324
Total emission factor of national grid [76]	$EF_{nat.grid}$	kg CO_2/MWh	466

11.4 RESULTS AND DISCUSSION

The main results of the thermodynamic and economic analyses are reported in Table 11.4, where the same numbering used in Figure 11.3 is employed. As mentioned, the sludge outlet moisture content is lower than 10%, reducing the mass of sludge to be disposed by 73.2% compared to the inlet conditions. The geothermal fluid needed for thermal drying is significantly lower than that used to feed the ORC. The fraction of geothermal fluid used for electricity production allows covering the whole WWTP demand and it also generates an electrical surplus.

The economic results are illustrated in Table 11.5. The system appears to be economically feasible, with SPB and NPV around 5 years and higher than 1.45 M€ respectively.

In this specific case study, the total investment costs are mainly related to the dryer (around 35%) and the ORC module (around 45%). It is worth noticing that, in geothermal applications, drilling cost heavily affects the economic feasibility of the system, while in this specific case study it represents the 18% of the total investment, owing to low depth at which geothermal fluid temperature is suitable for such applications.

Considering the AES, the avoided electricity purchase is the most affecting parameter (around 60%), followed by the avoided sludge disposal cost (around 34%).

In order to investigate the developed methodology, a sensitivity analysis on the WWTP size is carried out, varying the population equivalent from 10,000 to 20,000 with a step of 5000. The main results are reported in Table 11.6.

TABLE 11.4
Main Results of Thermodynamic Analysis

Component/Stream		Parameter	SI Unit	Value
Dryer	Sludge	X_{12}	%	6.50
		$\dot{m}_{12}$	kg/h	20.4
		T_{12}	°C	101
	Fresh air	$\dot{m}_5$	kg/s	0.258
		t_6	°C	69.3
	Desiccant Flow	$\dot{m}_7$	kg/s	0.767
		T_9	°C	79.3
		$\dot{m}_9$	kg/s	0.782
		$\dot{m}_{9a}$	kg/s	0.508
		$\dot{m}_{9b}$	kg/s	0.274
		T_{10}	°C	46.3
	Evaporated water	$\dot{m}_{w,ev}$	g/s	15.6
	Electricity demand	$\dot{P}_{dryer}$	kW	4.56
	Geothermal fluid	T_2	°C	83.1
		T_3	°C	70
	HE1	A_{HE1}	m²	1.19
		$\dot{Q}_{HE1}$	kW	13.9
	HE2	A_{HE2}	m²	6.03
		$\dot{Q}_{HE2}$	kW	43.4
ORC	R245fa	p_{eva}	bar	9.63
		p_{cond}	bar	1.84
		$\dot{P}_{net}$	kW	150
		$\dot{m}_{13}$	kg/s	6.33
		T_{13}	°C	31.0
		T_{14}	°C	31.5
		T_{15}	°C	88
		T_{16}	°C	46.8
	Cooling water	$\dot{m}_{17}$	kg/s	51.0
	Evaporator	A_{eva}	m²	192
		$\dot{Q}_{eva}$	MW	1.43
	Condenser	A_{cond}	m²	341
		$\dot{Q}_{cond}$	MW	1.28
	Efficiency	η_{ORC}	%	10.5
	WWTP electricity demand	$\dot{P}_{WWTP}$	kW	78.6
	Geothermal fuid pumping	$\dot{P}_{pumping}$	kW	16.8
	Total internal electricity demand	$\dot{P}_{tot,int}$	kW	100
	Electricity surplus	$\dot{P}_{surplus}$	kW	50.0
	Avoided electricity purchase	$\dot{P}_{av,purch}$	kW	78.6

Electrical power required by the WWTP plant ($\dot{P}_{WWTP}$), total electrical power required by the whole plant ($\dot{P}_{req,tot}$) under investigation, and avoided power purchased from the grid ($\dot{P}_{av,purc}$) have been calculated. Once the simulations have been performed, the total thermal demand, $\dot{Q}_{HE,tot}$, and the net power output of the ORC, $\dot{P}_{net}$, can be determined.

TABLE 11.5
Main Results of Economic Analysis

Parameter	SI Unit		Value
Dryer cost	J_{dry}	€	350,000
ORC cost	J_{ORC}	€	450,000
Well cost	J_{well}	€	180,000
HE2	J_{HE2}	€	3887
HE1	J_{HE1}	€	1093
Auxiliary boiler	$J_{Aux.boil}$	€	5930
Total investment	J_{tot}	€	990,909
O&M	$J_{O\&M}$	€/y	49,545
Avoided sludge disposal	R_{sl}	€/y	67,248
Avoided electricity purchase	$R_{el,purch}$	€/y	117,900
Produced electricity selling	$R_{el,sell}$	€/y	61,500
Electricity purchase	$J_{el,purch}$	€/y	0
Annual economic saving	AES	€/y	197,103
Simple pay back	SPB	y	5.03
Net present value	NPV	€	1,464,991

As shown in Figure 11.4, the total thermal demand increases with the WWTP capacity, mainly owing to the higher desiccant flow needed for drying process (Figure 11.5). The higher demand of thermal energy affects ORC operation. In fact, the net power output of the ORC (Figure 11.6) decreases with the WWTP capacity owing the lower availability of thermal source of the geothermal fluid. In Figure 11.7, the geothermal fluid mass flow rate used for sludge drying is reported. The mass flow rate needed for sludge drying increases with WWTP size. Since the extractable mass flow rate of geothermal fluid is 5 kg/s, and thermal drying is considered as a priority, a lower heat source is available for the ORC. However, it is worth noticing that only in case of the largest WWTP capacity the system is not capable to cover all of the internal demand of electricity.

Only in the case of the smallest WWTP capacity, is electricity surplus $\dot{P}_{surplus}$ sufficient and suitable for selling to the national grid. Indeed, in the case of WWTP capacity equal to 10,000, an electricity surplus of around 50 kW is available. In the case where WWTP capacity is equal to 15,000, the electricity surplus is around 8 kW, then a possible sale to the national grid has not been taken into account and the corresponding revenue, $R_{el,sell}$, has been considered null.

Then, the 10,000 PE WWTP can be considered as a small producer of electricity, while the 15,000 PE WWTP can be considered as a self-producer of electricity and completely self-sufficient.

Regarding the largest capacity, power production covers most of the internal electricity demand, which corresponds to the avoided electricity purchased, $\dot{P}_{av.purch}$.

The sensitivity analysis has shown again as most of the AES (Figure 11.8) is represented by the revenues related to the avoided cost of the purchase of electricity, $R_{el,purch}$ and to the avoided cost of sludge disposal, R_{sl}; the revenue obtained from selling electricity to the grid only represents a small part of the total AES.

TABLE 11.6
Main Results of the Sensitivity Analysis

		$\dot{P}_{WWTP}$ (kW)	$\dot{Q}_{HE,tot}$ (kW)	$\dot{P}_{net}$ (kW)	$\dot{P}_{req,tot}$ (kW)	$\dot{P}_{surplus}$ (kW)	$\dot{P}_{av,purch}$ (kW)	J_{HE} (k€)	J_{boil} (k€)
WWTP capacity	10,000	78.6	57.2	150	100.0	50.0	78.6	4.98	5.93
	15,000	118	85.8	147	139	7.80	118	6.83	8.44
	20,000	157	114	144	178	0	123	8551	10.8

		J_{tot} (k€)	$J_{el,purch}$ (k€)	R_{SI} (k€)	$R_{el,purch}$ (k€)	$R_{el,sell}$ (k€)	AES (k€)	SPB (y)	NPV (M€)	CO2 (t/y)
WWTP capacity	10,000	991	0.00	67.2	118	61.5	197	5.03	1.46	1045
	15,000	995	0.00	101	177	0	228	4.36	1.85	1031
	20,000	999	51.0	134	181	0	218	4.58	1.72	1128

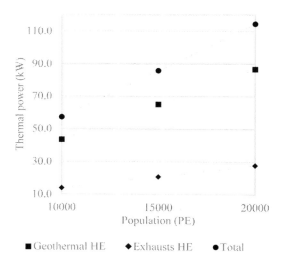

FIGURE 11.4 Thermal power required as a function of the WWTP capacity.

This result, together with the obtained values of the SPBs (Figure 11.9), leads to an interesting consideration about the economic convenience of such system. In fact, despite the possibility of selling electricity to the network, higher profitability has been obtained in case of higher WWTP capacity, characterized by higher revenue related to the avoided cost for plant electricity and to the avoided cost of sludge disposal.

Finally, the proposed system reduces the environmental impact of wastewater and sludge treatment, as shown by the avoided tons of CO_2, reported in Table 11.6.

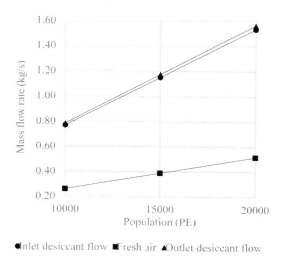

FIGURE 11.5 Mass flow rate of the desiccant flow as a function of the WWTP capacity.

Assessing the Feasibility of Renewable Energy Sources for Treatment

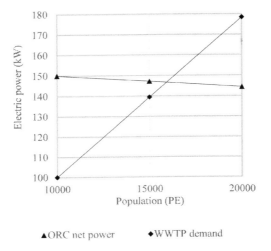

FIGURE 11.6 Net ORC power out and WWTP power demand as a function of the WWTP capacity.

The system under investigation is advantageous under specific conditions, which can be summarized as follows:

- Temperature of the geothermal source suitable for engineering applications available at low depth.
- Geothermal reservoir capacity large enough to guarantee an injection mass flow rate suitable for engineering applications and for long periods of time, without incurring in depletion of the reservoir.
- Calibration of the ORC module such as to make the plant self-sufficient, without over producing for electricity selling to the grid.

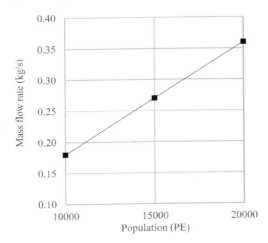

FIGURE 11.7 Mass flow rate of geofluid as a function of the WWTP capacity.

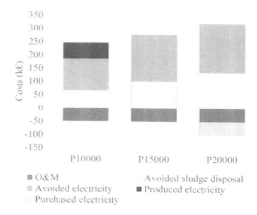

FIGURE 11.8 AES as a function of the WWTP capacity.

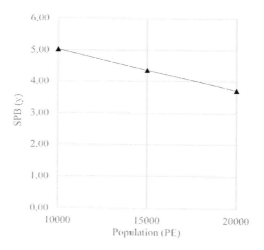

FIGURE 11.9 SPB as a function of the WWTP capacity.

11.5 CONCLUSIONS

Sludge thermal drying, aiming at reducing the water content, is a fundamental process for energy recovery from sludge, since it increases the calorific value of sludge. Owing to its high thermal demand and economic cost, sludge thermal drying is not a common practice yet. The key for environmental and economic sustainability in sludge thermal treatment is the integration of available renewable sources, employing conventional sources only when necessary.

The technologies of renewables exploitation, the issues associated with them, and the variables of the economic analysis to assess their profitability, strongly differ depending on the particular resource used. Moreover, renewables' availability is highly "site-specific" and therefore an in-depth "ad hoc" analysis is needed.

Assessing the Feasibility of Renewable Energy Sources for Treatment

In this chapter, a methodology to integrate renewable energy systems with wastewater and sludge treatment, along with an approach to estimate the economic feasibility of such integration, is proposed.

In order to investigate the proposed methodology, a case study is analyzed. A combined heat and power plant powered by geothermal source is considered to supply heat for sludge thermal drying and electricity by means of an ORC unit to the facilities of the wastewater treatment plant. The developed system is analyzed by thermodynamic, environmental, and economic point of view.

The economic analysis is carried out by considering specific indexes as investment and operational costs, avoided costs with respect to a reference technology, revenues, simple payback, and net present value.

The proposed system appears to be profitable for high WWTP capacity and when self-production of electricity is considered, the high investment costs of technologies do not justify a surplus production of electricity to be sold in the national grid.

LIST OF ABBREVIATIONS

Roman Symbol	SI Unit	Description
a	–	Discounting rate
A	m²	Surface
AES	€/y	Annual economic saving
AF	–	Annual factor
c	€/t - €/kWh - €/Sm³	Specific cost
CO_2	t/y	CO_2 emissions
c_p	J/kg K	Specific heat
D	m	Diameter
E	J	Energy
EF	CO_2 t/kg - CO_2/Sm³	Emission factor
F	–	Correction factor
G	m/s	Specific dry desiccant flow
h	J/kg	Enthalpy
h_{shell}	W/(mK)	Shell heat transfer coefficient
h_{tube}	W/(mK)	Tube heat transfer coefficient
ΔH_V	J/kg	Latent heat term for vaporization
ΔH_W	J/kg	Mean enthalpy of sorption
J	€	Cost
$\dot{m}$	kg/s	Mass flow rate
N	y	Service life
NPV	€	Net present value
p	bar	Pressure
$\dot{P}$	W	Power
PE	J	Primary energy
q_L	W	Rate of heat loss from the system
$\dot{Q}$	W	Heating demand
R	€	Revenue
SPB	y	Simple pay back
T	°C	Temperature
X	–	Moisture content on wet basis
Y	–	Humidity on dry basis

Greek Symbol	SI Unit	Description
η	%	Efficiency

Acronyms	Description
DM	Dry matter
HE	Heat exchanger
LHV	Lower heating value
O&M	Operation and maintenance
ORC	Organic Rankine cycle
VM	Volatile matter
WWTP	Wastewater treatment plant

Subscripts/Superscripts	Description
av	Avoided
boil	Boiler
cold	Cold
cond	Condenser
DF	Desiccant flow
dryer	Dryer
el	Electrical
eva	Evaporator
G	Dry gas
HE	Heat exchanger
hot	Hot
in	Inlet
is	Isoentropic
LW	Liquid water
net	NET
ORC	ORC
out	outlet
pinch	Pinch
pump	Pump
purch	Purchased
ref	Reference
req	Required
S	Saturation condition
sell	Selling
Sl	Sludge
surplus	Surplus
th	Thermal
tot	Total
turb	Turbine
v	Vapor
well	Well

REFERENCES

1. Kacprzak, M. et al., Sewage sludge disposal strategies for sustainable development. *Environmental Research.* 156, pp 39–46, 2017.
2. Fytili, D. and A. Zabaniotou, Utilization of sewage sludge in EU application of old and new methods—a review. *Renewable and Sustainable Energy Reviews.* 12(1), pp 116–140, 2008.
3. Bianchini, A. et al., Sewage sludge management in Europe: A critical analysis of data quality. *International Journal of Environment and Waste Management.* 18(3), pp 226–238, 2016.
4. Council, E., Directive 1999/31/EC on the landfill of waste. *Official Journal of the European Union L.* 182, pp 1–19, 1999.
5. Turunen, V., J. Sorvari, and A. Mikola, A decision support tool for selecting the optimal sewage sludge treatment. *Chemosphere.* 193, pp 521–529, 2018.
6. Kelessidis, A. and A.S. Stasinakis, Comparative study of the methods used for treatment and final disposal of sewage sludge in European countries. *Waste Management.* 32(6), pp 1186–1195, 2012.
7. Tyagi, V.K. and S.-L. Lo, Sludge: A waste or renewable source for energy and resources recovery? *Renewable and Sustainable Energy Reviews.* 25, pp 708–728, 2013.
8. Syed-Hassan, S.S.A. et al., Thermochemical processing of sewage sludge to energy and fuel: Fundamentals, challenges and considerations. *Renewable and Sustainable Energy Reviews.* 80, pp 888–913, 2017.
9. Houillon, G. and O. Jolliet, Life cycle assessment of processes for the treatment of wastewater urban sludge: Energy and global warming analysis. *Journal of Cleaner Production.* 13(3), pp 287–299, 2005.
10. Scholz, M., Chapter 21 - Sludge treatment and disposal, in *Wetlands for Water Pollution Control (Second Edition)*, M. Scholz, Editor. 2016, Elsevier. p. 157–168.
11. Chai, L.H., Statistical dynamic features of sludge drying systems. *International Journal of Thermal Sciences.* 46(8), pp 802–811, 2007.
12. Uggetti, E. et al., Sludge treatment wetlands: A review on the state of the art. *Bioresource Technology.* 101(9), pp 2905–2912, 2010.
13. Collard, M., B. Teychené, and L. Lemée, Comparison of three different wastewater sludge and their respective drying processes: Solar, thermal and reed beds – Impact on organic matter characteristics. *Journal of Environmental Management.* 203, pp 760–767, 2017.
14. Nazari, L. et al., 3 - Recent advances in energy recovery from wastewater sludge, in *Direct Thermochemical Liquefaction for Energy Applications*, L. Rosendahl, Editor. 2018, Woodhead Publishing. p. 67–100.
15. Chae, K.-J. and J. Kang, Estimating the energy independence of a municipal wastewater treatment plant incorporating green energy resources. *Energy Conversion and Management.* 75, pp 664–672, 2013.
16. Di Fraia, S., N. Massarotti, and L. Vanoli, A novel energy assessment of urban wastewater treatment plants. *Energy Conversion and Management.* 163, pp 304–313, 2018.
17. Batstone, D. et al., Platforms for energy and nutrient recovery from domestic wastewater: A review. *Chemosphere.* 140, pp 2–11, 2015.
18. Gu, Y. et al., The feasibility and challenges of energy self-sufficient wastewater treatment plants. *Applied Energy.* 204, pp 1463–1475, 2017.
19. Silvestre, G., B. Fernández, and A. Bonmatí, Significance of anaerobic digestion as a source of clean energy in wastewater treatment plants. *Energy Conversion and Management.* 101, pp 255–262, 2015.
20. Venkatesh, G. and R.A. Elmi, Economic–environmental analysis of handling biogas from sewage sludge digesters in WWTPs (wastewater treatment plants) for energy recovery: Case study of Bekkelaget WWTP in Oslo (Norway). *Energy.* 58, pp 220–235, 2013.

21. Patterson, T. et al., An evaluation of the policy and techno-economic factors affecting the potential for biogas upgrading for transport fuel use in the UK. *Energy Policy*. 39(3), pp 1806–1816, 2011.
22. Stasta, P. et al., Thermal processing of sewage sludge. *Applied Thermal Engineering*. 26(13), pp 1420–1426, 2006.
23. Chen, G., P.L. Yue, and A.S. Mujumdar, Dewatering and drying of wastewater treatment sludge, in *Handbook of Industrial Drying*. 2006, CRC Press. pp 912–929.
24. Trendewicz, A.A. and R.J. Braun, Techno-economic analysis of solid oxide fuel cell-based combined heat and power systems for biogas utilization at wastewater treatment facilities. *Journal of Power Sources*. 233, pp 380–393, 2013.
25. Panepinto, D. et al., Evaluation of the energy efficiency of a large wastewater treatment plant in Italy. *Applied Energy*. 161, pp 404–411, 2016.
26. Shi, C.Y., *Mass Flow and Energy Efficiency of Municipal Wastewater Treatment Plants*. IWA Publishing, 2011.
27. Gude, V.G., Energy and water autarky of wastewater treatment and power generation systems. *Renewable and Sustainable Energy Reviews*. 45, pp 52–68, 2015.
28. Nowak, O., P. Enderle, and P. Varbanov, Ways to optimize the energy balance of municipal wastewater systems: Lessons learned from Austrian applications. *Journal of Cleaner Production*. 88, pp 125–131, 2015.
29. Gao, H., Y.D. Scherson, and G.F. Wells, Towards energy neutral wastewater treatment: Methodology and state of the art. *Environmental Science: Processes & Impacts*. 16(6), pp 1223–1246, 2014.
30. Shen, Y. et al., An overview of biogas production and utilization at full-scale wastewater treatment plants (WWTPs) in the United States: Challenges and opportunities towards energy-neutral WWTPs. *Renewable and Sustainable Energy Reviews*. 50, pp 346–362, 2015.
31. Pfluger, A. et al., Anaerobic digestion and biogas beneficial use at municipal wastewater treatment facilities in Colorado: A case study examining barriers to widespread implementation. *Journal of Cleaner Production*. 206, pp 97–107, 2019.
32. Müller, J., M. Stüben, and R. Tyagi. Innovative energy concept for sludge treatment using renewable resources. in *Proceedings of the IWA Wastewater Biosolids Sustainability Conference: Technical, Managerial, and Public Synergy*. 2007.
33. Bianchini, A. et al., Sewage sludge drying process integration with a waste-to-energy power plant. *Waste Management*. 42, pp 159–165, 2015.
34. Murashko, K. et al., Techno-economic analysis of a decentralized wastewater treatment plant operating in closed-loop. A Finnish case study. *Journal of Water Process Engineering*. 25, pp 278–294, 2018.
35. Li, S. et al., Integrated drying and incineration of wet sewage sludge in combined bubbling and circulating fluidized bed units. *Waste Management*. 34(12), pp 2561–2566, 2014.
36. Verstraete, W. and S.E. Vlaeminck, ZeroWasteWater: Short-cycling of wastewater resources for sustainable cities of the future. *International Journal of Sustainable Development & World Ecology*. 18(3), pp 253–264, 2011.
37. Flaga, A. Sludge drying. in *Proceedings of Polish-Swedish Seminars, Integration and optimization of urban sanitation systems*. Cracow, March 2005.
38. Buonocore, E. et al., Life cycle assessment indicators of urban wastewater and sewage sludge treatment. *Ecological Indicators*. 94, pp 13–23, 2018.
39. Kollmann, R. et al., Renewable energy from wastewater-Practical aspects of integrating a wastewater treatment plant into local energy supply concepts. *Journal of Cleaner Production*. 155, pp 119–129, 2017.
40. MosayebNezhad, M. et al., Techno-economic assessment of biogas-fed CHP hybrid systems in a real wastewater treatment plant. *Applied Thermal Engineering*. 129, pp 1263–1280, 2018.

41. Chacartegui, R. et al., Molten carbonate fuel cell: Towards negative emissions in wastewater treatment CHP plants. *International Journal of Greenhouse Gas Control.* 19, pp 453–461, 2013.
42. Colzi Lopes, A. et al., Energy balance and life cycle assessment of a microalgae-based wastewater treatment plant: A focus on alternative biogas uses. *Bioresource Technology.* 270, pp 138–146, 2018.
43. Malcolm, J.W. and D.E. Cassel, *Use of Solar Energy to Heat Anaerobic Digesters: Part I, Technical and Economic Feasibility Study: Part II, Economic Feasibility Throughout the United States.* Vol. 1. Environmental Protection Agency, Office of Research and Development, Municipal Environmental Research Laboratory, 1978.
44. El-Ariny, A. and H. Miller, Utilization of solar energy for sludge drying beds. *Journal of Solar Energy Engineering.* 106(3), pp 351–357, 1984.
45. Bennamoun, L., Solar drying of wastewater sludge: A review. *Renewable and Sustainable Energy Reviews.* 16(1), pp 1061–1073, 2012.
46. Bux, M. et al., Volume reduction and biological stabilization of sludge in small sewage plants by solar drying. *Drying Technology.* 20(4–5), pp 829–837, 2002.
47. Salihoglu, N.K., V. Pinarli, and G. Salihoglu, Solar drying in sludge management in Turkey. *Renewable Energy.* 32(10), pp 1661–1675, 2007.
48. Mathioudakis, V. et al., Extended dewatering of sewage sludge in solar drying plants. *Desalination.* 248(1–3), pp 733–739, 2009.
49. Di Fraia, S. et al., An integrated system for sewage sludge drying through solar energy and a combined heat and power unit fuelled by biogas. *Energy Conversion and Management.* 171, pp 587–603, 2018.
50. Rubio-Maya, C. et al., Techno-economic assessment for the integration into a multi-product plant based on cascade utilization of geothermal energy. *Applied Thermal Engineering.* 108, pp 84–92, 2016.
51. Ambriz-Díaz, V.M. et al., Analysis of a sequential production of electricity, ice and drying of agricultural products by cascading geothermal energy. *International Journal of Hydrogen Energy.* 42(28), pp 18092–18102, 2017.
52. Luo, X. et al., Grey relational analysis of an integrated cascade utilization system of geothermal water. *International Journal of Green Energy.* 13(1), pp 14–27, 2016.
53. Lund, J.W. and M. Rangel. Pilot fruit drier for the Los Azufres geothermal field, Mexico. in *Processing of The World Geothermal Congress,* Florence, Italy. 1995.
54. Kinyanjui, S. Cascaded use of geothermal energy: Eburru case study. *Geothermal Policy in the US,* pp 21, 2012.
55. Keey, R.B., *Drying: Principles and Practice.* Vol. 13. Elsevier, 2013.
56. Kim, D., K. Lee, and K.Y. Park, Hydrothermal carbonization of anaerobically digested sludge for solid fuel production and energy recovery. *Fuel.* 130, pp 120–125, 2014.
57. Calise, F. et al., A geothermal energy system for wastewater sludge drying and electricity production in a small island. *Energy.* 163, pp 130–143, 2018.
58. Alves, L.M.M., A.L. Costa, and M. da Graça Carvalho, Analysis of potential for market penetration of renewable energy technologies in peripheral islands. *Renewable Energy.* 19(1–2), pp 311–317, 2000.
59. Maria, E. and T. Tsoutsos, The sustainable development management of RES installations. Legal aspects of the environmental impact in small Greek island systems. *Energy Conversion and Management.* 45(5), pp 631–638, 2004.
60. Carlino, S. et al., The geothermal system of Ischia Island (southern Italy): Critical review and sustainability analysis of geothermal resource for electricity generation. *Renewable Energy.* 62, pp 177–196, 2014.
61. Paoletti, V. et al., A tool for evaluating geothermal power exploitability and its application to Ischia, Southern Italy. *Applied Energy.* 139, pp 303–312, 2015.

62. Calise, F. et al., A novel solar-geothermal trigeneration system integrating water desalination: Design, dynamic simulation and economic assessment. *Energy.* 115, pp 1533–1547, 2016.
63. Karimi, S. and S. Mansouri, A comparative profitability study of geothermal electricity production in developed and developing countries: Exergoeconomic analysis and optimization of different ORC configurations. *Renewable Energy.* 115, pp 600–619, 2018.
64. Rubio-Maya, C. et al., Cascade utilization of low and medium enthalpy geothermal resources – A review. *Renewable and Sustainable Energy Reviews.* 52, pp 689–716, 2015.
65. Calise, F. et al., Thermoeconomic analysis and off-design performance of an organic Rankine cycle powered by medium-temperature heat sources. *Solar Energy.* 103, pp 595–609, 2014.
66. Fiaschi, D. et al., Exergoeconomic analysis and comparison between ORC and Kalina cycles to exploit low and medium-high temperature heat from two different geothermal sites. *Energy Conversion and Management.* 154, pp 503–516, 2017.
67. Bennamoun, L., P. Arlabosse, and A. Léonard, Review on fundamental aspect of application of drying process to wastewater sludge. *Renewable and Sustainable Energy Reviews.* 28, pp 29–43, 2013.
68. Aspen PLUS V8.8 [Computer Software], A.T., Inc. (2017).
69. de Andrés, J. M., M. Vedrenne, M. Brambilla, and E. Rodríguez. Modeling and model performance evaluation of sewage sludge gasification in fluidized-bed gasifiers using Aspen Plus. *Journal of the Air & Waste Management Association*, 69(1), pp 23–33, 2019.
70. Raibhole, V.N. and S. Sapali, Simulation and Parametric Analysis of Cryogenic Oxygen Plant for Biomass Gasification. *Mechanical Engineering Research.* 2(2), p 97, 2012.
71. Huang, Y. and M. Chen, Thin-layer isothermal drying kinetics of municipal sewage sludge based on two falling rate stages during hot-air-forced convection. *Journal of Thermal Analysis and Calorimetry.* 129(1), pp 567–575, 2017.
72. AspenTech, AspenPlus V8.8. https://www.aspentech.com/en/products/engineering/aspen-plus. Last accessed: 29th November 2018. 2018.
73. Manara, P. and A. Zabaniotou, Towards sewage sludge based biofuels via thermochemical conversion–a review. *Renewable and Sustainable Energy Reviews.* 16(5), pp 2566–2582, 2012.
74. Campanelli, M., P. Foladori, and M. Vaccari, *Consumi elettrici ed efficienza energetica del trattamento delle acque reflue.* Maggioli editore, 2013.
75. TERNA, http://www.terna.it/. Last accessed: 29th November 2018.
76. ISPRA, Istituto Superiore per la Protezione e la Ricerca Ambientale. http://www.isprambiente.gov.it/it. Last accessed: 29th November 2018.

12 Biomass-Based Low-Capacity Gas Generator as a Fuel Source

Tomasz Chmielniak, Aleksander Sobolewski, Joanna Bigda and Tomasz Iluk

CONTENTS

12.1 Introduction 271
12.2 EKOD Biomass Gasification Reactor 272
12.3 GazEla Gasifier 275
    12.3.1 Pilot Gasification Plant with GazEla Fixed Bed Reactor 275
    12.3.2 Demonstration Plant for Biomass Gasification with a GazEla Fixed Bed Reactor 280
12.4 Summary 281
References 282

12.1 INTRODUCTION

One of the basic methods to reduce anthropogenic CO_2 emissions is increasing the share of renewable energy sources in the balance of fuels used in energy and chemical industry among others, which can be implemented by thermochemical conversion of biomass. The use of biomass for energy purposes can be achieved by using three basic technologies: combustion, gasification, and pyrolysis. The products of biomass conversion processes are heat, electricity, chemicals, and fuels [1,2].

At present, gasification technology that will be decisive in the medium and long term seems to be of particular interest for biomass conversion.

An important advantage of the gasification process is, among others, multidirectional use of process gas, which is the main product of the process for the production of heat, electricity, or ecologically clean fuels such as methanol and hydrogen.

At present, gasification installations cooperating with power boilers in heat production systems have reached market maturity. Currently, works are being carried out on the development of small-scale, distributed systems integrated with cogeneration units for the production of electricity and heat. These systems are characterized by high efficiency often exceeding 90%, low emissions of harmful substances, and the possibility of using low-quality waste fuels. In Europe, several dozen systems of this type have been created at various levels of technological readiness [3]. Their further development requires improvement of gas cleaning systems and increase of operational reliability and, as a consequence, an increase of annual availability, which is of key importance for increasing production economics. Dissemination of this type of solution can significantly increase the share of biomass in the production of electricity and heat in Europe and the world.

Examples of such solutions in Poland are EKOD [4] gasification systems and its improved QM12 construction, as well as the GazEla reactor developed at the Institute for Chemical Processing of Coal (Zabrze, Poland), which is dedicated to small and medium cogeneration systems [5].

12.2 EKOD BIOMASS GASIFICATION REACTOR

The EKOD biomass gasification reactor is a construction operating in a countercurrent system. Currently operated devices cooperate with water gas boilers in heat production systems with a capacity of 2.5–5 MW, and modified constructions are also offered in cogeneration systems [6]. Reactors are designed for gasification of biomass (wood waste) and organic waste.

The general view of the EKOD reactor and the outline of the gasification system integrated with the water boiler is shown in Figure 12.1.

Lump fuel of larger sizes or irregular shapes, stored in containers, is supplied to the reactor by means of a fuel feeding system consisting of a transport and feeding

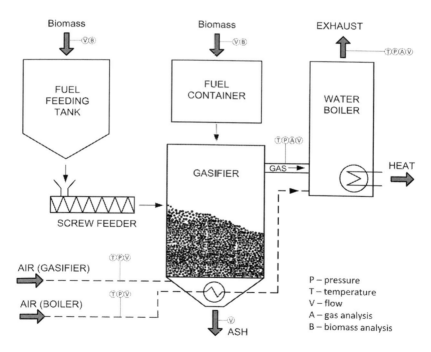

FIGURE 12.1 Outline of the biomass gasification installation (EKOD type) integrated with a gas boiler. (Based on Sobolewski, A. et al. *Zgazowanie biomasy. Nowoczesne technologie pozyskiwania i energetycznego wykorzystania biomasy.* Monografia. Warszawa 2010, 280–309.)

unit and a feeding lock. The task of the lock is to prevent the uncontrolled release of process gas from the reactor to the outside of the device while loading the fuel. In the case of powering the generator with shredded fuel (e.g., sawdust or woodchips), the feeding system is a container and a screw feeder dispensing fuel directly to the reactor. The gasifying agent is fed circumferentially.

The gasification process is carried out at a temperature of 600–800°C and atmospheric pressure. The residence time of raw fuel in the reactor chamber was set at the level of 2–3 h. The gasification process is controlled by introducing an appropriate amount of gasifying agent. Adjustment of the amount of air supplied to the system takes place by automatically maintaining a constant, low value of negative pressure in the upper part of the reactor.

Reactor tests were carried out in a system integrated with a water-steam boiler (heat production). During the gasification tests, waste wood was used as the fuel in the form of lumps with a maximum grain size of 300–400 mm. Gasification tests were also carried out with the use of other raw materials, that is, woodchips, fiberboards, particle boards, straw, tire, and wood mixtures [4,7]. The characteristics of the fuels used are presented in Table 12.1.

The scheme of the research installation together with the approximate location of the points and the scope of the conducted measurement works are presented in Figure 12.1. The scope of the measurement work included analysis and measurement of the fuel stream, measurements of temperature, pressure and the flow rates of both air for gasification and generated gas, as well as the determination of gas composition (gaseous components, content of dust and tar).

The fuel stream was measured by determining the weight of the wood feed at specific time intervals. The measurements included determination of the average weight of one fuel container and strict record of feeding cycles during

TABLE 12.1
Fuel Characteristics (as Received)

Parameter	Unit	Waste Wood	Woodchips	Wood Boards	Tires/Waste Wood
Form	–	Lumps	Lumps	Lumps	Lumps
Size	mm	<300	30–50	<300	<300
Total moisture, W_t^r	%	7.5	15	15	8
Ash, A^r	%	0.4	0.4	0.5	2.7
Volatiles, V^{daf}	%	79.6	71.9	69.0	68.2
C_t^r	%	48.7	44.1	42.9	63.1
H_t^r	%	5.9	5.2	5.0	4.8
N^r	%	0.1	0.05	0.8	0.1
Calorific value, Q_i^r	MJ/kg	17.6	16.1	15.6	25.5

Source: Chmielniak, T. et al. *Przemysł Chemiczny* 2006, 85, 8–9, 1247–1251; Sobolewski, A. et al. *Zgazowanie biomasy. Nowoczesne technologie pozyskiwania i energetycznego wykorzystania biomasy*. Monografia. Warszawa 2010, 280–309.

the measurements (about 10 h). Fuel analysis included proximate and ultimate analysis. The measurement of the gas flow rate was carried out by specifying the temperature, pressure, and gas velocity distribution in the channel (the EMIO Digital Micromanometer (E.M.I.O. Przedsiębiorstwo Innowacyjno-Wdrożeniowe Sp. z o.o., Poland). To determine the gas composition, gaseous samples were collected in tedlar bags. Determinations of gaseous components were carried out in accordance with the Polish standard PN-93/C-96012 using a Siemens gas chromatograph (Sichromat 2-8) equipped with FID (Flame Ionization Detector) and TCD (Thermal Conductivity Detector) detectors. The determination of organic impurities (tar substances) was carried out according to our own Institute for Chemical Processing of Coal (IChPW) procedure Q/ZF/P/15/04/B. The method is based on the weight of a solvent extract of tar substances in the gas in the form adsorbed on the surface of the dust and in the gas phase. Solvent extraction was performed in Tecator extractor, model Avanti (Foss Tecator, Sweden). Dust concentration measurements were carried out using a standard dust gas measurement kit, ZAM Kęty (Zakład Urządzeń Przemysłowych, ZAM Kęty sp. z o.o, Poland). The measurements were carried out following the recommendations of the Polish standard PN-Z-04030-7 regarding the measurement and concentration of dust in waste gases using the gravimetric method.

During the measurement, the generator gas was sampled repeatedly for composition analysis (gaseous components). Each sampling was accompanied by measurement of temperature, pressure, and flow rate of gas and air for gasification. Gas sampling was carried out between successive feeds.

The results of the measurements carried out are presented in Tables 12.2 and 12.3. During testing, the fuel stream to the reactor, depending on the fuel used, fluctuated within 360–680 kg/h. In all cases, the ratio of the stream fed to the air reactor (gasifying agent) to the fuel stream (m_a/m_f) was within the range of 1.8–2.1. The calorific values of the gas generated were in the range of 4770–9250 kJ/m_n^3 (Table 12.2). The highest calorific value was obtained for a mixture of waste wood and tires. In the case of biomass gasification, that is, woodchips and waste wood, the calorific

TABLE 12.2
Parameters of the Biomass Gasification Process—EKOD Gasification Reactor

Fuel	t	Fuel Stream	m_a/m_f	Dry gas Stream	W_d (dry gas)		Process Efficiency
	°C	kg/h	kg/kg	m_n^3/kg_{pal}	kJ/kg	kJ/m_n^3	%[a]
Waste wood	760	490	2.1	2.5	4700	5660	80
Woodchips	685	580	1.9	2.3	4220	5200	75
Wood boards	685	680	1.8	2.2	3840	4770	68
Tires/waste wood	690	360	2.1	2.4	7400	9250	86

Source: Chmielniak, T. et al. *Przemysł Chemiczny* 2006, 85, 8–9, 1247–1251.

Abbreviations: t, temperature at the reactor outlet; m_a/m_f, air to fuel ratio; W_d, calorific value.

[a] Efficiency defined as the ratio of the chemical enthalpy of the gas produced to the chemical enthalpy of the fuel supplied ("cold gas efficiency").

TABLE 12.3
Characteristics of Product Gas

Fuel	\multicolumn{7}{c}{Gas Compounds, % v/v (Dry Condition)}	Dust, mg/m_n^3	Tar Compounds, mg/m_n^3
	H_2	$N_2 + O_2$	CO	CH_4	CO_2	C_2H_4	Others[a]		
Waste wood	7.4	59.3	18.9	4.4	8.6	1.1	0.3	1055	643
Woodchips	6.8	59.1	17.3	3.7	11.8	0.9	0.4	350	406
Wood boards	6.2	60.9	16.7	2.8	12.0	1.0	0.4	1970	2934
Tires/waste wood	3.0	58.0	20.0	3.5	8.0	6.5	1.0	2870	1210

Source: Chmielniak, T. et al. *Przemysł Chemiczny* 2006, 85, 8–9, 1247–1251.

[a] C_2H_6, C_3H_6, C_3H_8, C_4H_{10}, C_5H_{12}, C_6H_6

value of the obtained gas amounted to 5200 and 5660 kJ/m_n^3, which corresponds to the typical results obtained in the case of biomass air gasification in fixed bed [8–12]. The calorific value of the generated gas increased with the increase of the calorific value of the fuel.

The gasification process in the tested device proceeds with relatively high efficiency, in the range of 68%–86% (so-called "cold gas efficiency," Table 12.2). The obtained results are higher than those typical for countercurrent reactors and comparable with cocurrent structures [10].

The composition of the produced gas, depending on the fuel used, is presented in Table 12.3. In the case of gaseous components and dust, the obtained values are similar to the literature data [8–12]. Noteworthy is the very low tar content in the generated gas, as for this type of construction solutions, in the range of 400–3000 mg/m_n^3 (literature data: 10,000–150,000 mg/m_n^3). During biomass gasification, the tar content corresponds to the results obtained in the case of cocurrent reactors [8–12].

12.3 GazEla GASIFIER

The GazEla fixed bed biomass gasification reactor was developed at the Institute for Chemical Processing of Coal in Zabrze (Poland) [13]. Cooperation with Syngaz S.A., which is an industrial partner of IChPW, in the development of this technology has resulted in the construction and testing of two biomass gasification systems for production in cogeneration of electricity and heat on both pilot (60 kWt) and demonstration scale (1.5 MWt).

The GazEla gasifier with the fixed bed was developed for the production of process gas from biomass. Owing to the type of the bed, it is dedicated to small and medium power systems based on gas piston engines.

12.3.1 PILOT GASIFICATION PLANT WITH GAZELA FIXED BED REACTOR

The pilot installation is located in the IChPW. Its main element is reactor with a capacity of approximately 60 kW$_t$. It is a vertical cylindrical generator with an internal diameter of 400 mm and a technological height of approximately 900 mm. In the axis of the device, there is a riser pipe with an adjustable position, aimed at collecting the process gas directly from the gasification zone. Appropriate selection of the technological height H_r (distance between the air distributor and the inlet to the riser pipe, Figure 12.2) allows obtaining gas with a lower amount of tar impurities and of higher temperature [14,15].

Fuel is supplied to the top of the reactor. The gasifying agent is air supplied by the fan in three points of the reactor: under the grate, in the middle part and above the fuel bed. The reactor works at low overpressure. In the drying and pyrolysis zone, fuel and air move cocurrently, in the gasification zone there is mixed current, while the combustion process takes place in a countercurrent. The reactor capacity and the load of individual process zones are regulated by the amount of air supplied to individual air nozzles of the device. Figure 12.3 shows pictures of the pilot scale reactor.

This device combines the advantages of both a countercurrent and cocurrent reactor and allows for the possibility of using fuel with different granulation and increased humidity. The outlet of the process gas directly from the gasification zone, that is, to a place with high temperatures causing tar compounds decomposition, allows for the opportunity of reducing a significant amount of tar.

During the technological tests of the reactor, lump and fibrous woodchips, as well as pellets made of wood waste and wheat straw, were used as fuel. Table 12.4 presents an analysis of the properties of the biomass used.

The results of reactor tests for selected types of biomass are presented in Table 12.5. They confirm the possibility of using different biomass raw materials in the GazEla

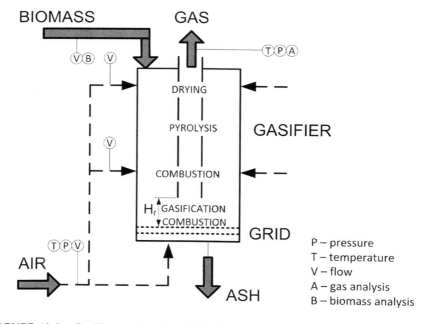

FIGURE 12.2 GazEla gasifier—the principle of operation. (Based on Kotowicz, J. et al. *Energy* 2013, 52, 265.)

Biomass-Based Low-Capacity Gas Generator as a Fuel Source

FIGURE 12.3 GazEla gasifier—general view of the bottom and upper part of the gasifier.

reactor. The calorific value of the (dry) gas obtained ranged from 3.82 to 4.72 MJ/m_n^3. For woodchips, the dry gas composition was in the following range: $H_2 = 6.1–7.5\%$, $CH_4 = 2.1–2.5\%$, $CO = 19.5–25.0\%$, $CO_2 = 9.5–11.8\%$, and $N_2 = 55.9–60.1\%$. In the case of pellets, the main gas components were, respectively: $H_2 = 6.7–8.9\%$, $CH_4 = 0.9–2.6\%$, $CO = 20.3–22.0\%$, $CO_2 = 12.8–12.9\%$, and $N_2 = 55.4–57.5\%$. Variations in the main components of the gas are small, and the concentration changes of individual components are stable (Figure 12.4).

The basic gas composition was measured using a mobile SIEMENS Ultramat 23 gas analyzer station and SIEMENS Oxymat (Siemens, Germany). The extended composition of the process gas was determined in IChPW laboratories by chromatographic analysis of gas samples collected in tedlar according to IChPW procedures.

TABLE 12.4
Characteristics of the Fuel

Parameter	Unit	Woodchips I	Woodchips II	Wood Pellets	Straw Pellets
Form	–	Lumps	Fibrous	Peletts	Peletts
Size	mm	<8	20–50	Ø6	Ø6
Total moisture, W_t^r	%	21.4	14.6	4.4	9.3
Ash, A^a	%	1.3	0.5	0.3	5.5
Volatiles, V^{daf}	%	81.11	83.48	83.61	81.36
C_t^a	%	49.4	48.5	49.6	45.0
H_t^a	%	5.56	5.75	5.88	5.47
N^a	%	<0.05	0.19	0.27	0.53
Calorific value, Q_i^r	MJ/kg	14.2	15.7	17.8	15.4

Source: Quaak, P. et al. World Bank Technical Paper No. 422, Energy Series, 1999.

TABLE 12.5
Change in the Composition of the Process Gas Depending on the Type of Biomass Used

Gas Component	Unit	Woodchips I	Woodchips II	Wood Pellets I	Straw Pellets II
H_2	% v/v	6.1	7.5	6.7	8.9
CH_4		2.5	2.1	0.9	2.6
CO		19.5	25.0	22.0	20.3
CO_2		11.8	9.5	12.9	12.8
O_2		0.0	0.0	0.0	0.0
N_2		60.1	55.9	57.5	55.4
W_d	MJ/m^3_n	4.01	4.72	3.82	4.45

The sampling of the process gas pollutants for laboratory tests was carried out using a set consisting of a dust separator (filter), allowing separation of solid particles and ensuring sampling temperature >150°C and system of scrubbers containing isopropanol for extraction of compounds present in the liquid phase under standard conditions. The first scrubber was kept at ambient temperature and its task was to collect tar and water. The second scrubber was immersed in a bath at −15°C and was responsible for the thorough purification of gas from light hydrocarbons. At the outlet of the system, a filter was placed which stopped the condensation fog. The aspirator enabled constant values of the flow of collected gases, along with the measurement of

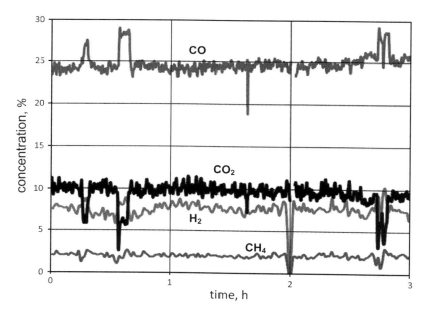

FIGURE 12.4 Changes in the main fuel components during the gasification process. (Based on Kotowicz, J. et al. *Energy* 2013, 52, 265.)

their volume. The pollutants obtained this way were subjected to laboratory analyses, including chromatography.

The nominal fuel stream for the tested reactor was determined at the level of approximately 14–15 kg/h. The experimental generator has a rated power of approximately 62 kW$_t$.

It should also be emphasized that for some types of biomass, it is necessary to modify the method of conducting the gasification process. This applies especially to the demanding raw material, which is straw and sewage sludge, for which it is necessary to lower the temperatures in the lower part of the reactor to eliminate sintered ash.

The process gas (raw) obtained from the reactor is characterized by moderate content of tar and dust. The average content of dust pollutants for gasification of woodchips was about 600 mg/m$_n^3$, while organic pollutants were at the level of 1200 mg/m$_n^3$.

An important feature of the reactor is also the stability of its operation expressed in a stable gas composition over a long period of time (Figure 12.4).

The process gas was obtained from the GazEla gasifier, with a temperature of approximately 550–650°C and is subjected to the cleaning system for preparation for combustion in the piston engine. The gas wet cleaning system presented in Figure 12.5 is the result of the IChPW long-term work on dry and wet methods of process gas purification. The basic tasks of the gas cleaning system include process gas dedusting, removal of organic substances (in particular from tars), and drying and removal of acidic and alkaline substances. In the system proposed by the IChPW the above objectives are achieved through a high temperature filter, an oil scrubber, and a gas cooler with a dedicated demisting system. During the tests, the possibility of removing dust from process gas to the value of 5 mg/m$_n^3$ (at the outlet after the high temperature filter) was confirmed, while organic substances were reduced to the value of approximately 100–150 mg/m$_n^3$ [16,17].

As a part of the research work on the biomass gasification installation with the use of GazEla gas generator, combustion tests of the produced process gas in a dual-fuel piston engine were also carried out in order to confirm the possibility of efficient use of process gas to supply piston engines (Figure 12.6).

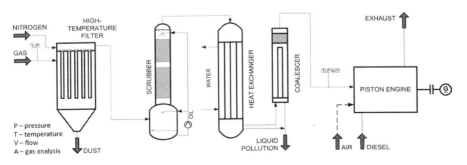

FIGURE 12.5 Process gas wet cleaning system. (Based on Iluk, T. et al. *Przemysł Chemiczny* 2015, 94, 464–468.)

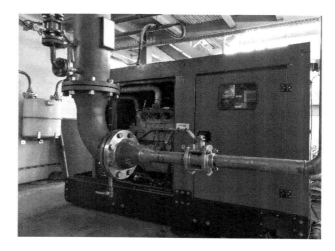

FIGURE 12.6 Work of a piston engine on diesel oil and process gas for a power of 15.5 kW$_{el}$.

12.3.2 Demonstration Plant for Biomass Gasification with a GazEla Fixed Bed Reactor

The next stage in the development of biomass gasification technology with the GazEla gas generator was to increase the scale of the plant. The obtained results from the pilot plant allowed to design a biomass gasification plant at a scale of 1.5 MW$_t$. The demonstration gasification plant was located in a wood processing plant (Opole voivodeship, Poland), which produces pellets and briquettes from sawdust and straw for energy purposes, among others. An important aspect when choosing a location was that the plant has a high demand for electricity and heat, which is needed for technological purposes (i.e., drying sawdust and straw). Figure 12.7 shows a view of the demonstration plant.

The investor at the plant is Syngaz S.A. from Swietochlowice, which is a partner of IChPW in the development of biomass gasification technology based on the GazEla gas generator.

The demonstration plant consists of four main technology blocks. The first block is the biomass feeding system for the gas generator. The second one is a three-zone fixed bed gasifier, the construction of which has been developed at the Institute for Chemical Processing of Coal.

Another technological system associated with further use of the process gas in the piston engine is the purification system. As a concept, a dry gas cleaning system was chosen, which is characterized by the absence of additional pollution and low operating costs.

Basic parameters of the demonstration gasification plant include:

- Fuel stream: 450 kg/h,
- Gas stream: 1200 kg/h,
- Gas calorific value: 4.5–5 MJ/m$_n^3$,
- Gas generator efficiency (cold): 60–65%.

FIGURE 12.7 Demonstration plant for biomass gasification of 1.5 MW$_t$.

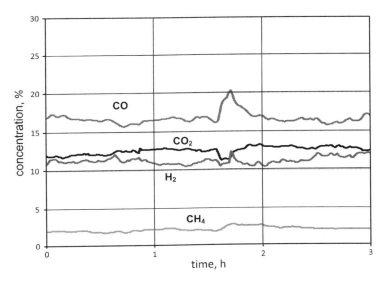

FIGURE 12.8 Changes in the main fuel components during the gasification process. (Based on Sobolewski, A. et al. *Rynek Energii* 2013, 106, 3.)

The products are about 350 kW of electricity, as well as about 800 kW of heat. Figure 12.8 shows the timeline of concentration changes of the main components of the process gas [18].

12.4 SUMMARY

Biomass is widely regarded as a significant potential fuel and energy source. According to the analyses of the International Energy Agency (IEA), the share of biomass in electricity generation will increase as the sources of oil and natural gas become depleted. Among the methods of thermochemical conversion of biomass, the largest developmental potential has gasification processes, which is confirmed by research and demonstration projects supported by, among others, the IEA and the European Commission (Energy and Transport Directorate). The gasification technologies in heat production systems have reached market maturity and their further development is related to electricity generation systems, chemical synthesis, or the production of clean fuels, including hydrogen. An important direction in the development of biomass gasification technology is also micro-cogeneration systems.

The technological solutions presented in this chapter enable the effective use of biomass for the production of heat, as well as electricity and heat in cogeneration units.

REFERENCES

1. Chmielniak, T. *Technologie energetyczne.* Gliwice. Politechnika Śląska; 2004.
2. Al-Mansour, F., Zuwala, J. An evaluation of biomass co-firing in Europe. *Biomass and Bioenergy* 2010, 34, 620–9.
3. http://www.ieatask33.org/
4. Chmielniak, T., Ściążko, M., Zawistowski, J., Dudyński, M. Badania technologii zgazowania biomasy w złożu stałym w skali 3,5 MW. *Przemysł Chemiczny* 2006, 85, 8–9, 1247–1251.
5. Kotowicz, J., Iluk, T.: Instalacja zgazowania biomasy zintegrowana z silnikiem spalinowym. *Rynek Energii* 2011, 3, 94, 47–52.
6. http://qenergy.pl/
7. Sobolewski, A., Ilmurzyńska, J., Iluk, T., Czaplicki, A. *Zgazowanie biomasy. Nowoczesne technologie pozyskiwania i energetycznego wykorzystania biomasy.* Monografia. Warszawa 2010, 280–309.
8. Bridgwater, A.V. The technical and economic feasibility of biomass gasification for power generation. *Fuel* 1995, 74, 5, 631.
9. Quaak, P., Knoef, H., Stassen, H. World Bank Technical Paper No. 422, Energy Series, 1999.
10. Stassen, H.E.M., Knoef, H.A.M. *Small Scale Gasification Systems.* The Netherlands. Biomass Technology Group BV. (http://www.gasnet.uk.net)
11. Hasler, P., Nussbaumer, T. Gas cleaning for IC engine applications from fixed bed biomass gasification. *Biomass and Bioenergy* 1999, 16, 385–395
12. Neeft, J.P.A., Knoef, H.A.M., Zielke, U., Sjöström, K., Hasler, P., Simell, P.A., Dorrington, M.A. et al. Guideline for Sampling and Analysis of Tar and Particles in Biomass Producer Gases. Energy project ERK6-CT1999–2002, (Tar protocol).
13. Billig, P., Ściążko, M., Sobolewski, A. Pat. pol. PL-208616, 2010.
14. Kotowicz, J., Sobolewski, A., Iluk, T., Matuszek, K. Zgazowanie biomasy w reaktorze ze złożem stałym. *Rynek Energii* 2009, 2, 81, 52–58.
15. Kotowicz, J., Sobolewski, A., Iluk, T. Energetic analysis of a system integrated with biomass gasification. *Energy* 2013, 52, 265.
16. Iluk, T., Sobolewski, A., Stelmach, S. Oczyszczanie gazu procesowego pochodzącego ze zgazowania biomasy pod kątem możliwości wykorzystania w silnikach tłokowych. *Przemysł Chemiczny* 2015, 94, 464–468.
17. Iluk, T., Sobolewski, A., Szul, M. Zgazowanie osadów ściekowych, SRF oraz biomasy w generatorze gazu ze złożem stałym – porównanie wyników prac eksperymentalnych w skali pilotowej. *Przemysł Chemiczny* 2016, 95, 8, 1634–1640.
18. Sobolewski, A., Iluk, T. Doświadczenia eksploatacyjne z rozruchu demonstracyjnej instalacji zgazowania biomasy o mocy 1,5 MWT. *Rynek Energii* 2013, 106, 3.

13 Production of Generator Gas from Biomass and Fuels from Waste on a Small Scale

Danuta Król and Sławomir Poskrobko

CONTENTS

13.1 Introduction 284
13.2 Production of Generator Gas with Enhanced Methane Share:
 Theoretical Bases..286
 13.2.1 Methane Formation Mechanism..286
13.3 Examples of the Application..290
 13.3.1 Gasification of Waste Wood Biomass..290
 13.3.1.1 Results and Discussion...291
 13.3.2 Gasification Process of Olive Pits..294
 13.3.2.1 Results and Discussion...295
 13.3.3 Gasification of Wood Biomass with the Addition of
 Post-Extraction Rapeseed Meal..297
 13.3.3.1 Results and Discussion...300
 13.3.4 Gasification of RDF and Bio-CONOx..304
13.4 Micro Cogeneration: Rich-Methane Gasifier and Micro Gas Turbine.........309
 13.4.1 Modeling of a Syngas Microturbine-Based Cogeneration Plant......309
 13.4.2 Gas Microturbine Model..309
 13.4.2.1 Air Compressor Model 310
 13.4.2.2 Combustion Chamber Model...310
 13.4.2.3 Gas Turbine Model 311
 13.4.2.4 Gas Microturbine Model Flue Gas Heat Exchanger..........311
 13.4.2.5 Thermodynamic Performance 311
 13.4.3 Results and Discussion 311
13.5 Summary 313
References..314

13.1 INTRODUCTION

Biomass fuels, waste (biomass waste), and fuels formed from waste containing biomass are thermochemically converted to generate energy, or (as in the case of waste and fuels from waste), to their disposal with simultaneous recovery of the chemical energy stored in them.

Biomass and fuels formed from waste belong to the so-called "difficult fuels." "Difficult fuels," owing to the specificity of their physicochemical properties (e.g., high humidity, heavy load of alkali metals, heavy metals, sulfur, chlorine), pose particular problems in the processes of their thermal treatment. The organization of these processes is associated with many difficulties and the need to solve many emissions, corrosion, and technical problems. The degree of difficulty during combustion is determined by the so-called "fuel ratio," defined as the ratio of coke residue to volatile matter and with values up to 10. For biomass, it is within the range of ~ 0.8 to $\ll 0.1$, and for solid fuels, it can reach values as high as 10.

Combustion is the main process used for their thermal transformation. An alternative to combustion is the gasification process, which enables processing of biomass fuels or waste into a combustible generator gas (often called syngas), which is widely used. This gas can be burned directly in the boiler or, after cleaning, burned in the engine, for example, a Stirling engine or gas turbine. In power boilers, it can be used as a reburning gas to reduce NOx. The gas can also be a source of substrates for the synthesis of second-generation fuels, for example, dimethyl ether (DME).

Gasification of fuel (in this case biomass or solid waste) is the transformation of a combustible substance into a gas fuel, that occurs as a result of the fuel being treated with a gasifying agent at high temperature, at atmospheric pressure or at elevated pressure (Figure 13.1). The gasifying agents used are air (21% O_2 + 78% N_2), oxygen (O_2), steam (H_2O), carbon dioxide (CO_2), hydrogen (H_2), or their mixtures. Biomass fuels or waste

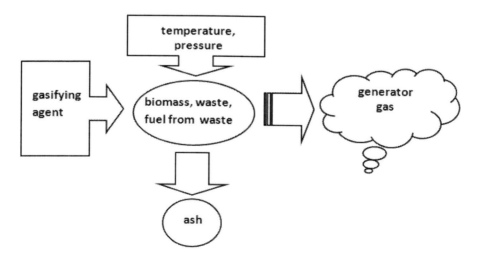

FIGURE 13.1 Model of the gasifying process.

Production of Generator Gas from Biomass

containing significant moisture content do not requiresteam provided into the gasifier (sometimes it is even necessary to dry them) at the temperature in the gasifier chamber; the moisture in the fuel becomes steam, therefore, being a reactive agent.

Partial oxidation in the gasification process converts fuel—biomass, biomass waste, and fuels from waste containing biomass, into gas consisting mainly of CO, CO_2, H_2, CH_4, H_2O, and other C_nH_m hydrocarbons. In the case of gasifying the fuel with air, nitrogen (N_2) is also present in the generator gas.

The gasification process includes three main stages:

I. Moisture loss

   fuel (biomass) $_{wet}$ → fuel (biomass) $_{dry}$ + H_2O

II. Pyrolysis—dry fuel thermal decomposition under anaerobic conditions, degassing of fuel, resulting in gas and tar mixtures when coke is formed. The mutual proportions of products depend on the type of fuel, temperature and heating rate.

   fuel (biomass)$_{dry}$ → coke + tar + wH_2O + cCO + bCO_2 + hH_2 + aCH_4

III. Gasification—thermal decomposition products react with the gasifying agent, combustible products of CO, H_2, CH_4, C_nH_m, and CO_2 are formed.

   The basic gasification process is described in a simplified way by the chemical equations:

Oxidation reactions

$C(fuel) + O_2 = CO_2$ (the exothermic reaction) − 406 kJ/mol
$C + 1/2\ O_2 = CO$ (the exothermic reaction) − 123 kJ/mol

Secondary reactions

$CO + 1/2\ O_2 = CO_2$ (the exothermic reaction) − 283 kJ/mol
$C_nH_m + n/2\ O_2 = nCO + m/2\ H_2$ (the exothermic reaction)

Boudouard reaction

$C + CO_2 = 2CO$ (the endothermic reaction) + 172.6 kJ/mol

Reactions with steam

$C + H_2O(steam) = CO + H_2$ (the endothermic reaction) + 131.4 kJ/mol
$C + 2H_2O(steam) = CO_2 + 2H_2$ (the endothermic reaction) + 87.2 kJ/mol
$C_nH_m + nH_2O = nCO + (n + m/2)\ H_2$ (the endothermic reaction)

Methanation

$C + 2H_2 = CH_4$ (the exothermic reaction) − 74.9 kJ/mol
$2C + 2H_2O = CH_4 + CO_2$ (the exothermic reaction) − 8.79 kJ/mol
$CO + 3H_2 = CH_4 + H_2O$ (the exothermic reaction) − 205 kJ/mol
$CO_2 + 4H_2 = CH_4 + 2H_2O$ (the exothermic reaction) − 167.3 kJ/mol
$2CO + 2H_2 = CH_4 + CO_2$ (the exothermic reaction) − 247 kJ/mol

Gas conversion, steam reforming

$CO + H_2O = CO_2 + H_2$ (the exothermic reaction) − 41.2 kJ/mol
$CH_4 + H_2O = CO + 3H_2$ (the endothermic reaction) + 206.4 kJ/mol
$CH_4 + CO_2 = 2CO + 2H_2O$ (the endothermic reaction) + 248.4 kJ/mol
$CH_4 + 1/2\ O_2 = CO + 2H_2$ (the endothermic reaction) + 35.5 kJ/mol

Thermal cracking reactions

$p\ C_xH_y = q\ C_nH_m + r\ H_2$ (the endothermic reaction)
$C_nH_m = n\ C + m/2\ H_2$ (the endothermic reaction)

The course of these reactions has an obvious effect on the gasification process, but the factors such as fuel (biomass) properties, its fragmentation, fuel contact time (biomass) with a gasifying agent, heat and mass exchange, gas flow rate, and pressure cannot be neglected. The greater the reactivity of the fuel (biomass), the faster the gasification reactions. The resulting gas contains more CO and less CO_2. The high reactivity of fuel to CO_2 has a positive effect on the ability to reduce CO_2 to CO. The high content of volatile parts in fuel (biomass) is beneficial due to the increased calorific value of the gas produced; however, it causes problems owing to the need to remove tar from the generator gas if it is not used in a hot state. Higher pressure values lead to higher equilibrium concentrations of CO and H_2. Higher pressure is beneficial for the methanation process and less favorable for the formation of CO and H_2, for example. This pressure effect is obvious—higher pressure promotes the thermodynamic existence of larger molecules—CH_4 methane contains five atoms, and in total CO, carbon monoxide and H_2, hydrogen, only four atoms.

In the following sections, the results of the gasification process carried out in a counter-current gas tube generator (authors' project) are presented. The process was aimed at obtaining increased CH_4 content in the generator gas, at the expense of reducing H_2, CO, or CO_2 concentrations. Gasified materials were various types of woody biomass (in various physical forms—pellets, sawdust, and bark, and with different moisture contents, including highly wet fuels), olive seeds, waste fuel containing biomass waste, Bio-CONOx [1]—waste biomass of agricultural origin. The gasifying factor in each case was air. The process was always carried out without elevated pressure. The construction of the gasifier and the temperature conditions used enabled the autothermal operation of the process, and the composition of the generator gas and the shares of its individual components directly enabling calculation of its calorific value. The calorific value of the syngas obtained, above 10 MJ/m³, makes it possible to use it for the supply of gas microturbines, for which the minimum value of 10 MJ/m³ is a barrier.

13.2 PRODUCTION OF GENERATOR GAS WITH ENHANCED METHANE SHARE: THEORETICAL BASES

13.2.1 METHANE FORMATION MECHANISM

A characteristic feature of the gasification process in a tubular reactor, where synthesis gas with high methane content is obtained, is the temperature distribution in the

Production of Generator Gas from Biomass

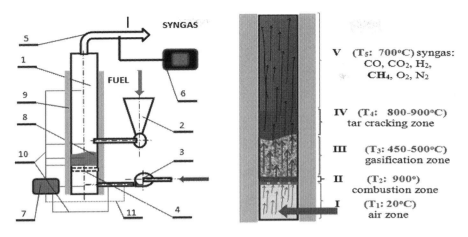

FIGURE 13.2 Scheme of the laboratory gasifier 1.5 kW, experimental setup where: 1—reaction chamber, 2—fuel feeder, 3—air blower and regulation system, 4—grate, 5—syngas outlet, 6—syngas pipeline (syngas sampling point), 7—monitoring of flows and temperatures, 8—biomass fuel, 9—thermal insulation, 10—temperature sensors, 11—air flow measurement m³/h. (From Król D., Poskrobko S.: *Energy & Fuels* 2017, 31, 3935–3942.)

reaction chamber of the gasifier, which fosters the methane creation processes [2]. In the following areas, combustion (II)—$T_2 = 900°C$, gasification (III)—$T_3 = 450$–$500°C$, cracked tars (IV)—$T_4 = 800$–$900°C$ (Figure 13.2), the temperature of the air gasification medium was (I) $T_1 = 20°C$. The laboratory test stand tubular reactor is shown in Figure 13.3.

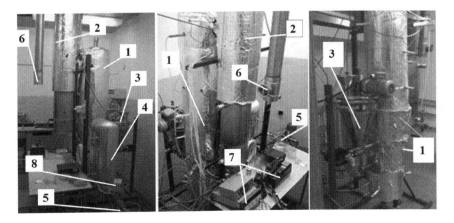

FIGURE 13.3 Laboratory test stand 4–5 kW, tubular gasifier, where: 1—gasifier, 2—combustion chamber and gas channel, 3—fuel tank and dosing system, 4—air heater, 5—air flow channel and fan, 6—control chimney draft, 7—control panel, 8—measurement of airflow. (From Król D. et al.: Micro cogeneration - rich-methane gasifier and micro gas turbine, *E3S Web of Conferences 2017*, Vol. 14, 01024, Energy and Fuels 2016.)

The basis of the methane creation process of thermal decomposition of biomass is to maintain a stable temperature, $T_3 = 450$–$500°C$, in the gasification zone and a temperature of 800–900°C over the gasification zone (over a layer of fuel) permitting thermal cracking of heavy hydrocarbon pairs, that is, tar. Producing and maintaining a high temperature in zone IV is promoted by partial combustion of flammable condensable organic vapors, for example, according to the kinetic model of Brayton [4]:

$$CH_{1.522}O_{0.0228} + 0.867O_2 \rightarrow CO + 0.761H_2O. \tag{13.1}$$

The remaining pairs not burned are cracked according to the reaction:

$$TAR_{(gas)} \rightarrow \vartheta_{H_2}H_{2(gas)} + \vartheta_{CH_4}CH_{4(gas)} + \vartheta_{CO_2}CO_{2(gas)} \\ + \vartheta_{CO}CO_{(gas)} + \vartheta_{tar}tar_{inert}, \tag{13.2}$$

where: ϑ_i is the stoichiometric coefficient.

In temperature conditions of the degassing zone (III), in the gas phase in the fuel layer there occurs, apart from conventional reactions typical for generator processes, exothermic methane forming Fischer–Tropsch and Sabatier–Senderens reactions [5]:

$$CO + 3H_2 \leftrightarrow CH_4 + H_2O \quad \Delta H = -206.4 \text{ kJ/mol.} \tag{13.3}$$

The water forming in the reaction (13.3) also reacts with carbon monoxide according to the equation:

$$CO + H_2O \leftrightarrow CO_2 + H_2 \quad \Delta H = -41.2 \text{ kJ/mol.} \tag{13.4}$$

As the concentration of CO in the reaction system is reduced, the reaction equilibrium shifts and the reaction takes place in the opposite direction (from right to left). Carbon monoxide, which is forming, and is subject to methanation by Equation 13.3.

By adding the reaction (13.4) extending from right to left to reaction (13.3), an equation of carbon dioxide methanation reaction (13.5) is obtained [5]:

$$CO_2 + H_2 \leftrightarrow CO + H_2O$$

$$CO + 3H_2 \leftrightarrow CH_4 + H_2O$$

$$CO_2 + 4H_2 \leftrightarrow CH_4 + 2H_2O \quad \Delta H = -164.9 \text{ kJ/mol.} \tag{13.5}$$

Methane is also formed in methanization reactions:

$$2CO + 2H_2 \rightarrow CH_4 + CO_2 \quad \Delta H = -254.1 \text{ kJ,} \tag{13.6}$$

$$CO + 4H_2 \rightarrow CH_4 + 2H_2O. \tag{13.7}$$

Methanation of CO and CO_2 are exothermic reactions and run with reduced volume. Therefore, lowering the temperature and increasing the pressure favors the shift of the equilibrium to the increase of methane in equilibrium gas mixture.

In the presented experiment, the pressure was not increased and the temperature was maintained in the range $T_3 = 450–500°C$.

The effectiveness of the above reactions in zone III (Figure 13.2) (at $T_3 = 450–500°C$), shown in Figure 13.4, indicate dependences making the equilibrium constant K_p dependent of the temperature, which is considered due to the simple way of simplification up the calculations according to the estimate Nernst equation, which provides good accuracy of the calculations in the gas phase:

$$\lg K_p = \frac{Q}{2.3RT} + \frac{14.7}{R}\sum \gamma_j \lg T + (\sum \gamma''_j i - \sum \gamma'_j i) \qquad (13.8)$$

where Q is the thermal effect of reaction at standard conditions; $\sum \gamma = \sum \gamma''_j i - \sum \gamma'_j i$ is the difference between the number of moles of reaction products and numbers of moles of the starting materials; and i is the contractual chemical constant. These are determined on the basis of empirical equations of resilience curve of $\lg p_j$ pairs:

$$i = \lg p_j + \frac{r_j}{2.3RT} - \frac{14.7}{R} \lg T, \qquad (13.9)$$

where r_j is the vaporization heat (J/mol) with pressure $p = 1.013$ bar. The agreed chemical constant for mononuclear gases can take: $i = 1.5$, polynuclear $i = 3$.

Figure 13.4 indicates that the above methanation reactions occur with high intensity at process temperatures in the gasification layer III. The results of the gasification

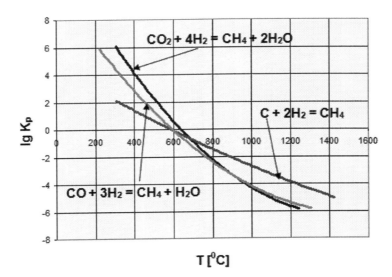

FIGURE 13.4 Logarithmic dependence of the equilibrium constant K_p of methanation reaction to temperature. (From Król D., Poskrobko S.: *Energy & Fuels* 2017, 31, 3935–3942.)

experiment showed that the process was maintained within the specified temperature range, despite increasing the gasification air flow. In such cases, the temperature in the layer of fuel increased up to 500°C. The increase in the temperature was recorded over a layer of fuel to 1000°C with an increase in the share of O_2 in the gaseous products of the process. Considering the methanation process, one cannot overlook the reaction of carbon with hydrogen:

$$C + H_2 \leftrightarrow CH_4 \quad \Delta H = -74.9 \text{ kJ/mol} \qquad (13.10)$$

In this case, the kinetics of the gasification is directed by the correlation of chemical and physical phenomena. The reaction substrate of gasification, that is, fuel (biomass and post-extraction rapeseed meal), after the first step—degassing (which takes place at elevated temperature), changes its form (structure) and becomes a porous solid. Heterogeneous gasification reaction occurs at the interface of gas–solid.

Grains of gasified fuels are surrounded by a thin (film) coating. Through this coating, the molecules of the gaseous reactants must firstly penetrate oxygen and water vapor, and which further diffuse into the pores. On the walls of the pores, they react with carbon. The speed of heterogeneous gasification reaction depends therefore on the rate of chemical reaction and diffusion rate of the gasifying agent into the pores of the fuel. However, in low reactive temperatures, at which the methanation process is carried out, the reaction equilibrium (13.10) is determined slowly.

In the gasification of biomass, in addition to gaseous products and solid residue, condensable substances are formed, mainly tar. In the cracking zone at a temperature of 850–900°C, thermal cracking of fat vapor and condensing substances takes place. Cracking products are CO, CH_4, H_2, and CO_2.

13.3 EXAMPLES OF THE APPLICATION

13.3.1 Gasification of Waste Wood Biomass

The research material was a waste forest biomass in the form of bark, sawdust, and wood pellets: (1) pine sawdust, wood sawdust of deciduous trees (except oak), pellets from coniferous trees, and sawdust came from the forest, from the area of the Podlasie Province; and (2) combined wet sawdust, bark pine, and wood pellets made from mixed sawdust came from TRAK Timber Production Plant in Garbatka Długa, in the Mazowieckie Province [6].

Samples of the research material were collected according to the procedures used for sampling solids. After homogenization (in order to obtain representative samples) fuel analysis was carried out. For this purpose, the following were determined: humidity, flammable and noncombustible parts, heat of combustion, in accordance with PN-ISO standards (PN-Z-15008-02:1993; PN-ISO 1171:2002; PN–ISO 1928:2002), elemental composition of the flammable substance—carbon, hydrogen, nitrogen, and sulfur by means of CHNS elemental analyzer model 2400 series II from Perkin Elmer, and chlorine according to PN-ISO 587/2000. Mixed wet sawdust and coniferous trees bark contain a substantial amount of moisture (42% and 47%, respectively). The combustion of these biofuels causes technical difficulties, and the fuel can be called "difficult" (the explanation of the term "difficult fuels" is presented in the chapter Introduction). In the gasification process, the fuel moisture can be a factor advantageously influencing the process, but more than 40% of water in the fuel significantly decreases its calorific value. The ash content and elemental composition of the combustible substance of the analyzed forest residues (sawdust, wood pellets, and bark) do not differ from typical values for wood biomass. Fuels from waste pulp having the characteristics of fuel, as in Table 13.1, were converted into heat in the gasification process, in order to obtain the combustible gas generator. Gasification was carried out in a laboratory tubular reactor with a fixed bed (Figure 13.3). The maximum heat output of the gasifier was 5 kW. The gasifying agent was air (0.56–0.93 g/s), which was fed to the bed through the sieve grate from below the gasifier. Fuel was supplied to a feeder and then transported to a reactor with a screw conveyor. The air was heated by a system of electric heaters. Self-ignition of the fuel layer was initiated by heating the reactor chamber with air to a temperature of 320°C. Gasification was carried out at 375–435°C in the fuel layer temperature. In contrast, above the bed in the gas phase, the temperature was 870–940°C. At a temperature of 375–435°C, in the fuel layer the autothermal process occurred. Intensive evaporation of high-calorific condensing substances took place, including tar. It is suggested that these high-calorific compounds were burned over the layer of fuel, in the atmosphere (in the lower layer) of unreacted oxygen, which resulted in a temperature increase to a value $T = \sim 940°C$. The lack of high-calorific C_nH_m hydrocarbons in the generator gas seems to be evidence of this. Once the temperature in the reaction chamber was stabilized, the gas was collected from the upper part of the chamber by an aspirator. Once passing through the scrubber system, gas was transported to the analyzer. Measurement of concentrations of gaseous gasification products was made using a synthesis gas analyzer GAS 3000.

The ash was removed during the process by being blown off from the gasification chamber with a string of exhaust fans. The ash was deposited in the bottom part of the secondary combustion chamber. The resulting gas was burned in a secondary combustion chamber and before the exhaust fan, the exhaust gases were cooled to a temperature of about 180°C. For cooling, ambient air was sucked through the flue exhaust duct. The gas generator and the channel blower are equipped with an electric heating system that serves to provide a hot process zone in the gasifier start-up phase. The stabilization process was performed by regulating the flow of fuel and the gasification air flow regulation.

13.3.1.1 Results and Discussion

The results of wood gasification varied depending on the change in intake air flow.. The change of the air flow in conducted experimental studies was obtained by adjusting the air blow. Figure 13.5(a–f) show the % share of flammable components: carbon monoxide CO, hydrogen H_2, and methane CH_4, as a function of the oxygen content ratio $O_2[\%_{vol}]$ in the gas to carbon C share in the fuel. The quality (calorific value) of generator gas is decided by the share of flammable components.

TABLE 13.1
Fuel Properties of Wood Waste

Properties	Pine Sawdust	Deciduous Tree Sawdust	Mixed Wet Sawdust	Coniferous Trees Pellets	Mixed Wood Pellets	Coniferous Bark
Moisture content [%]	11.7	9.3	42.1	9.1	9.0	47.4
Ash content[a] [%] (inflammable substance in fuel)	1.5	3.0	1.6	0.5	0.6	1.1
Flammable fraction content[a] [%] (flammable substance in fuel)	98.5	97.0	98.4	99.5	99.4	98.9
LHV^a [kJ/kg]	19,050	19,101	17,958	18,151	18,145	18,680
LHV [kJ/kg]	17,433	17,542	8794	16,170	16,165	8031
Carbon c^{daf} [%]	48.54	51.70	45.07	49.1	48.93	47.86
Hydrogen h^{daf} [%]	6.10	6.10	6.26	6.32	6.48	6.16
Nitrogen n^{daf} [%]	0.32	0.21	1.44	0.20	0.93	1.27
Sulfur s^{daf} [%]	0.05	0.10	0.01	0.04	0.02	0.03
Chlorine cl^{daf} [%]	0.11	0.10	0.16	0.07	0.01	0.12
Oxygen O^{daf} [%]	44.88	41.79	47.06	44.27	43.54	44.56

Source: Poskrobko S. et al.: *Drewno* 2016, 59, 197, 241–248.
[a] Expressed on a dry basis.

Production of Generator Gas from Biomass 293

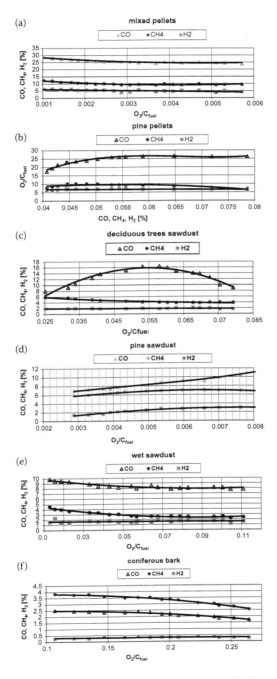

FIGURE 13.5 [%$_{vol}$]CH$_4$, H$_2$, CO=f(O$_2$/C$_{fuel}$); (a) mixed pellets, (b) pine pellets, (c) deciduous trees sawdust, (d) coniferous trees sawdust, (e) wet sawdust, (f) coniferous bark. (From Poskrobko S. et al.: *Drewno* 2016, 59, 197, 241–248.)

The calorific value of the generator gas was calculated using the following formula (its source is found in the instruction manual of the analyzer, [2,7]):

$$LHVsyngas = 126\,[\%\,CO] + 108\,[\%\,H_2] + 359\,[\%CH_4] \\ + 665\,[\%\,C_nH_m\,[kJ/Nm^3], \qquad (13.11)$$

where [%CO], [%H$_2$], [%CH$_4$], and [%CnHm] are the measured percentage share of the combustible components in the generator gas.

The greatest amount of combustible components were found in the generator gases resulting from the gasification of pellets, both mixed and pine (Figure 13.5a and b): carbon monoxide CO at up to about 28%–30%, 11%–14% methane CH$_4$, and hydrogen H$_2$ 5%–6%. This gas composition was reflected in their calorific value (Figure 13.6) - 7600–8700 kJ/Nm³. Gasification of sawdust, regardless of their type (Figure 13.5c–e), resulted in gas production which was, by more than half of the flammable components share and calorific value, twice lower. A significant amount of methane in the gas (up to 14% with gasified pellets) should be duly noted. It should also be noted that in the gasification of wood biomass in a compact fixed bed, the share of methane in the synthesis gas is approximately 2%–3% and the heating value of synthesis gas typically reaches the level of 2000–4000 kJ/Nm³ [8]. The bark of conifers (Figure 13.5f) turned out to be the worst fuel for gasification. Several percentages of methane and carbon monoxide and a fraction of hydrogen in the gas, which resulted in its low calorific value of about 1600 kJ/Nm³. Here, the process occurred at the highest values of O$_2$/C$_{fuel}$ (>0.1). Gasification of mixed pellets and pine sawdust was carried out under conditions of least oxygenation—O$_2$/C$_{fuel}$ 0.001–0.008. Research on waste wood materials gasification enabled the production of low-calorific gases (Figure 13.6).

13.3.2 Gasification Process of Olive Pits

Biomass of crushed olive pits was used for the gasification process (Table 13.2) [9].

Table 13.2 shows the fuel and the physicochemical properties—PN-EN-ISO 18134-2:2015-11; PN-EN-ISO 9029:2005; PN-ISO 1171:2002 and chlorine gram

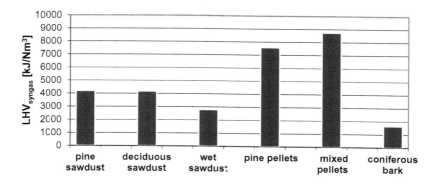

FIGURE 13.6 The calorific value of gas generated from the gasification of wood. (From Poskrobko S. et al.: *Drewno* 2016, 59, 197, 241–248.)

TABLE 13.2
Elemental Composition of Dry Mass of Surveyed Fuel and its LHV Calorific Value

Moisture [%]	4.97
Ash [%$_{dry\ m.}$]	6.35
Flammable fraction [%$_{dry\ m.}$]	93.65
Carbon c [%]	52.09
Hydrogen h [%]	3.67
Nitrogen n [%]	1.21
Sulfur s [%]	0.03
Chlorine cl [%]	0.42
Oxygen o [%]	41.58
LHV [MJ/kg$_{dry\ m.}$]	24.04

Source: Król D., Poskrobko S.: The influence of the conversion medium on the gasification process of olive pits, *Proceedings of the International Conference Experimental Fluid Mechanics* 2018, Prague, Czech Republic, 314–319.

fraction was denoted in compliance with Polish standard PN-ISO 587/2000, LHV-PN-ISO 1928:2002. Elemental composition of the flammable substances—carbon (C), hydrogen (H), nitrogen (N), and sulfur (S) has been determined with the use of an elemental analyzer, CHNS, model 2400, series II by Perkin Elmer.

Gasification was carried out in a 1.5 kW tubular gasifier (Figure 13.2) with the following dimensions: internal diameter of the reaction chamber $d_{r1} = 70$ mm, external diameter $d_{r2} = 75$ mm and total height $h = 800$ mm, insulation layer thickness 60 mm, sieve grid with holes of diameter $\varphi = 2$ mm, grid thickness $g = 5$ mm, and height of the air chamber $ha = 70$ mm. The gasifying agent was air fed to the sieve grate from below of the air chamber. The air flow for gasification in subsequent tests was 0.5 m^3/h, 0.85 m^3/h, 1.1 m^3/h, 1.4 m^3/h, and 1.7 m^3/h. The fuel flow was 0.37 kg/h. The temperature of the blowing air was 20°C. Gasification took place on the grate, in the fuel layer. Above the gas phase (over the fuel layer), in the atmosphere of unreacted oxygen, tar, and oil vapors from the gasified olive pits were partially burned. This ensured autothermality of the process and caused the temperature to rise to T = ~800°C. At this temperature, unburned organic vapors were cracked to form CH_4 [2,7]. The fuel was fed into the reaction chamber in a continuous manner by means of a screw feeder. The ash from the grate was removed during the process by blowing the grate with air. The measurement of concentrations of gaseous gasification products (CO_2, CO, O_2, H_2, CH_4) was made using the GAS 3000 gas analyzer. The calorific value of syngas was determined according to the dependence 11.

13.3.2.1 Results and Discussion

The results of research on the olive oil gasification process are shown in Figures 13.7 and 13.8. The basic test results indicating the efficiency of the process are shown in

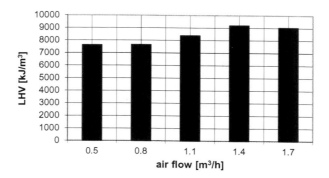

FIGURE 13.7 The calorific value of the syngas, characteristic for different gas streams. (From Król D., Poskrobko S.: The influence of the conversion medium on the gasification process of olive pits, *Proceedings of the International Conference Experimental Fluid Mechanics* 2018, Prague, Czech Republic, 314–319.)

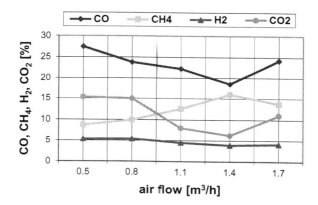

FIGURE 13.8 The effect of changes in the gasifier stream value on the value of gaseous products. (From Król D., Poskrobko S.: The influence of the conversion medium on the gasification process of olive pits, *Proceedings of the International Conference Experimental Fluid Mechanics* 2018, Prague, Czech Republic, 314–319.)

Figure 13.8, where the concentration values of gasification products (CO, CH_4, H_2, CO_2) depend on the flux gasifier stream value; this is the air supplied to the reactor.

The experiment confirmed the results of previous research carried out by the authors [6,7,10], with the use of other biomass fuels and refuse-derived fuel (RDF). A characteristic feature of the methanation process is hydrogenation reactions, occurring efficiently at temperatures of 300–500°C:

$$CO + 3H_2 \rightarrow CH_4 + H_2O, \qquad (13.12)$$

$$CO_2 + 4H_2 \rightarrow CH_4 + 2H_2O, \qquad (13.13)$$

$$C + 2H_2 \rightarrow CH_4. \qquad (13.14)$$

TABLE 13.3
Temperature Distribution in the Gasifier

Stream of gasification air [m³/h]	Temp. in zone II [°C]	Temp. in zone III [°C]	Temp. in zone IV [°C]	Temp. in zone V [°C]
0.5	865	456	742	763
0.8	878	462	751	768
1.1	841	437	815	834
1.4	834	428	823	842
1.7	856	542	862	871

Source: Król D., Poskrobko S.: The influence of the conversion medium on the gasification process of olive pits, *Proceedings of the International Conference Experimental Fluid Mechanics* 2018, Prague, Czech Republic, 314–319.

The formation of CH$_4$ was accompanied by a drop in the shares of CO and CO$_2$ as well as H$_2$.

These processes occurred in the fuel layer (area III, Figure 13.2) under the temperature conditions characteristic for zone III—the gasification zone (Table 13.3). In this zone, the methanation (hydrogenation) process took place.

The most favorable working conditions of the gasifier were related to the gas flow rate of 1.4 m³/h. Then, the generator gas was created with the highest calorific value of 9.2 MJ/nm³. The increase in the air stream directed the process toward the combustion reaction. This was evidenced by an increase in CO and CO$_2$ and a decrease in the concentration of CH$_4$ in the gas fraction (Figure 13.8). In zone II—above the grate, in contact with the excess blowing air, there occurred a process of fuel combustion, 0.5–1.7 m³/h. Temperatures relevant for the methanation process concern zone III [2], where kinetic conditions favor the processes of CO and CO$_2$ hydrogenation, in the direction of obtaining CH$_4$. The indicator characteristics obtained in the experiment relate to the gasification factor of air and temperatures in the range of zone III (Figure 13.2) (according to the data in Table 13.3) and the process taking place under atmospheric pressure.

13.3.3 Gasification of Wood Biomass with the Addition of Post-Extraction Rapeseed Meal

The research of the gasification process was conducted for biofuels such as wooden biomass and rapeseed meal [2]. Wood biomass was in the form of pellets, and rapeseed meal is a loose material (Figure 13.9).

Table 13.4 provides the partial elemental compositions of tested materials related to dry mass and their lower heating value (LHV), which were calculated based on measurements of elemental composition of flammable substances and measured values of higher heating value (HHV) combustion heats (enthalpy), dependent mainly on gram shares of elements such as carbon C and hydrogen H (and combustible sulfur). The fuel properties were marked as in Section 13.3.1.

FIGURE 13.9 Grain structure of fuel (a) post-extraction rapeseed meal, (b) wood pellet. (From Król D., Poskrobko S.: *Energy & Fuels* 2017, 31, 3935–3942.)

Post-extraction rapeseed meal is a waste from the production of rapeseed oil. As a combustible substance it contains amino acids, peptides, proteins, and as a result, a considerable proportion of nitrogen, so an element which in proper conditions of the combustion chamber, particularly in gasification, is released in the form of ammonia. Fat compounds contained in the rapeseed meal make the ignition temperature of the degassing substances higher than conventional biomass materials, which results in a time delay of the ignition. At the same time, the total time of thermal decomposition of the biopreparation [11,12] is longer than other fuels with the same high calorific value, comparable to wood biomass.

The requirement of trace amounts of alkali metals (Table 13.5) and a low chlorine content (Table 13.4) in a combustible material, allow for the use of the rapeseed meal at temperatures above 800°C, since it does not form low-melting chlorides. A high sulfur content (Table 13.4) in the flammable substance of rapeseed meal, also protects the fuel against the formation of low-melting salts involving chlorine. The rapeseed meal can thus be used not only as a supplement to coal, but it can be also successfully used in the process of their cocombustion or cogasification with

TABLE 13.4
Elemental Composition of Dry Mass of Surveyed Fuels and Its LHV Calorific Value (of Dry Mass [%$_{dry.m}$])

Sample type	Moisture [%]	ash [%]	c [%]	h [%]	n [%]	s [%]	cl [%]	o [%]	LHV [kJ/kg]
Softwood pellets	6.0	2.65	44.93	6.71	2.32	0.17	0.03	43.19	15,581
Rapeseed meal	8.76	7.05	49.30	6.96	6.76	1.38	0.01	35.59	16,836

Source: Król D., Poskrobko S.: *Energy & Fuels* 2017, 31, 3935–3942.

TABLE 13.5
Alkali Metal Concentration in Post-Extraction Rapeseed Meal

	Metal [ppm]		
Preparation	Calcium Ca	Potassium K	Sodium Na
Rapeseed meal	226.4	227.4	228.4

Source: Król D., Poskrobko S.: *Energy & Fuels* 2017, 31, 3935–3942.

agro-biomass. Dispensed in suitable proportions (i.e., with such content, so that the ratio of S/Cl>2.2 in the fuel; if S/Cl<2.2 the danger of chloride corrosion is very high; the molar ratio of S/Cl can be used as an indicator for the high-temperature corrosion risk), it also reduces the effects of high-temperature corrosion being the result of alkali metal chlorides presence, for example, if agricultural waste is used as fuel.

Environmental conditions of the methanation reaction is influenced by the fuel composition. In addition to the main fuel—wood biomass—an important supplement is a biomass preparation of rapeseed meal, with a high (376°C) flash point. Fat components contained in post-extraction rapeseed meal cause the ignition temperature of degasifying substances at a higher temperature than in case of typical biomass materials, which translates into a time delay of ignition. The presence of the oil compounds in the form of vapor isolates fuel particles from the oxygen, which forms the local, reductive reaction environment, conducive to the formation of the reduced form of carbon—CH_4. The addition of rapeseed meal supports the methane forming process because the meal is composed of approximately 4% of the fatty acids. These are primarily linolenic acid, linoleic, erucic, oleic, eicosenoic, palmitic, and stearic. In the temperature zone—in the layer of fuel $T_3 = 450–500°C$—apart from the gasification of fuel (wood biomass and rapeseed meal), evaporation of the above-mentioned fatty compounds takes place.

Biomass gasification studies were performed using the gas generator shown in Figure 13.2. The maximum thermal power of the gasifier was about 4 kW. Gasification was carried out in a fluidized compact bed, in an atmosphere of air shortage and continuous dosing of wood biomass mixed in different ratios with the product of rapeseed meal to the gasifier chamber. The ratio of the fuel stream (F) to the air (A), F/A = 0.76, where the efficiency of the gasifying agent, air A = 0.5 g/s, and fuel, that is, wood biomass loading capacity containing meal, F = 0.38 g/s.

In order to obtain synthesis gas having different contents of CH_4, the share of the rapeseed meal in the fuel was graded from 5% to 30%. The initiation of the gasification process was carried out by means of resistance heating elements installed along the gasification chamber and the air supply channel. In the initial phase of the process, spontaneous combustion of fuel took place. Then, the -up heating elements were turned off and the chamber was heated up to a temperature of 900°C, burning biomass. At this temperature, the tested fuels were dosed, and then the gasification process was stabilized by adjusting the flow of air and fuel. Stabilization meant establishing a proper process temperature, which in this case was $T_2 = 900°C$ on the grate, $T_3 = 450–500°C$ in the fuel layer, $T_4 = 800–900°C$ above the fuel layer, and

$T_4 = 700°C$ in the convection zone and at the outburst of the tubular reactor (Figure 13.2). The air flow was controlled by changing the blower motor speed, changing the frequency of current supplying the engine with the inverter. The installation was featured with continuous measurement of the mass flow rate of air. Identification of the mass flow of fuel was performed using the previously formed characteristics of the dispensing system equipped with a stepping motor. The measurement of the composition of the gas was made by the synthesis gas analyzer type GAS 3100. The reaction chamber of the gasifier with a diameter of 80 mm and a length of 1800 mm was equipped with a perforated grate of perforation size, that is, diameter hole of 4 mm, which was backfilled with fuel (biomass containing rapeseed meal). Through the openings in the grate, the air was supplied into the reaction chamber. The tubular reactor is also adapted for use in a fluidized bed and in a compact double layer [13]. The aim of the experiment was to determine optimal technological parameters for the process of identified fuels gasification, that is, the mixture of wood biomass and rapeseed meal. In order to identify the properties for the production of the reducing zone, the formulation of rapeseed meal was dosed in the amounts from 5% to 30%, referring to the weight of biomass. The results of the experiment are shown in the form of shares of the synthesis gas composition (CO, CH_4, H_2, CO_2) and the oxygen share in the syngas. Identification of the adjustment process to the conditions set was made by the indication of O_2 share in the synthesis gas, for which the process happened automatically.

13.3.3.1 Results and Discussion

The test results of the gasification of wood biomass with the addition of preparation rapeseed meal are shown in Figures 13.10 through 13.13. In each succeeding attempt, the preparation was added to wood biomass in amounts from 5% to 30% in relation to the biomass. The process took place at 800–850°C. Figure 13.10 shows the percentages of each of the gaseous products of gasification (CO_2, CO, CH_4, H_2) forming a synthesis gas. Combustible gas components, that is, CO, H_2, and CH_4 were

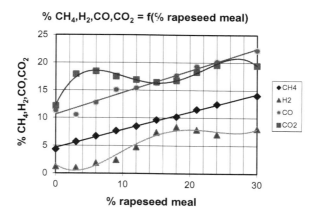

FIGURE 13.10 Synthesis gas composition in function of the rapeseed meal share in the fuel. (From Król D., Poskrobko S.: *Energy & Fuels* 2017, 31, 3935–3942.)

Production of Generator Gas from Biomass   301

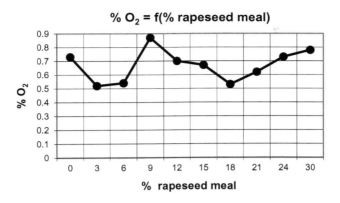

FIGURE 13.11 The share of O_2 in the synthesis gas for different shares of rapeseed meal. (From Król D., Poskrobko S.: *Energy & Fuels* 2017, 31, 3935–3942.)

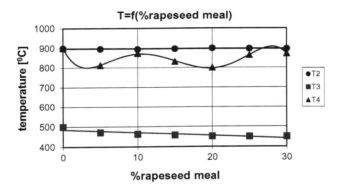

FIGURE 13.12 Temperature distribution in the reaction chamber of the gasifier, based on the proportion of rapeseed meal. (From Król D., Poskrobko S.: *Energy & Fuels* 2017, 31, 3935–3942.)

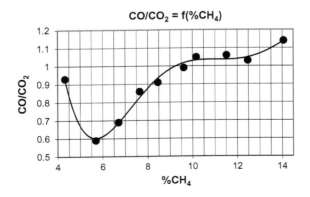

FIGURE 13.13 Dependence CO/CO_2 for the process of CH_4 formation. (From Król D., Poskrobko S.: *Energy & Fuels* 2017, 31, 3935–3942.)

gradually rising with the increase of the share of rapeseed meal in the fuel, which contributed to an increase in the calorific value of the synthesis gas. The proportional increase in CO and CH_4 was followed by an increase in the share of rapeseed meal in the fuel. The addition of rapeseed meal to wood biomass results in the formation of local reduction zones within the grains of fuel (it relates to the presence of oil fractions). When the amount of rapeseed meal in the fuel increases from 15% to 25%, a reduction zone expands. This results in a proportional increase of the gaseous forms of carbon (with insufficient oxygen) in the syngas: CO and CH_4. In contrast, the amount of its oxidized form, CO_2, decreases. The process achieves a balance, maintaining the $CO/CO_2 >1$ ratio. In such conditions a proportional increase of CO and CH_4 in the synthesis gas takes place, but a weakening growth in the share of H_2 can be noted, as shown in Figure 13.10. As regards the participation of rapeseed meal at 0%–10% (Figure 13.10), the increase of CO_2 suggests rapid combustion of wood biomass. The addition of rapeseed meal in amounts up to 5% of the wood biomass (with the fuel to air ratio, $F/A = 0.76$), does not particularly affect the process of its gasification in the direction of methanation. Under such conditions, the composition of the generator gas is dominated by CO_2—the oxidizing zone is maintained (Figure 13.10). This is confirmed by the decreasing share of O_2 in the gas. That means that oxygen in the gasifier reaction chamber (Figure 13.11), was actively involved in oxidation reactions—the temperature in the bed was 500°C. Increasing the share of rapeseed meal in the fuel increases the efficiency of the gasification process through an increase in the content of CO and CH_4. The conversion of the fuel into gas takes place at a preferred ratio of $CO/CO_2 > 1$. The reductive nature of the process ($CO/CO_2 > 1$) stabilizes from approximately 20% of rapeseed meal share in the wood fuel. The growth of the meal share makes the O_2 content in the gas increase (Figure 13.11) and the temperature of the bed decreases and is maintained at 450°C. A slight decrease in the temperature preserves the reductive nature of the process and improves the kinetic conditions of CO and CO_2 hydrogenation, that is, methanation (which will be discussed below).

The recorded changes of O_2 (Figure 13.11), depending on the meal share in the fuel, can be affected by inaccurate mixing of the meal with wood biomass. The quality of the homogenization of fuel is significantly affected by the difference in the size of the particles and the density of wood biomass and meal.

Fuel composed of ingredients—pelletized wood biomass with the addition of rapeseed meal—was thoroughly mixed. However, owing to the difference in weight of the two components, when dosing the fuel, the meal flaked. No homogenization of the fuel on the grid (changes of rapeseed meal content in the entire layer of fuel) causes variable temporary oxygen demand. In the absence of a homogeneous mixture on the grate of the reactor, from a practical point of view, we found this result (a difference of 0.3% O_2 in the generator gas) very satisfactory.

The presence of H_2 in the generator gas (Figure 13.10) has its origins in the fact that this is hydrogen which is produced at a high temperature (850°C) in the cracking zone (cracking occurs over the layer of fuel). Hydrogen formed in radical reactions $H + H = H_2$ in the layer of fuel (at $T = 450°C$), is almost exclusively involved in the methanation processes by hydrogenation of C, CO, and CO_2. It is a

Production of Generator Gas from Biomass

possible hypothesis, for which at this stage of the research, there is no conclusive evidence.

Temperature distribution in zone I (oxidative), II (reduction), and III (above the bed, in the cracked tars zone) is shown in Figure 13.12. Temperature distribution indicates that the process, principally in the reducing layer (gasification layer), is run stably.

The effect of the CO/CO_2 ratio on the methane formation process is shown in Figure 13.13. The process is disrupted by the changes in the concentration of CO_2 in the synthesis gas, so the most sensitive gas to oxygen, as indicated by the test results obtained. This graph also shows that, irrespective of the atmosphere in the gasification chamber of the reactor, the proportion of CH_4 always increases with the addition of rapeseed meal.

Figure 13.14 shows the fuel characteristics of the obtained synthesis gas for various bulk quantities of fuel added to the formulation, the operation of which, naturally tends to produce a reducing atmosphere. The diagram has an essential technological meaning, as on its basis (for a given calorific value) the composition of the synthesis gas is estimated.

Figure 13.15 illustrates, in turn, the effect of the formulation rapeseed meal on the calorific value of the synthesis gas obtained in the gasification. It is worth noting that the 30% share of the rapeseed meal in wood biomass increases the calorific value of gas by about three times. This has an impact on increasing the efficiency of the chemical conversion process of solid fuel into generator gas, in relation to the gasification technology with a low share of CH_4 in gas. Chemical conversion (chemical efficiency) is defined as the ratio of the calorific value of the gas to the calorific value of the fuel (density of the generator gas obtained is 1.019 kg/m³ in standard conditions), $LHV_{(g)}/LHV_{(f)}$. This factor indicates the development of gasification technology. In the present case of the gasification of wood biomass with the extracted rapeseed meal, the ratio is $LHV_{(g)}/LHV_{(f)} = 0.65$ (30% addition of rapeseed meal), and referring to the quality of the gas in the literature [14] $LHV^*_{(gas)}/LHV^*_{(fuel)} = 0.29$. The calorific value of the generator gas is calculated from Reference [7].

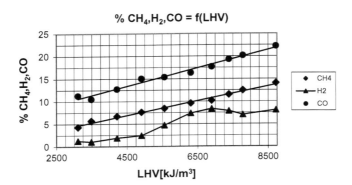

FIGURE 13.14 Fuel characteristics of synthesis gases. LHV of synthesis gas for various gas shares of flammable components (CH_4, H_2, CO). (From Król D., Poskrobko S.: *Energy & Fuels* 2017, 31, 3935–3942.)

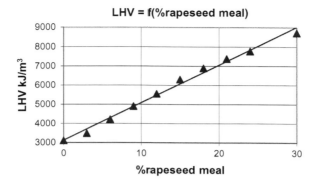

FIGURE 13.15 LHV of synthesis gases depending on the share of rapeseed meal in the fuel. (From Król D., Poskrobko S.: *Energy & Fuels* 2017, 31, 3935–3942.)

13.3.4 Gasification of RDF and Bio-CONOx

The RDF pellet fuel used in these studies is generated from the combustible fraction of municipal waste [3]. The preparation of Bio-CONOX is composed of combustible biomass waste materials. Tables 13.6 and 13.7 show RDF and Bio-CONOx preparation fuel properties. RDF fuel and mixtures of the fuel with Bio-CONOx 13%, 23%, and 33% were applied to the gasification process.

Gasification was carried out on a laboratory scale, in the tubular gasifier with fluidized bed (Figure 13.3). The gasifying agent was air (0.64 g/s) which was fed to

TABLE 13.6
Pellets' Fuel Properties of RDF on a Dry Weight Basis and Elemental Composition of Flammable Substance Expressed on a Dry Ash-Free Basis

Properties	Symbol	Value
Moisture content	[%]	5.4
Ash content[a] (inflammable substance in fuel)	[%]	12.9
Flammable fraction content[a] (flammable substance in fuel)	[%]	87.1
LHV[a]	[kJ/kg]	21,798
Carbon c^{daf}	[%]	66.7
Hydrogen h^{daf}	[%]	7.5
Nitrogen n^{daf}	[%]	4.1
Sulfur s^{daf}	[%]	1.8
Chlorine cl^{daf}	[%]	0.3
Oxygen o^{daf}	[%]	19.6

Source: Król D. et al.: Micro cogeneration - rich-methane gasifier and micro gas turbine, *E3S Web of Conferences* 2017, Vol. 14, 01024, Energy and Fuels 2016.

[a] Expressed on a dry basis.

TABLE 13.7
Physical, Chemical, and Fuel Properties of Bio-CONOx Expressed on a Dry Basis and Elemental Composition of Flammable Substance Expressed on a Dry Ash-Free Basis

Properties	Symbol	Value Acc. [15]	Value
Moisture content	[%]	<12	7.3
Flammable fraction content (flammable substance in fuel)	[%]	>90	93.2
Ash content (inflammable substance in fuel)	[%]	<10	6.8
LHV	[kJ/kg]	>17,000	17,745
HHV	[kJ/kg]	>15,500	18,942
Protein content	[%]	24–40	31.0
Plant fats content	[%]	2.5–5.0	4.4
Carbon c^{daf}	[%]	49.00–52.00	50.2
Hydrogen h^{daf}	[%]	6.00–8.00	6.3
Nitrogen n^{daf}	[%]	6.00–8.00	5.9
Sulfur s^{daf}	[%]	1.00–1.5	1.2
Chlorine cl^{daf}	[%]	<0.05	0.02
Oxygen o^{daf}	[%]	30.45–38.00	36.38

Source: Król D. et al.: Micro cogeneration - rich-methane gasifier and micro gas turbine, *E3S Web of Conferences* 2017, Vol. 14, 01024, Energy and Fuels 2016.

the bed by a perforated grate from below the gasifier. The diameter of the perforation hole in the grate was d = 1 mm. During the experiment, the air was not heated—the air temperature was 22°C. The maximum thermal power of the gasifier was 5 kW. The temperature of the process was defined in the range of 380–410°C, and above the layer of fuel approximately 900°C. In a fluid bed (in the layer of fuel)—at a temperature of 380–410°C—the gasification process occurred autothermally. In the gaseous phase (above the bed), the temperature reached 880–930°C. An intense evaporation of high-calorific condensable substances, including tar, occurred. It is suggested that these high-calorific compounds were partially burned over the fuel layer in an atmosphere of unreacted oxygen, which resulted in the temperature increase up to the value of ~900°C. That is evidenced by the lack of high-calorific hydrocarbons, CnHm, in the generator gas.

The fuel was fed into the reaction chamber continuously with a feed screw driven by a stepper motor, with precise speed control, that is, 0.41 g/s fuel flow. The ash was removed in the process, exhausting from the gasification chamber by revolving (vacuum and performance) the exhaust fan rotor. The resultant gas was cooled by a stream of air sucked from the environment to a temperature of approximately 180°C. The gas generator and blowing air channel are fitted with an electric heating system, which is used to heat the reaction zone, where spontaneous combustion of fuel took place. The stabilization of the process was carried out by regulating the flow of fuel and the gasification air flow.

The test results of RDF fuel gasification containing the formulation Bio-CONOx are shown in Figures 13.16 and 13.17. Figure 13.16 shows the dependence of the generator gas composition CH_4, CO, H_2, and CO_2 from the amount of oxygen. Taking into account the fact that the air flow during the gasification process was maintained at a constant level, the demand of the oxygen is significantly reduced, with increasing participation of the additive Bio-CONOx in the fuel (Figure 13.16). This is expressed in increasing participation of oxygen in the generator gas. These results show that the addition of Bio-CONOx clearly influences the formation of a strong reducing zone in the reaction environment. Under these conditions, the temperature in the bed was maintained in the range of 380–410°C, the temperature over the bed gradually grew from 880°C (gasification without the addition of Bio-CONOx) to 930°C (gasification with 33% addition of Bio-CONOx). Changes in the temperature suggest the increased participation of condensing vapor cracking (including tars), which favored methane forming processes according to the scheme (13.15):

$$(TAR)gas \rightarrow V_{CO}CO + V_{CH4}CH_4 + V_{H2}H_2 + V_{CO2}CO_2. \quad (13.15)$$

The high temperature above the bed was assured by the process of partial combustion, mainly high calorific tars, according to the Brayton kinetic model: $CH_{1.522}O_{0.0228} + 0.867\ O_2 \rightarrow CO + 0{,}761\ H_2O$. Gasification of RDF fuel with air made it possible to obtain a synthesis gas, having a composition of combustible gases up to 10.3%CH_4, 11.6%CO, and 7 3%H_2, and calorific value at a maximum of $\sim$6000 kJ/Nm^3.

Gasification kinetics is directed by a complex interdependence of chemical and physical phenomena. The speed of heterogeneous gasification reactions depends both on the speed of chemical reactions, and the speed of diffusion of the gasifying agent into the pores of fuel:

$$Vr = -dnc/dt = f(k_{ef}, O_{,mr}, C, w, p, d), \quad (13.16)$$

where k_{ef} is the effective reaction speed constant; O is the specific surface area of the fuel; m_r is the fuel mass in the reaction space; C is the the concentration of the gasification agent in the reaction space; w is the the flow rate of the reactants; p is pressure; and d is the diameter of the fuel grains.

The results shown in Figure 13.16 indicate that use of the preparation Bio-CONOx in amounts of 13%, 23%, and 33% of the fuel mixture causes expansion of the reducing zone. The result is a continuous increase in the proportion of methane, CH_4 (even double with the 33% share of Bio-CONOx in the fuel) and hydrogen, H_2 in the syngas. Attention is drawn to a significant drop in CO_2 content from 16% (for the gasification of RDF fuel only) to a very low level of 1–2.5% when Bio-CONOx was added. In addition, the amount of CO_2 in the process remains unchanged (Figure 13.16—points are arranged in parallel to the *x*-axis). This reflects the dominant reducing zone in the area of the gasification reaction, favoring the increase of the CH_4 share. The primary

Production of Generator Gas from Biomass

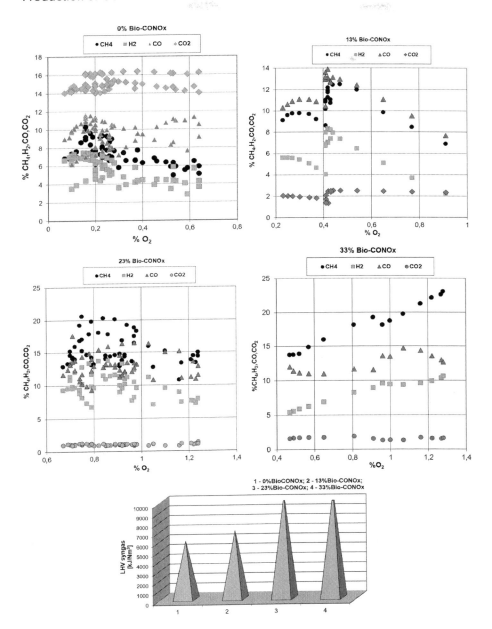

FIGURE 13.16 %CH$_4$, H$_2$, CO, CO$_2$ = f (O$_2$) for different shares of Bio-CONOx in the fuel. (From Król D. et al.: Micro cogeneration - rich-methane gasifier and micro gas turbine, *E3S Web of Conferences* 2017, Vol. 14, 01024, Energy and Fuels 2016.)

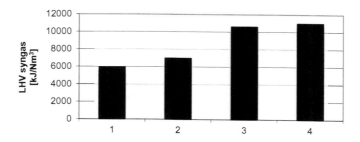

FIGURE 13.17 Calorific value of the synthesis gas depending on the share of Bio-CONOx in the fuel: 1–0%Bio-CONOx; 2–13%Bio-CONOx; 3–23%Bio-CONOx; 4–33%Bio-CONOx. (From Król D. et al.: Micro cogeneration - rich-methane gasifier and micro gas turbine, *E3S Web of Conferences* 2017, Vol. 14, 01024, Energy and Fuels 2016.)

objective of the processes of the second and third generation is, among others, the preparation of higher calorific gas consisting mainly of methane. The obtained results clearly fit in with this trend. They show that in the area of the fuel layer, in favorable temperature conditions of 380–410°C, exothermic methanation reactions occurred. Basic relations oriented toward methanation of synthesis gases are (13.17):

$$CO + 3H_2 \leftrightarrow CH_4 + H_2O \quad (\Delta H = -206 \text{ kJ/mol}). \tag{13.17}$$

The water formed during the methanation reaction also reacts with carbon monoxide according to Equation 13.18:

$$CO + H_2O \leftrightarrow CO_2 + H_2 \quad (\Delta H = -41.5 \text{ kJ/mol}). \tag{13.18}$$

As in the reaction system the concentration of CO decreases, the balance of the reaction Equation (13.18) moves. This reaction changes direction. It begins to occur from the right side to the left. Carbon monoxide, which is being formed, undergoes the methanation according to Equation 13.17. By adding reaction (13.17) to reaction (13.18)—with its course from right to left—we get the sum equation of the CO_2 methanation reaction (13.19):

$$CO_2 + 4H_2 \rightarrow CH_4 + 2H_2O \quad (\Delta H = -164.9 \text{ kJ/mol}). \tag{13.19}$$

Hydrogen in the low-temperature thermal conditions of the deposit was created in the radical reactions $H* + H* \rightarrow H_2$ and hydrogen was involved in the methanation reaction of CO and CO_2. Methanation reactions of CO and CO_2 are exothermic reactions taking place with a reduction of volume, so reducing the temperature favors a shift of the equilibrium in the direction to increase methane concentration in the equilibrium gas mixture. The methane forming process is more widely described in Reference [7]. The intensification of the methanation process for the gasification of fuel with 23% and 33% shares of Bio-CONOx, where the highest share of CH_4 and calorific value of LHVsyngas ~10 MJ/Nm^3 were recorded. The calorific value of the generator gas was calculated on the basis of the following formula (its source is the instruction manual of the analyzer [2,7]).

13.4 MICRO COGENERATION: RICH-METHANE GASIFIER AND MICRO GAS TURBINE

13.4.1 Modeling of a Syngas Microturbine-Based Cogeneration Plant

The chapter presents a conceptual, small-scale cogeneration plant (30 kW) that combines a gasifier, a single shaft microturbine fed by syngas, and a flue gas heat exchanger (FGHEX) [3]. It is assumed that the synthesis gas provided to the microturbine is clean enough in the scrubbing unit segment (Figure 13.18). The basic scheme of a cogeneration plant is shown in Figure 13.18. The part of the installation responsible for producing syngas are described in detail in above. In further discussion, we consider a theoretical Brayton cycle (we assume that in the microturbine combustion chamber and in the FGHEX gas side the losses of inflow and outflow air are negligent, and so are the losses of the pressure).

13.4.2 Gas Microturbine Model

The computational part of the work undertaken mainly concerns the thermodynamic aspects of a microturbine design [3]. The turbine consists of four components: compressor, combustion chamber, turbine, and generator. It was assumed that the thermodynamical cycle of the microturbine (Brayton) is realized without heat regeneration.

In the cycle considered, atmospheric air (state 1—inlet GM) is continuously drawn into the compressor and then after compression (state 2—outlet GM) is delivered to the combustion chamber. Syngas fuel is used to increase the temperature of

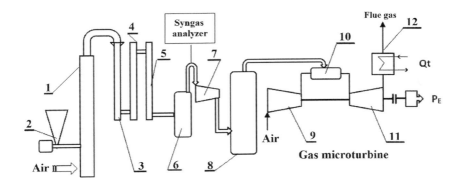

FIGURE 13.18 Simplified scheme of a GM-based cogeneration plant: 1—gasifier, 2—feed system, 3,4—scrubbing unit, 5—carbon (or ceramic) filter, 6—compensatory syngas container, 7—compressor for syngas, 8—container for compressed syngas, 9—GM (air compressor), 10—GM (combustion chamber), 11—GM (expansion turbine), 12—FGHEX (flue gas heat exchange), PE-electric power, Q_t-heat power.) (From Król D. et al.: Micro cogeneration—rich-methane gasifier and micro gas turbine, *E3S Web of Conferences* 2017, Vol. 14, 01024, Energy and Fuels 2016.)

compressed air through a combustion process. Hot gases (combustion products) leaving the combustion chamber (state 3) expand in the turbine, which produces power to drive the compressor and generator, and finally discharges flue gas (state 4 for flue gas, also subscript g1).

13.4.2.1 Air Compressor Model

Knowing the air inlet temperature, T_1, to the compressor, pressure ratio, π_c, and isentropic efficiency, $\eta_{i(c)}$ for the compressor, can be determined using the following parameters:

$$\pi_c = p_2/p_1, \qquad (13.20)$$

where p_1 and p_2 are compressor inlet and outlet air pressure, respectively:

$$\eta_{i(c)} = (T_{2s} - T_1)/(T_2 - T_1). \qquad (13.21)$$

where T_1 and T_2 are the compressor inlet and outlet air temperature, respectively, and T_{2s} is the compressor isentropic outlet temperature, so the final temperature of the compressor can be obtained from the equation:

$$T_2 = T_1 \cdot \left(1 + \left(\pi_{i(c)}^{pc} - 1\right)/\eta_c\right); \quad pc = (k_c - 1)/k_c. \qquad (13.22)$$

Assuming that blade cooling is not taken into account, the specific work of the compressor, w_c and consequently compressor power, N_C can be calculated by:

$$w_c = c_{p(a)} \cdot T_1 \cdot (\pi_c^{p_c} - 1)/(\eta_{m(c)} \cdot \eta_{i(c)}) \quad \rightarrow \quad N_C = \dot{m}_a \cdot w_c, \qquad (13.23)$$

where $c_{p(a)}$ is the specific heat of air and $\eta_{m(c)}$ is the mechanical efficiency of the compressor.

13.4.2.2 Combustion Chamber Model

From energy balance in the combustion chamber:

$$\dot{m}_a \cdot c_{p(a)} \cdot T_2 + \dot{m}_f \cdot W_d + \dot{m}_f \cdot c_{p(f)} \cdot T_f = (\dot{m}_a + \dot{m}_f) \cdot c_{p(g)} \cdot T_3, \qquad (13.24)$$

where, $\dot{m}_f$ is the fuel mass flow rate (kg/s), $\dot{m}_a$ is the air mass flow rate (kg/s), W_d is the low heating value, T_3 is the turbine inlet temperature, T_f is the temperature of fuel, $c_{p(f)}$ is the specific heat of the fuel, and $c_{p(g)}$ is the specific heat of the flue gas. After manipulating (13.5), the fuel air ratio can be expressed as

$$FAR = \dot{m}_f/\dot{m}_a = (c_{p(g)} \cdot T_3 - c_{p(a)} \cdot T_3 \cdot (1/\pi_t^{pg}))/(W_d + c_{p(f)} \cdot T_f - c_{p(g)} \cdot T_3). \qquad (13.25)$$

13.4.2.3 Gas Turbine Model

The exhaust gases' temperature from the gas turbine is given by

$$T_4 = T_3 \cdot (1 - \eta_{i(t)} \cdot (1 - 1/\pi_t^{pg})); \quad pg = (k_g - 1)/k_g, \quad (13.26)$$

where T_4 is also the inlet gases' temperature ($=T_{g1}$) to the FGHEX.

The specific shaft work, w_t of the turbine and turbine output power are given by

$$w_t = c_{p(g)} \cdot T_3 \cdot \eta_{i(t)} \cdot (1 - (1/\pi_t^{pg}))/\eta_{m(t)} \quad \rightarrow \quad N_T = (\dot{m}_a + \dot{m}_f) \cdot w_t. \quad (13.27)$$

The net turbine output power [7, 16] can be expressed as

$$N_{GT} = (N_T - N_C) \cdot \eta_g. \quad (13.28)$$

If the heat supplied is expressed as

$$E_{ch} = \dot{m}_f \cdot W_d, \quad (13.29)$$

then the gas turbine efficiency η_{th} can be determined by the equation

$$\eta_{th} = N_{GT} / E_{ch}. \quad (13.30)$$

13.4.2.4 Gas Microturbine Model Flue Gas Heat Exchanger

For recovered waste heat from the gas turbine, exhaust gas flows into the FGHEX (it is assumed with T_{g1}, T_{g2} inlet, and outlet flue gas temperature, respectively). Usually, the FGHEX represents a single pressure heat recovery steam generator (HRSG) or water-gas heat recuperator.

13.4.2.5 Thermodynamic Performance

There are three known and widely used thermodynamical methodologies for the performance evaluation of a plant, including gas-fired systems. Two of them are about the energetic evaluation, which enables calculation of the fuel utilization efficiency and power to heat ratio. The third methodology operates on an exergetic approach and allows for system evaluation on the basis of the second law, exergetic efficiency. In the aforementioned case, we limited our attention to the usage of the first approach (named also as the energy utilization factor) and defined as:

$$\eta_{ov} = EUF = \frac{N_T + Q_{ex}}{E_{ch}} \quad \text{where} \quad Q_{ex} = (\dot{m}_a + \dot{m}_f) \cdot \tilde{c}_{p(g)} \cdot (T_{g1} - T_{g2}). \quad (13.31)$$

13.4.3 Results and Discussion

All calculations have been performed using EES software. For better reliability of the results, they are compared with the technical data available for a commercial microturbine, the Capstone C30 (Tables 13.8 through 13.11).

TABLE 13.8
Microturbine Performance Data (Capstone Model C30)

Electrical power N_T [kW]	30
Total heat extraction Q_{ex} [kW]	56
Heat added E_{ch} [kW]	115
Flue gas temperature T_4 [K]	271[°C]/544.15[K]
Electrical efficiency η_{el} [−]	26
Overall efficiency EUF [%]	77
Rotational speed [r/min]	96,000
Air flow rate at compressor inlet [kg/s]	0.31

Source: Król D. et al.: Micro cogeneration - rich-methane gasifier and micro gas turbine, *E3S Web of Conferences* 2017, Vol. 14, 01024, Energy and Fuels 2016.

Comments: Designed for ambient conditions (ISO-standard reference conditions): temperature 288.15 [K], pressure 101.325 [kPa], relative humidity 60%.

TABLE 13.9
Design Parameters for Simulation of the Model Microturbine

Compressor	$\pi_c = 4,5, \eta_{i(c)} = 0,8, \eta_{m(c)} = 0,95, k_a = 1,4$
Turbine	$\pi_{i(t)} = 4.5, \eta_{i(t)} = 0.88, \eta_{m(t)} = 0.95, k_g = 1.326$
Combustion chamber	$\eta_{cc} = 1,0$
Generator: generator efficiency	$\eta_g = 1,0$
Mass flow rate of air [kg/s]	0.31
Outlet temperature [K]	544.15
Output power N_{GT} [kW]	30.0

Source: Król D. et al.: Micro cogeneration - rich-methane gasifier and micro gas turbine, *E3S Web of Conferences* 2017, Vol. 14, 01024, Energy and Fuels 2016.

TABLE 13.10
Operating Parameters of the Model GM Results of Calculation Using EES

State of Cycle	Pressure [kPa]	Temperature [K]
1. Inlet air to the compressor	101.3	288.15
2. Outlet air from the compressor	456	481.53
3. Inlet gas to the turbine	456	(see Table 13.6)
4. Gas exhaust from the turbine	101.3	544.15
Compressor input power N_C [kW]		63.8
Turbine output power N_T [kW]		94.4
Net output power N_{GT} [kW]		30.0

Source: Król D. et al.: Micro cogeneration - rich-methane gasifier and micro gas turbine, *E3S Web of Conferences* 2017, Vol. 14, 01024, Energy and Fuels 2016.

TABLE 13.11
Calculated Cases for Different Fuels Including Design Parameters[a]

	Fuel I	Fuel II	Fuel III	Fuel IV	Methane
Temperature of inlet gas to turbine T_3 [K]	811.2	815.3	821.1	821.3	828.3
Mass flow rate of fuel [kg/s]	0.02216	0.0172	0.01027	0.01007	0.002182
Ratio air/fuel AFR [-]	13.99	18.03	30.18	30.78	142.1
Heat extracted [kW][b]	58.229	59.112	56.996	56.961	55.557
Heat added [kW]	107.395	107.827	108.431	108.449	109.136
Efficiency η_{th} [%]	28.5	28.39	28.23	28.23	28.05
Efficiency EUF [%]	82.98	81.82	80.23	80.19	78.39

Source: Król D. et al.: Micro cogeneration - rich-methane gasifier and micro gas turbine, *E3S Web of Conferences* 2017, Vol. 14, 01024, Energy and Fuels 2016.

[a] Calorific values W_d: 4.847 MJ/kg—Fuel I, 6.270 MJ/kg—Fuel II, 10.558 MJ/kg—Fuel III, 10.769 MJ/kg—Fuel IV and for pure methane 50 MJ/kg (all W_d based on lower heating values)

[b] Heat extracted rates have been calculated taking reference temperature (T_{g2} = 368,15[K]/95[°C]—the temperature of the exhaust gases exit from the FGHEX).

13.5 SUMMARY

The presented technologies for the gasification of biomass fuels, biomass waste, and fuels from waste (in a gasifier designed by the authors), aimed at producing a generator gas with an increased share of methane, can be implemented in dispersed energy (including prosumer) systems for electricity and heat generation, and in low power facilities (e.g., up to 0.5 MW).

Another application for these technologies and gas generators can also be off-grid systems.

An important element worth emphasizing is the obtained composition of the generator gas, in which the methane content is increased, but the hydrogen content is reduced (in relation to technologies in which the combustible gas contains mainly carbon monoxide and hydrogen, and to a minimal extent methane or other light hydrocarbons). Its use in dispersed energy facilities, where it is burnt in low-power boilers, minimizes operational problems. Low hydrogen content guarantees flame stability during combustion. The increased methane content has a positive effect on its calorific value. Therefore, gas generators based on the presented simple structure (in which the efficiency of the hydrogenation reaction of carbon monoxide, CO and carbon dioxide, CO_2 to CH_4, methane increases), using air as a converter, operating at atmospheric pressure conditions, converting into flammable gas, renewable, readily available solid fuels, which are biomass or waste fuels, and are part of the needs of low-emission technologies.

The generator gas produced in such conditions can also be burnt in gas microturbines, in which the calorific value of gas with the lowest value of 10 MJ/m³ is the preferred value of the producers, for example, the Capstone. It is therefore often a barrier value to this method of syngas application, which is generated in many known devices of other structures. In the end, it comes down to the fact that the calorific value of the generated syngas does not meet the imposed criteria.

REFERENCES

1. Król D., Poskrobko S., Łach J.: Use of the "Bio-CONOX" additive in combustion, gasification and degassing of solid fuels, resulting in a reduction of nitrogen oxide emissions and an increase of the efficiency of generator processes, 2014, patent No. P.390001.
2. Król D., Poskrobko S.: Experimental study on co-gasification of wood biomass and post-extraction rapeseed meal – rich-methane gasification, *Energy & Fuels* 2017, 31, 3935–3942.
3. Król D., Poskrobko S., Gościk J.: Micro cogeneration - rich-methane gasifier and micro gas turbine, *E3S Web of Conferences* 2017, Vol. 14, 01024, Energy and Fuels 2016.
4. Radmanesh R., Chaouki J., Guy Ch.: Biomass gasification in bubbling fluidized bed reactor: Experiments and modeling. *Environmental and Energy Engineering* 2006, 52, 4258–4271.
5. Rönsch S., Schneider J., Matthischke S., Schlüter M., Götz M., Lefebvre J., Prabhakaran P., Bajohr S.: Review on methanation – From fundamentals to current projects. *Fuel* 2016, 166, 276–296.
6. Poskrobko S., Król D. Borsukiewicz-Gozdur A.: Gasification of waste wood biomass, *Drewno* 2016, 59, 197, 241–248.
7. Król D., Poskrobko S.: High-methane gasification of fuels from waste – Experimental identification, *Energy* 2016, 116, 592–600.
8. Saravanakumar A., Haridasan T.M., Reed T.B., Bai R.K.: Experimental investigations of long stick wood gasification in a bottom lit updraft fixed bed gasifier, *Fuel Processing Technology* 2007, 88, 617–622.
9. Król D., Poskrobko S.: The influence of the conversion medium on the gasification process of olive pits, *Proceedings of the International Conference Experimental Fluid Mechanics* 2018, Prague, Czech Republic, 314–319.
10. Król D., Gałko G.: Studies on sewage sludge gasification, *Przemysł Chemiczny* 2017, 96, 2, 341–342.
11. Król, D., Poskrobko, S.: Waste and fuel from waste - Analysis of thermal decomposition - Part I, *Journal of Thermal Analysis and Calorimetry* 2012, 109, 619–628.
12. Poskrobko S., Król D.: Biofuels - Thermogravimetric research of dry decomposition - Part II, *Journal of Thermal Analysis and Calorimetry* 2012, 109, 629–638.
13. Van de Steene L., Tagutchou J.P., Mermoud F., Martin E., Salvador S.: A new experimental Continuous Fixed Bed Reactor to characterise wood char gasification, *Fuel* 2010, 89, 3320–3329.
14. Bhaduri S., Contino F., Jeanmart H., Breuer E.: The effects of biomass syngas composition, moisture, tar loading and operating conditions on the combustion of a tar-tolerant HCCI (Homogeneous Charge Compression Ignition) engine, *Energy* 2015 87, 289–302.
15. Poskrobko S., Król D., Łach J.: A primary method for reducing nitrogen oxides in coal combustion through addition of Bio-CONOx, *Fuel Processing Technology* 2012, 101, 58–63.
16. Granatstein D.L.: Case study on Lahden Lampovoima gasification project Kymijarvi Power Station, Lahti Finland. Natural Resources Canada/CANMET Energy Technology Centre (CETC), *IEA BIOENERGY AGREEMENT—TASK* 36, November 2002.

Part V

Computational Fluid Dynamics as a Modern Tool in Studies of Biomass-Based Small-Scale Energy Devices

14 Computational Fluid Dynamics as a Modern Tool in Studies of Biomass-Based Small-Scale Energy Devices

Mateusz Szubel, Maciej Kryś and Karolina Papis

CONTENTS

Symbols.. 317
14.1 What Is Computational Fluid Dynamics? 319
14.2 Structure of a Classic CFD Project 320
14.3 Simulating the Combustion Process Using CFD 323
14.4 Challenges in CFD Modeling of Thermochemical Treatment of Biomass 326
14.5 Features of Reacting Flows and Combustion Modeling 326
14.6 Role of Multiphase Transport in CFD Modeling of Small-Scale Biomass-Based Energy Devices 329
14.7 Thermochemical Conversion of Solid Fuels in Fixed-Bed Devices 331
14.8 Selected Popular Combustion Models for Fast and Slow Chemistry 333
14.9 Eddy Dissipation Model 333
14.10 Finite Rate Model and Finite Rate/Eddy Dissipation Model 334
14.11 Eddy Dissipation Concept Model 335
14.12 Summary 336
References 337

SYMBOLS

ARABIC LETTERS

A:	preexponential factor
C_{MK}:	mole concentration of M_K compound
D:	diffusion coefficient
Da:	Damköhler number
$E[\phi]$:	expected value of variable
h:	mixture enthalpy per unit mass
h_k:	specific "k" compound enthalpy

k:	turbulence kinetic energy
k_j:	reaction rate coefficient
L:	characteristic dimension
m:	total number of reactions
M_{wi}:	molar mass of "i" substrate in "r" reaction
M_{wj}:	molar mass of "j" product in "r" reaction
$\mathbf{n}$:	unit vector
N:	total number of compounds in the chemical process
p:	pressure
$\dot{q}_{kj}$:	reaction rate
R_{ir}:	rate of production of "i" species in "r" reaction
R_{slow}:	rate of the slowest reaction in stoichiometric conditions
R_u:	universal gas constant
S:	external integration area
S_{CK}:	Schmidt number
S_{rad}:	radiation heat gain or loss
S_φ:	source term in transport equations
$\overline{S}_\varphi$:	averaged source value in transport equations
t:	time
T:	absolute temperature
$\mathbf{u}$:	velocity vector of fluid element
U:	characteristic velocity
$\mathbf{u}_i$:	velocity vector of the "i" particle
$\mathbf{u}'$:	velocity fluctuation vector
V:	total volume
x:	space coordinate of fluid particle
Y_i:	mass fraction of the "i" substrate
Y_K:	mass fraction of the "k" species
Y_K^*:	species mass fraction in small scale
Y_p:	mass fraction of the "p" product

Greek Letters

α:	temperature exponent
γ:	length fraction
Γ:	dynamic diffusion coefficient
ε:	turbulence dissipation rate
μ:	dynamic viscosity
υ:	kinematic viscosity
υ_{ir}:	stoichiometric coefficients of "i" substrates in "r" reaction
υ_{jr}:	stoichiometric coefficients of "j" substrates in "r" reaction
υ'_{kj}:	stoichiometric coefficients of "M_k" substrates in "j" reaction
υ''_{kj}:	stoichiometric coefficients of "M_k" products in "j" reaction
ρ:	density
ρ_{ad}:	density of the adiabatic flame

σ_k: Prandtl number
τ: reaction time scale
$\bar{\phi}$:   time averaged variable
ϕ':   variable fluctuation
$\dot{\omega}$:   volumetric production or consumption rate

14.1 WHAT IS COMPUTATIONAL FLUID DYNAMICS?

Computational fluid dynamics (CFD) is an advanced tool that allows for the solution of systems of differential transport equations. Although the name of this group of methods suggests that it is dedicated to studies of fluid flow, it is also suitable for simulating the phenomena in solids and multiphase systems. Numerical methods are an effective supplementation to methods in analytical and experimental fluid mechanics (Figure 14.1). Many simple problems relating to laminar fluid flow may be solved analytically, which means that it is not necessary to apply advanced numerical methods. This way, the solution may be achieved by means of algebraic equations.

Carrying out an experiment is certainly the most instinctive way to study the problems of fluid mechanics, although it requires metrology and experimental planning experience; however, the drawback to this analytical method is the limited possibility for installing measurement sets in actual fluid flow systems. Even in the case of a dedicated experimental stand, implementation of a large number of various sensors may result in problems owing to the requirement to maintain the flow characteristics. Certainly, most of the studies in fluid dynamics (but also in many other fields) are nowadays supported by numerical tools [1–10]. In general, the basic system is founded on conservation principles, such as mass conservation (represented by the continuity equation), momentum conservation (momentum equation—"N–S" or momentum equations), and the first law of thermodynamics (energy equation).

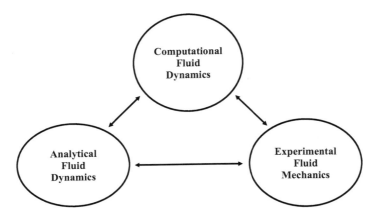

FIGURE 14.1 Relations between the complementary approaches to fluid dynamics studies.

14.2 STRUCTURE OF A CLASSIC CFD PROJECT

From the engineering practice point of view, CFD allows for the description of a problem, its solution, and obtaining the required set of results through three groups of methodological elements:

- Preprocessing
- Solution process
- Postprocessing

The stages are interrelated and constitute the main line process. Its structure is presented in Figure 14.2.

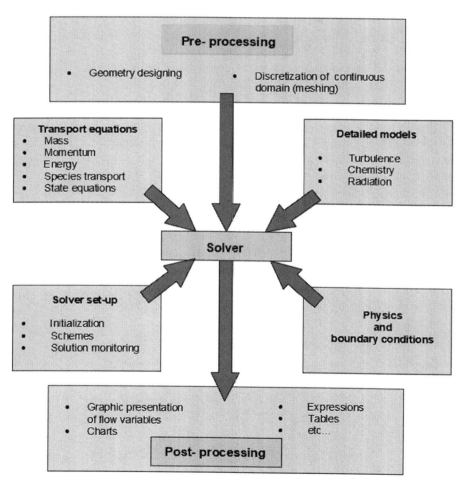

FIGURE 14.2 General structure of a CFD project.

CFD as a Tool in Studies of Biomass-Based Devices

Preprocessing is the first step in defining the problem. It allows for the description of the geometry (using internal tools of the software or an external computer aided design [CAD] environment), generation of the computational grid, determining the boundary conditions, and all the physical and chemical phenomena that are significant in view of the case under consideration.

An appropriate level of simplification of the physical model during the stages of designing the geometry and generating a sufficiently dense and high-quality mesh (computational grid) is decisive for achieving reasonable results with the simulation.

Meshing of the geometry is a crucial stage of CFD model design, as the idea behind the method assumes the substitution of an analytical solution obtained in theoretical fluid dynamics by the approximated numerical solution in the so-called nodes, which are selected points in space. This approach is called discretization. The final effect of discretization is a system of differential equations and its solution allows for a set of dependent variables that belong to the discrete set of points to be obtained. The area that is the closest to one of these points is the so-called computational cell. Furthermore, vertices of the cell build the mesh—a computational grid that covers the entire domain (model) considered within the study [11]. During computation, the so-called governing equations are solved in discretized form to reach the solution in the computational domain. To be more precise, it has to be noted that depending on the selected discretization method, individual solutions may refer to the finite difference, finite element, or the finite volume. Thus, the finite difference method (FDM), finite element method (FEM), and the finite volume method (FVM) may be distinguished.

An analysis of the model parameters' sensitivity to the mesh cells size and quality is the preliminary stage of many studies, especially when complex simulations are performed (e.g., [12,13]). The evaluation of mesh quality should be carried out based on dedicated indicators. These have been widely described in numerous papers [11,14–18]. Owing to the complex structure of the model, in the case of thermochemical conversion simulations it is crucial to find a compromise between the number of mesh nodes and elements and the time of computation.

Before the computation can be commenced, a series of solution parameters has to be set. CFD solvers can operate based on different calculation methods of velocity and pressure fields. Generally, the velocity field is obtained from the momentum equation using a pressure-based solver (pressure-based Navier-Stokes—PBNS). However, in the case of compressible fluid flows, where the Mach number is relatively high, it is reasonable to use the so-called density-based approach (density-based Navier-Stokes—DBNS), where the density field is calculated based on the continuity equation assuming constant pressure. Subsequently, the pressure is obtained from the equation of state [11,19].

Apart from the FDM and the FEM, the FVM of the CFD problem discretization and solution should be mentioned. It is currently the most widely used method in the numerical analyses of fluid flow. The method assumes that the neighboring nodes in a mesh create solids with a volume of ΔV. The shape of these solids depends on the mesh structure. Usually, in the case of structural grids, the assumed shape is cuboid (hexahedron), while in nonstructural grids it is a tetrahedron. These separated areas of the domain may be called computational cells, control volumes, or finite volumes.

The generalized transport equation in the FVM can be expressed as the integral form referring to the control volume (CV) that is surrounded by the control surface (CS):

$$\frac{\partial}{\partial t}\int_{CV}\rho\phi\,dV + \int_{CS}\rho\phi\boldsymbol{u}\ast\boldsymbol{n}\,dS = \int_{CS}\Gamma grad\phi\ast\boldsymbol{n}\,dS + \int_{CV}S_\circ\,dV. \quad (14.1)$$

If the transport of the intensive variable takes place in a transient system, the above equation should be integrated to achieve the following relation:

$$\int_{t}^{t+\Delta t}\left(\frac{\partial}{\partial t}\int_{CV}\rho\phi\,dV\right)dt + \int_{t}^{t+\Delta t}\left(\int_{CS}\rho\phi\boldsymbol{u}\ast\boldsymbol{n}\,dS\right)dt$$
$$= \int_{t}^{t+\Delta t}\left(\int_{CS}\Gamma grad\phi\ast\boldsymbol{n}\,dS\right)dt + \int_{t}^{t+\Delta t}\left(\int_{CV}S_\circ\,dV\right)dt. \quad (14.2)$$

The above equation serves as a basis for the computation of a discretized form of the transport equations. Some simplifications, however, are applied to reduce the case to a simple set of derivative equations [20].

Solving N–S and other governing equations directly can be very challenging, since most fluid flows, especially turbulent, are not stationary and fluctuate considerably over time. To tackle this issue, the RANS (Reynolds averaged Navier-Stokes) time averaging method can be applied to all transport equations. For a steady state process, the current value of the intensive variable can be expressed as the sum of fluctuations and the time average value as given below:

$$\phi(\boldsymbol{x},t) = \overline{\phi}(\boldsymbol{x}) + \phi'(\boldsymbol{x},t). \quad (14.3)$$

In the case of a transient process, the ensemble average is applied:

$$\langle\phi(\boldsymbol{x},t)\rangle = \lim_{N\to\infty}\left[\frac{1}{N}\sum_{n=1}^{N}\phi^n(\boldsymbol{x},t)\right]. \quad (14.4)$$

In general, the statistical definition of the expected value is applied in order to generalize the combination of fluctuation and ensemble average:

$$\overline{\phi}(\boldsymbol{x},t) = E[\phi(\boldsymbol{x},t)]. \quad (14.5)$$

The value of the average fluctuation is equal to 0. This averaging of linear expressions does not introduce any new expressions into the system; however, it is different in the case of nonlinear expressions, where additional nonzero terms are created. These new elements in the equations represent the contribution of the turbulent transport in the mean flow and result in new variables in the system of equations. Thus, additional equations that represent the phenomena of turbulent flow are applied in order to balance the system of equations [21–23].

CFD as a Tool in Studies of Biomass-Based Devices

The expression that derives from the application of RANS averaging in the general transport equation is given below:

$$\frac{\partial(\rho\bar{\phi})}{\partial t} + div(\rho\bar{\phi}\bar{\boldsymbol{u}} + \overline{\rho\phi'\boldsymbol{u}'}) = div(\Gamma grad\bar{\phi}) + \bar{S}_{\phi}. \qquad (14.6)$$

Independently from the mathematical bases of discretization and solution in the model, some discretization errors occur and influence the results of the simulation in a different manner. Usually, where the diffusion is much less significant than convection, the most important discretization error type is the so-called numerical diffusion [25].

Discretization errors result from the approximation of the advection terms in the transport equations in the CFD model [25]. Stream components for the transport equation are usually approximated on the surfaces dividing the cells in the computational grid, thus the general distribution of the variable depends on the method of this approximation.

To avoid an excessive discretization error, it is necessary to select an appropriate spatial discretization method. The description of the impact of the spatial discretization scheme on the computation results may be found in References [26–28].

Besides the spatial discretization, time discretization occurs in transient systems where time has to be considered as the fourth dimension. Just like the cell is the basic element of the computational grid used in spatial discretization, the time step is the basic element of time discretization. An appropriate selection of the time step in the actual time simulation is the key factor that influences the accuracy of the results and has to be set depending on the type of simulated phenomena [29–33].

The postprocessing of the simulation results is the next, near-final step of the study. Although it could be carried out using a number of independent tools, most modern CFD platforms are equipped with dedicated modules that allow a complex simulation report to be created, containing isolines and vectors distribution, charts, tables, etc.

After processing the data obtained in the simulation, it is necessary to perform a model validation. The CFD result needs to be compared with other available data to verify the agreement level and ultimately to confirm the accuracy of the simulation. Depending on the available resources, results obtained by other researchers, data from validation experiments, or analytical calculations may be used at this point. However, it is always recommended to apply as many validation methods as necessary to eliminate potential mistakes in the CFD model. The validation process allows to conclude whether the CFD model is worthwhile—and that CFD actually stands for computational fluid dynamics rather than colors for deciders.

14.3 SIMULATING THE COMBUSTION PROCESS USING CFD

From a practical point of view, combustion is one of the most important processes in view of thermochemical biomass conversion. It combines a number of physical and chemical phenomena, such as turbulent flow and its interactions with the chemistry, heat transfer, and chemical processes. Owing to this fact, reaching the chemical equilibrium requires some time that may be approximated based on the mathematical description of the chemical kinetics of the process, including the reaction mechanisms and their rates.

According to the simplest definition of combustion, the fuel and the oxidizer react and result in a product. In reality, however, even the simplest reaction includes a series of elementary reactions that result in the presence of some transitional substrates (such as radicals) in the system. The radicals are unstable and very reactive. Usually, the radicals that take part in such processes are H, O, OH, CH, HO_2, or C_2. To exemplify this, simple hydrogen combustion can be presented as [19]:

$$H_2 + \frac{1}{2}O_2 \rightarrow H_2O. \tag{14.7}$$

Even in the case of such a basic process, a series of elementary reactions may be listed:

$$H_2 + O_2 \leftrightarrow 2OH, \tag{14.8}$$

$$H_2 + OH \leftrightarrow H_2O + H, \tag{14.9}$$

$$H + O_2 \leftrightarrow OH + O, \tag{14.10}$$

$$H_2O + O \leftrightarrow OH + OH, \tag{14.11}$$

$$H + OH + M \leftrightarrow H_2O + M, \tag{14.12}$$

where M stands for a random additional compound that takes part in the reaction.

In the case of CFD-simulated combustion, the approach that assumes a detailed description of the chemistry (including the detailed description of mechanisms) results in a cost inefficiency of the entire study, owing to the long computation time and—in most cases—the application of a high power computer. Most of the studies related to biomass combustion, or thermochemical conversion in general, need to give consideration to dynamic, time-dependent processes, which additionally increases the time required to obtain the final result. Furthermore, it is always difficult to achieve a satisfying convergence of the governing equations' solution, thus the result of the CFD analysis might be of unacceptable quality.

The efficient way to solve the issues of combustion in complex reactors is the application of reduced reaction mechanisms. This approach allows for good quality results in academic-level research to be obtained. Thanks to this method, it is possible to significantly reduce the complexity of the combustion model by removing selected stages of reactions or even entire reactions from the system. Usually, when the goal of the study is to find a relationship between the phenomena or to optimize the reactor (e.g., a boiler or gasifier), the process has little meaning as compared to the effect. In such a case, the reduced reaction mechanisms become an effective tool for providing balance between the computation time and the accuracy of the simulation results. This factor becomes crucial in the case of industrial-scale studies.

CFD as a Tool in Studies of Biomass-Based Devices 325

Each chemical process exhibits certain rates of product creation and consumption of reagents. The correct definition of these parameters is crucial in CFD modeling of biomass thermochemical conversion. Owing to the complexity of the process, the same compound or chemical element may be both a substrate and a product at different stages of the chemical reaction, as well as in different reactions.

Chemical processes such as combustion can be expressed as [19]:

$$\sum_{k=1}^{N} v'_{kj} M_k \rightarrow \sum_{k=1}^{N} v''_{kj} M_k \quad \text{for } j = 1, 2, \ldots, m. \tag{14.13}$$

The reaction rate can be expressed as [19]:

$$\dot{q}_{kj} = \frac{dC_{M_k}}{dt} = (v''_k - v'_k) k_f \prod_{k=1}^{N} (C_{M_k})^{v_{kj}}. \tag{14.14}$$

The most useful and common way of expressing the reaction rate coefficient is the Arrhenius equation:

$$k_f = A T^\alpha \exp\left(-\frac{E_a}{R_u T}\right). \tag{14.15}$$

To solve the problem of describing combustion, the reaction rate coefficient has to be applied in order to calculate the volumetric rate of the substrate consumption or product creation. This allows the changes of concentrations during the chemical process to be expressed. In case of the CFD reacting flow simulation, where volumetric species transport occurs, a separate transport equation is required for each compound and chemical element present in the system. The aforementioned volumetric rate is implemented in each reagent transport equation as the source term that is positive for the production and negative in the case of reagent consumption.

The transport equation for the Y_k component of the reacting mixture can be expressed in a general form as [19]:

$$\frac{\partial}{\partial t}(\rho Y_k) + \frac{\partial}{\partial x_i}(\rho u_i Y_k) = \frac{\partial}{\partial x_i}\left(\rho D \frac{\partial Y_k}{\partial x_i}\right) + \dot{\omega}_k. \tag{14.16}$$

Of course, the chemical energy that is related to the considered process has to be taken into account in the energy balance. The basic form of the equation that allows for the enthalpy to be calculated is given below [19]:

$$\frac{\partial}{\partial t}(\rho h) + \frac{\partial}{\partial x_i}(\rho u_i h) = \frac{\partial}{\partial x_i}\left[\frac{\mu}{\sigma_h}\frac{\partial h}{\partial x_i} + \mu\left(\frac{1}{Sc_k} - \frac{1}{\sigma_h}\right)\sum_{k=1}^{N} h_k \frac{\partial Y_k}{\partial x_i}\right] + \frac{\partial p}{\partial t} + S_{rad}. \tag{14.17}$$

Of course, as in the CFD model of nonreacting fluid flow, mass and momentum conservation principles still apply.

14.4 CHALLENGES IN CFD MODELING OF THERMOCHEMICAL TREATMENT OF BIOMASS

Although, even on a basic level, CFD heat and mass transfer simulations are still considered quite complex, they are comparatively simple as compared to the modeling of reactive flows. As far as the combustion, gasification, or other similar models do not just serve as examples in an academic course, the correct design, solution, and interpretation thereof may pose an actual challenge for the researcher.

Numerous different barriers may occur in the case of CFD models aimed at studying energy devices at an actual, commercial scale. In fact, some difficulties may be noted even at the very beginning of the design process, as the shape of the simulated device is usually quite complex. This may result in a time-consuming and complicated process aimed at generating a mesh characterized by a sufficient quality.

Nevertheless, it is not the only one problem. To achieve a reliable reflection of the simulated process, it is crucial to correctly determine the boundary conditions, which are often difficult or even impossible to measure. Moreover, these conditions may change during the process—solid biomass combustion is strongly time-dependent. It is recommended to use as much experimental data as possible as inputs and to adapt the same experimental stand in the subsequent validation of the model.

A third group of barriers refers to the physical and chemical characteristics of the simulated phenomena. As was previously mentioned, the detailed simulation of chemical processes requires good knowledge of the subject matter, including reaction kinetics, chemical relations between reagents, chemistry-turbulence dependences, the impact of radiation on the process, and other issues.

Aside from the stage of designing the model (geometry, grid, setting physics, and chemistry), the effective compromise between the accuracy of the result and the model weight is crucial for the final achievement of the assumed goal. Very often this compromise is reached at the expense of accuracy of reflecting a given phenomenon. As long as this approach allows for the goal to be achieved, its application is adequate, as the simplest models usually provide the best answers. Such models may provide a lot of information regarding the process, which can be helpful in the reduction of the time or range of the experiment and, broadly speaking, support the designing and prototyping processes.

14.5 FEATURES OF REACTING FLOWS AND COMBUSTION MODELING

A large number of classic and modern technologies for utilization of biomass in small-scale devices for energy production purposes result in the lack of a single approach that can be applied universally. The process of designing a CFD model of biomass combustion or related processes has to begin with determining the manner in which the fuel and the oxidizer are supplied into the system. Depending on the device type, these reagents can be fed individually or as a mixture. Owing to this

CFD as a Tool in Studies of Biomass-Based Devices

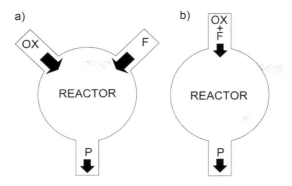

FIGURE 14.3 Graphic representation of premixed and non-premixed combustion: "OX"—oxidizer, "F"—fuel, "P"—product.

feature, two basic types of flame, or generally combustion, may be distinguished (Figure 14.3) [34]:

- Non-premixed combustion, which is related to the so-called diffusion flame that is formed in the conditions of two individual inlets of fuel and oxidizer to the system; reagents are transported to the reaction area from each side of the flame.
- Premixed combustion, which causes the formation of a kinetic flame, where both reagents are completely mixed and the reaction front is directly related to the temperature front in the flame.

Of course, an indirect case, in which none of above described cases can be clearly indicated is also possible. In such a case, it can be said that partially-premixed combustion occurs.

Apart from the moment when the mixing takes place, it is necessary to determine the characteristics of this phenomenon. In most industrial-scale biomass-fired energy devices, the transport phenomena are turbulent. Thus, the mixing process is very intensive and, in general, the chemical process takes place in turbulent conditions. The correct type of interaction between the turbulence and the chemistry in the CFD model always has to be determined and it has a significant impact on the accuracy of the result.

The sequence of the phenomena occurring, for example, in the combustion chamber of a biomass-fired boiler, where a turbulent flow field is observed, resemble a closed loop (Figure 14.4). In this loop, the turbulences and flame have a mutual impact and form system that may even have an influence on the entire reactor. Such a situation occurs in a certain area until the reagents have been exhausted.

Based on the phenomena referred to above, it may be concluded that it is necessary to determine if the flow is turbulent or not prior to deciding which approach to the simulation design is adequate for a given device and process model. In systems in which only the laminar fluid flow occurs, the range of possible modeling methods of combustion or related processes is significantly reduced. However, the presence of

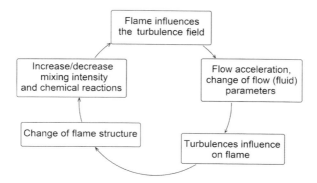

FIGURE 14.4 Relationship between the turbulence field and the flame.

turbulences in the system makes it unsubstantiated to apply certain models that are dedicated to simulating reactions in laminar cases. The Reynolds number allows the applicability of turbulence models to be determined:

$$Re = \frac{inertial\ force}{viscous\ effects} = \frac{\rho U L}{\mu}. \tag{14.18}$$

Another useful indicator for determining the character of combustion in the gas phase is the Damköhler number. It allows the transport phenomena and the reaction timescales to be correlated [35–37]:

$$Da = \frac{Mixing\ timescale}{Chemical\ timescale} = \frac{L/U}{\rho_{ad}/R_{slow}} = \frac{k/\varepsilon}{\rho_{ad}/R_{slow}}. \tag{14.19}$$

If the slowest reaction is still fast enough, the case may be simplified to one where substrates react in the moment of first contact. Then, Da ≫ 1 and the so-called "fast chemistry" takes place. Otherwise, when Da ~ 1, the so-called "slow chemistry" processes occur. From the CFD simulation point of view, the statement that the system is based on fast or slow chemistry assumptions results in the requirement of assuming a different approach to the description of the simulated process.

In the case of small-scale boilers, stoves, fireplaces, biomass burners, or incinerators, as well as biomass gasifiers, Da usually significantly exceeds 1 and thus, the reaction kinetics may be neglected and it can be assumed that the chemical process is limited by the turbulent mixing of the reagents. However, owing to this fact, the selection of the correct approach to turbulence modeling is a key issue that needs to be considered very carefully.

Notwithstanding the above, in many cases such as the research regarding pollution, soot decomposition, or ignition and extinction of the flame, the Da value may be close to 1 and it may be reasonable to design the model based on the slow chemistry (finite rate chemistry). In that case, contrary to the fast chemistry, reactions are limited by chemical kinetics. Owing to this fact, reliable values of kinetic parameters have to be applied and a correct definition of the reaction mechanisms has to be carried out. The time relationships between the turbulence field and the reactions are the key issues.

14.6 ROLE OF MULTIPHASE TRANSPORT IN CFD MODELING OF SMALL-SCALE BIOMASS-BASED ENERGY DEVICES

Among the wide range of problems related to combustion, one should mention multiphase transport with solid or fluid particles. Although the majority of these kinds of issues are related to the utilization of fossil fuel-based devices (oil burners, pulverized coal-fired boilers, fluidized bed boilers, and gasifiers, etc.), the CFD simulations of multiphase biomass combustion cannot be omitted in this discussion.

For example, transportation of the solid particles with the exhaust gas is undesirable in view of the heat balance of the biomass boiler owing to the content of combustible organic components. The amount of energy in the material leaving the combustion chamber with the exhaust gas is called the loss of fly ash and it depends on the physical properties of the solid particle components (specific heat, conductivity, density, etc.). However, while the organic part of the particulate matter (PM) causes problems, the inorganic ones are also problematic due to the possible presence of potassium, sulfur, sodium, and zinc, which could lead to increased emissions. The greatest amount of PM emissions from biomass combustion consists of particles with an aerodynamic diameter smaller than 2.5 [38].

Of course, the inclusion of PM in the simulation may have a different impact on the reliability of the results depending on the technologies applied in the device concerned. Discrete phase transport would constitute yet another physical value in wood-fired stoves or fireplaces, while it would constitute a substantial issue in the case of fluidized bed gasifiers (full multiphase modeling is then applied). Returning to solid biomass combustion and gasification, the decision to expand the model by discrete phase transport needs to be based on the energy and environmental impact analysis of the considered device. It should be determined if a given fuel and the solid phase transport phenomenon would result in a significant increase of PM emissions and the reduction of efficiency (fly ash loss and heat accumulation in the particulate matter).

The inclusion of two-phase transport (especially in the case of full multiphase models) always results in a longer computation time as well as some difficulties in achieving the required convergence level in some transport equations. Usually, the greater the content of the second phase (where fluid is the continuous, primary phase) in the reactor, the bigger the problem with achieving a quick computation and high convergence.

Generally (based on the ANSYS Fluent solver terminology and functionalities [34]), different characteristics of the mass and heat transfer can be related to some specific cases of discrete phase transport:

- Mass-less transport without drag forces (only for very simple problems)
- Inert transport, where a particle may exchange heat with the surroundings, however disregarding chemistry or phase-change phenomena
- Droplet, where both the heat transfer and the phase-change (evaporation and boiling) are possible

- Multicomponent, similar to the above but for multicomponent processes
- Combusting, with heat and mass transfer related to the heterogeneous chemical reactions occurring between solid and fluid phases

The thermochemical conversion of the solid particle that is transported with the gas phase through the reactor, however, has to include all the important physical and chemical stages, namely:

- Inert heating of the particle material
- Moisture release
- Devolatilization
- Phase-change phenomena besides the moisture release (if required)
- Char combustion
- Further inert heating

General approaches that are usually applied to define the mass transfer laws in case of solid fuel combustion is presented in Figure 14.5.

Of course, in the gas phase, the devolatilization products that are no longer directly related to the solid particle conversion have to be included as an additional chemical equation or equations.

The ANSYS Fluent solver offers a mathematical apparatus that allows for multiphase simulations using different approaches, which are aimed at specific issues, to be run:

- The discrete phase model or DPM, is a classic method of simulating discrete phase transport in a Lagrangian reference frame in case of fluid droplets and solid particles. This approach is limited to a maximum of 10% of discrete phase volume in relation to the continuous phase. This is due to the lack of interaction between the particles.

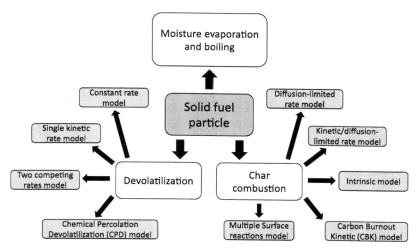

FIGURE 14.5 Common approaches to modeling of devolatilization and char combustion in solid fuel burning (as exemplified by the ANSYS Fluent functionality). (Based on ANSYS Fluent 17.2 Help (ANSYS Workbench 17.2 help files).)

CFD as a Tool in Studies of Biomass-Based Devices

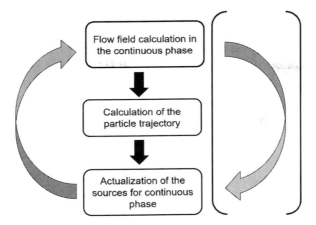

FIGURE 14.6 The coupling of continuous and discrete phases, based on the example of a Lagrangian reference frame.

- Multiphase models that are based on the Eulerian reference frame. This includes three basic approaches, each for a different kind of phase configuration: Euler-Euler for gas–liquid flows; Euler-granular for gas–solid flows; volume of fluid for the gas–liquid configuration also (it is dedicated, however, to systems in which phases are not dispersed, and a clear phase boundary can be observed).
- The dense discrete phase model (DDPM) that consist of Lagrangian tracking of the phase transport in finite volume in the considered Euler reference frame.
- The discrete element method, allowing for DDPM to be combined with efficient modeling of particle collisions.

Figure 14.6 presents a schematic diagram of interactions between continuous and discrete phases in the area of the computational domain. Movement of the particles in the chemical reactions region affects the fluid flow field. Thus, in most cases the discrete phase transport (except the ones disregarding phase interactions) is coupled with the chemistry. In the case of the inert particles, the fields of turbulences and temperature are changed, while in other cases the discrete phase takes an active part in chemical reactions.

14.7 THERMOCHEMICAL CONVERSION OF SOLID FUELS IN FIXED-BED DEVICES

Despite the understood fact that the majority of the currently available CFD solvers provide for modeling of combustion in the gas phase or multiphase processes with discrete phase transport, it is not sufficient from the point of view of modeling the biomass-based energy-appliances.

In general, combustion or gasification of biomass (woodchip, log, straw bale, etc.) is not different from combustion of a small particle that is simulated using discrete phase transport methods—both cases concern thermochemical conversion of the solid fuel. However, the previously presented tools that are widely used to describe the chemistry of small particles are not appropriate for packed or fixed bed devices.

The physical and chemical properties of biomass may differ depending on many factors, which makes the description of the chemistry much more difficult than for example, coal. One of the important problems results from the highly diverse kinetics of devolatilization, depending on the type of biofuel (the content of volatiles and kinetics of fuel decomposition is difficult to predict for certain fuels—only general, estimated values are available) [39,40]. Moreover, the form of the fuel itself is problematic due to the requirement of a precise description of the bed geometry changes during the process (this, of course, also concerns coal fixed-bed appliances). Only several, very general issues need to be solved to gain the understanding of this problem, but a number of new ones appear when each of the cases is considered in more detail. Broadly speaking, it may be stated that owing to the very specific characteristics of biomass and the number of dedicated technologies, it is difficult to develop efficient modeling methods that could be universal and provide reliable results.

Currently, the most popular way to carry out CFD simulations in such cases is to extend the software's functionality using the designer's original script, which can be introduced into the CFD solver. In the ANSYS Fluent solver it is possible by means of the so-called user defined functions (UDFs) that allow, among other things, the boundary conditions, the physical and chemical properties, the reaction kinetics, and transport phenomena using macros created in the C programming language, to be described.

A common and reliable approach to defining the fuel bed assumes that it should be considered a porous media [41–44]. In such a case, the UDFs can be initially used to describe the relationship between the porosity and medium permeability and the thermal (or other) conditions in the reactor. Subsequently (and this is the most interesting issue), UDFs may be applied to modify the selected governing equation by implementing additional source terms. A simple example that may be provided is a function that sets the value of the source of water released from the moisture contained in the fuel, and the negative source that controls the decrease of this moisture (owing to the release to the gas phase). In the simplest case, the argument in this function is the temperature. Moreover, the sources control the process of devolatilization and char combustion (owing to the description of kinetics by means of the Arrhenius equation). Such an approach is well-substantiated as long as each devolatilization and combustion product is assigned its own transport equation in which the source can be applied. As in the example relating to moisture, functions describing the change of the mass of fuel and residues from the thermochemical conversion can be developed and applied in the solver to register the shrinking of the simulated fuel.

Owing to the highly time-dependent characteristics of the process, it is necessary to simulate the heterogeneous combustion or gasification with a fixed fuel bed in a dynamic-transient mode. Thus, some variables need to be set (or reset) at the beginning of each new time step of the simulation. Owing to this fact, UDFs are often applied to control this process in these kinds of models. A simplified summary of the design process of a model of combustion in a fixed-bed device using the methodology presented above is shown in Figure 14.7.

CFD as a Tool in Studies of Biomass-Based Devices

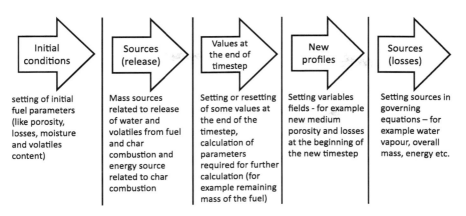

FIGURE 14.7 Proposal of a design process of a biomass-fixed-bed-combustion-CFD model with user defined functions.

14.8 SELECTED POPULAR COMBUSTION MODELS FOR FAST AND SLOW CHEMISTRY

As it was previously stated, performing time-efficient and reliable simulations of the combustion process is one of the biggest challenges in CFD. Although the chemistry characteristics (fast/slow, premixed/non-premixed, etc.) may require the application of different approaches, the so-called species transport approach can be used in the majority of cases.

Four different models encompassing fast, slow, and transition chemistry cases can be listed based on the example of ANSYS Fluent [34]:

- Eddy dissipation model (EDM)
- Finite rate model (FRM)
- Finite rate/Eddy dissipation model (FREDM)
- Eddy dissipation concept (EDC)

First of all, the selection of the appropriate approach needs to be based on the chemistry—the turbulence relation and the assessment of the significance of detailed reaction kinetics.

Of course, the evaluation of the significance of the kinetics is based on the Damköhler number (Equation 14.19).

14.9 EDDY DISSIPATION MODEL

For highly turbulent reacting flow, where the mixing rate is much higher than the chemical rate, it may be assumed that the reaction time is negligible. In that case, the reaction is limited by mixing and once the reagents reach contact, the reaction is completed. In this situation $Da \gg 1$ and the EDM approach is applicable.

However, as in other approaches, EDM has some drawbacks that need to be considered before using it in a given instance. The most important feature of the method is that the impact of the chemistry on the reaction is completely omitted due to the assumption of the fast chemistry conditions. In practice, this means that the reaction kinetics are not taken into account and the reaction rate is governed by large-eddy mixing time scale. It may thus be stated that the reaction is limited only by the mixing which does not occur in laminar flow. The original version of this model (eddy break-up model—EBUM) has been created by Spalding [45] and then elaborated on by Magnussen and Hjertager [46] to allow the application in non-premixed combustion.

Owing to the omission of the reaction kinetics, another method to express the reaction rates is required. The equation for the rate of species production is given below [46–49]:

$$R_{i_r} = MIN\left(R_{i_r(react)}, R_{i_r(prod)}\right). \tag{14.20}$$

$$R_{i_r(react)} = v_{i_r} M_{W_i} A \rho \frac{\varepsilon}{k} \min\left(\frac{Y_i}{v_{i_r} M_{W_i}}\right). \tag{14.21}$$

$$R_{i_r(prod)} = v_{i_r} M_{W_i} A B \rho \frac{\varepsilon}{k} \min\left(\frac{\sum_p (Y_p)}{\sum_j^N v_{j_r} M_{W_j}}\right). \tag{14.22}$$

Based on the mathematical description of the model's chemistry presented above, it may be noted that temperature does not impact the reaction rate. This may result in an over-prediction of the temperature distribution in the reaction area. Moreover, there is no possibility to introduce advanced reaction schemes with detailed chemistry effects. Although it is impossible to solve the latter problem here, the undesirable thermal effects can be reduced by a correct calibration of the A and B constants. However, this requires either experience in this kind of simulation, which may be a slightly subjective, or detailed experimental data.

14.10 FINITE RATE MODEL AND FINITE RATE/EDDY DISSIPATION MODEL

Although most of the commercially available biomass-based small-scale energy devices can be considered fast chemistry systems, in some instances this may not serve as a basis for the simulation. In some of the reactors with a relatively large reaction zone (straw-fired batch boilers, gasifiers, and incinerators), or a simplified air distribution system (as in fireplaces), it is possible to distinguish flow-stagnation regions besides the turbulent-flow area, where the temperature is high enough to allow certain chemical reactions to proceed owing to radiation (such as decomposition), even if no efficient mixing occurs.

In this case, the application of the model based on fast chemistry, such as the classic EDM, leads to seriously disturbed simulation results. On the other hand, the presence of turbulent-flow areas suggests that the methodology of modeling based on the mixing rate could provide good effects in these areas of the reactor.

In these complex cases, a combination of two different approaches that control the process in specific reaction conditions may be implemented. The FREDM allows the reaction rates between kinetics-dependent and mixing-dependent to be switched. The evaluation of conditions is based on the calculation of both the turbulence timescale and the Arrhenius equation (or Arrhenius chemical rate coefficient) that is related to the activation energy and temperature (Equation 14.15) and thus, the kinetics. It is also common to use one global reaction mechanism for the kinetic reaction when no detailed chemistry is required.

The rates of production and consumption of reagents calculated using these two methods are compared and the smaller value is selected as the one that controls the reaction in a given reactor area. In dynamic simulations, of course, besides the region of computational domain, the chemistry type may vary in time owing to the changes of the flow characteristics and the temperature field.

Every time a classic case of slow chemistry is considered (such as gasification in laminar flow), the FRM can be used separately. Then, only the detailed chemistry approach is applied. This approach is useful in each case where not just the result, but—most of all—the chemical process is considered. It is possible to create sets of transition reactions and study the advanced chemistry—for example, related to the formation of pollutants.

This advantage, however, may also pose a problem. The reliable representation of complex chemistry in certain processes requires the setting of kinetic parameters and rate exponents. The determination of these data may require running experiments—especially in the case of relatively complex processes (i.e., simultaneous thermal analysis—STA).

14.11 EDDY DISSIPATION CONCEPT MODEL

It is always a challenge to represent actual chemical phenomena in a turbulent, reacting mixture flow. In CFD simulations, the representation of actual chemical mechanisms proceeding in turbulent combustion conditions entails a high computational costs. However, this approach is also widely used in studies that require detailed analysis of the phenomena that take a place in the reactor. In these instances, it is necessary to employ methods fully encompassing the interactions between the chemistry and the turbulences in the considered system.

In 1996, Gran and Magnussen [50–54] developed an offshoot of the EDM that allows the considerations to the finite-rate chemistry to be extended. However, the new method has significantly differed from the original model, owing to, among other things, the scales of reactions.

To understand the idea behind the eddy dissipation concept model (EDCM), one may imagine that the reaction area is somewhat discretized to a certain number of very small, individual reactors where the chemistry occurs. It means that the analyses relate to small-scale phenomena. The processes in these "small-scale reactors" take place in constant pressure enthalpy conditions. Based on the above, it can be concluded that a certain length of scale is certainly important, but the timescale of the reaction occurring in a given reactor needs additionally to be considered.

Equations 14.23 and 14.24, provided below, describe the length fraction of small individual reactors (γ) and the timescale of the reaction (τ) that occurs inside, respectively.

$$\gamma = 2.1377 \left(\frac{v\varepsilon}{k^2}\right)^{1/4}. \qquad (14.23)$$

$$\tau = 0.4082 \sqrt{\frac{v}{\varepsilon}}. \qquad (14.24)$$

Based on these two parameters and species, the mean reaction rate can be expressed as provided below:

$$S_k = \frac{\rho \gamma^2}{\tau(1-\gamma^3)}(Y_k^* - Y_k). \qquad (14.25)$$

The EDCM provides the possibility to decide on how complicated the chemistry in a considered CFD simulation should be. Furthermore, it has to be noted that in case of the EDCM, there is no requirement to apply certain unintuitive constants (as in the EDM). However, it should always be remembered that the more complex the case is, the more unknowns will have to be considered.

14.12 SUMMARY

The information provided in this chapter was aimed at presenting selected methods that may be used to perform RANS-based CFD analyses of small-scale biomass-based energy devices. However, the context is general enough to be useful at the early stages of many different CFD projects. The authors' goal was not to cover the entire range of methods that are available for CFD modeling of combustion or gasification, but to indicate approaches that are common in current engineering studies and to inspire beginners to search for the best solutions to the engineering problems they are facing.

Irrespective of the method selected to solve the biomass thermochemical conversion issue, however, it is necessary to go through the main stages that are the same in the case of all software, as exhibited in Figure 14.8.

It should be remembered that all the steps need to be performed with equally great attention to detail. An excessively optimistic approach to the results of the simulation, without a careful consideration of the methodology, will most likely result in further difficulties. Despite seeming impressive and very convincing, the results of simulations without an in-depth validation are just a collection of colorful visualizations.

CFD as a Tool in Studies of Biomass-Based Devices

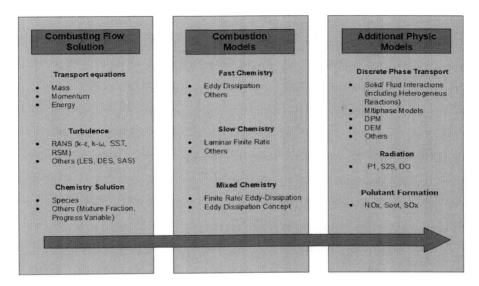

FIGURE 14.8 The main stages that need to be completed in CFD simulations of thermochemical treatment of biomass. (Based on ANSYS Fluent 17.2 Help (ANSYS Workbench 17.2 help files))

REFERENCES

1. M. S. Thordal, J. C. Bennetsen, H. Holger, H. Koss, Review for practical application of CFD for the determination of wind load on high-rise buildings, *Journal of Wind Engineering and Industrial Aerodynamics*, vol. 186, 2019, p. 155–168.
2. Y. Toparlar, B. Blocken, B. Maiheu, G. J. F. van Heijst, A review on the CFD analysis of urban microclimate, *Renewable and Sustainable Energy Reviews*, vol. 80, 2017, p. 1613–1640.
3. S. R. Shah, S. V. Jain, R. N. Patel, V. J. Lakhera, CFD for centrifugal pumps: A review of the state-of-the-art, *Procedia Engineering*, vol. 51, 2013, p. 715–720.
4. P. Hui, C. Xi-Zhong, L. Xiao-Fei, Z. Li-Tao, L. Zheng-Hong, CFD simulations of gas–liquid–solid flow in fluidized bed reactors—A review, *Powder Technology*, vol. 299, 2017, p. 235–258.
5. J. Silva, J. Teixeira, S. Teixeira, S. Preziati, J. Cassiano, CFD Modeling of combustion in biomass furnace, *Energy Procedia*, vol. 120, 2017, p. 665–672.
6. M. A. Gómez, R. Martín, S. Chapela, J. Porteirob, Steady CFD combustion modeling for biomass boilers: An application to the study of the exhaust gas recirculation performance, *Energy Conversion and Management*, vol. 179, 2019, p. 91–103.
7. H. Mätzing, H. J. Gehrmann, H, Seifert, D. Stapf, Modelling grate combustion of biomass and low rank fuels with CFD application, *Waste Management*, vol. 78, 2018, p. 686–697.
8. M. R. Karim, J. Naser, CFD modelling of combustion and associated emission of wet woody biomass in a 4 MW moving grate boiler, *Fuel*, vol. 222, 2018, p. 656–674.
9. U. Kumar, M. C. Paul, CFD modelling of biomass gasification with a volatile break-up approach, *Chemical Engineering Science*, vol. 195, 2019, p. 413–422.
10. S. Chapela, J. Porteiro, M. A. Gómez, D. Patiño, J. L. Míguez, Comprehensive CFD modeling of the ash deposition in a biomass packed bed burner, *Fuel*, vol. 234, 2018, p. 1099–1122.

11. T. Jiyuan, Y. Gaun-Heng, L. Chaoqun, *Computational Fluid Dynamics. A Practical Approach*, Butterworth-Heinemann, Oxford, 2013, pp. 33, 37–38, 160–173.
12. M. Szubel, M. Filipowicz, B. Matras, S. Podlasek, Air manifolds for straw-fired batch boilers— Experimental and numerical methods for improvement of selected operation parameters, *Energy*, vol. 162, 2018, p. 1003–1013.
13. E. Przenzak, G. Basista, P. Bargiel, M. Filipowicz, Research on high-temperature heat receiver in concentrated solar radiation system, *EPJ Web of Conferences*, vol. 143, 2017.
14. Z. A. Paul, G. Tucker, S. Shahpar, Optimal mesh topology generation for CFD, *Computer Methods in Applied Mechanics and Engineering*, vol. 317, 2017, p. 431–457.
15. A. Katz, V. Sankaran, Mesh quality effects on the accuracy of CFD solutions on unstructured meshes, *Journal of Computational Physics*, vol. 230, 2011, p. 7670–7686.
16. F. Gagliardi, K. C. Giannakoglou, A two-step radial basis function-based CFD mesh displacement tool, *Advances in Engineering Software*, vol. 128, 2019, p. 86–97.
17. G. Abbruzzese, M. Gómez, M. Cordero-Gracia, Unstructured 2D grid generation using overset-mesh cutting and single-mesh reconstruction, *Aerospace Science and Technology*, vol. 78, 2018, p. 637–647.
18. S. Menon, K. G. Mooney, K. G. Stapf, D. P. Schmidt, Parallel adaptive simplical re-meshing for deforming domain CFD computations, *Journal of Computational Physics*, vol. 298, 2015, p. 62–68.
19. H. K. Versteeg, W. Malalasekera, *An Introduction to Computational Fluid Dynamics. The Finite Volume Method*, Pearson Education Limited, Harlow, 2007, p. 179–211.
20. R. B. Bird, W. E. Stewart, E. N. Lightfoot, *Transport Phenomena*, John Wiley & Sons, New York, 2002, p. 10–25.
21. J. T. Davies, *Turbulence Phenomena. An Introduction to the Eddy Transfer of Momentum, Mass, and Heat, Particularly at Interfaces*, Academic Press, Cambridge, 1972, p. 79–120.
22. J. M. O'Brien, T. M. Young, J. M. Early, P. C. Griffin, An assessment of commercial CFD turbulence models for near wake HAWT modelling, *Journal of Wind Engineering and Industrial Aerodynamics*, vol. 176, 2018, p. 32–53.
23. L. Xiangdong, T. Jiyuan, Evaluation of the eddy viscosity turbulence models for the simulation of convection–radiation coupled heat transfer in indoor environment, *Energy and Buildings*, vol. 184, 2019, p. 8–18.
24. P. R. Spalart, Philosophies and fallacies in turbulence modeling, *Progress in Aerospace Sciences*, vol. 74, 2015, p. 1–15.
25. M. T. Odman, A quantitative analysis of numerical diffusion introduced by advection algorithms in air quality models, *Atmospheric Environment*, vol. 31, 1997, p. 1933–1940.
26. H. R. Thakare, A. D. Parekh, CFD analysis of energy separation of vortex tube employing different gases, turbulence models and discretisation schemes, *International Journal of Heat and Mass Transfer*, vol. 78, 2014, p. 360–370.
27. P. Cyklis, P. Młynarczyk, The influence of the spatial discretization methods on the nozzle impulse flow simulation results, *Procedia Engineering*, vol. 157, 2016, p. 396–403.
28. B. P. Leonard, S. Mokhtari, Beyond first-order upwinding: The ultra-sharp alternative for non-oscillatory steady-state simulation of convection, *International Journal of Numerical Methods in Engineering*, vol. 30, 1990, p. 729–766.
29. F. Tabet, V. Fichet, P. Plion, A comprehensive CFD based model for domestic biomass heating systems, *Journal of the Energy Institute*, vol. 89, 2016, p. 199–214.
30. R. Mikulandrić, D. Böhning, R. Böhme, L. Helsen, M. Beckmann, D. Lončar, Dynamic modelling of biomass gasification in a co-current fixed bed gasifier, *Energy Conversion and Management*, vol. 125, 2016, p. 264–276.
31. M. A. Gómez, J. Porteiro, D. De la Cuesta, D. Patiño, J. L. Míguez, Dynamic simulation of a biomass domestic boiler under thermally thick considerations, *Energy Conversion and Management*, vol. 140, 2017, p. 260–272.

32. M. A. Gómez, J. Porteiro, D. Patiño, J. L. Míguez, Eulerian CFD modelling for biomass combustion. Transient simulation of an underfeed pellet boiler, *Energy Conversion and Management*, vol. 101, 2015, p. 666–680.
33. F. Balduzzi, A. Bianchini, G. Ferrara, L. Ferrari, Dimensionless numbers for the assessment of mesh and timestep requirements in CFD simulations of Darrieus wind turbines, *Energy*, vol. 97, 2016, p. 246–261.
34. ANSYS Fluent 17.2 Help (ANSYS Workbench 17.2 help files).
35. M. J. Evans, C. Petre, P. R. Medwell, A. Parente, Generalisation of the eddy-dissipation concept for jet flames with low turbulence and low Damköhler number, *Proceedings of the Combustion Institute*, vol. 37, 2019, p. 4497–4505.
36. N. Chakraborty, A. N. Lipatnikov, Conditional velocity statistics for high and low Damköhler number turbulent premixed combustion in the context of Reynolds averaged Navier Stokes simulations, *Proceedings of the Combustion Institute*, vol. 34, 2013, p. 1333–1345.
37. F. Hampp, S. Shariatmadar, R. P. Lindstedt, Quantification of low Damköhler number turbulent premixed flames, *Proceedings of the Combustion Institute*, vol. 37, 2019, p. 2373–2381.
38. M. Szubel, W. Adamczyk, G. Basista, M. Filipowicz, Homogenous and heterogeneous combustion in the secondary chamber of a straw-fired batch boiler, *EPJ Web of Conferences*, vol. 143 (2017).
39. M. Szubel, A. Dernbecher, T. Dziok, Determination of kinetic parameters of pyrolysis of wheat straw using thermogravimetry and mathematical models, *IOP Conference Series: Earth and Environmental Science*, vol. 214, 2019.
40. M. Szubel, M. Filipowicz, W. Goryl, G. Basista, Characterization of the wood combustion process based on the TG analysis, numerical modelling and measurements performed on the experimental stand, *E3S Web of Conferences*, vol. 10, 2016.
41. S. M. Hashemi, S. A. Hashemi, Flame stability analysis of the premixed methane-air combustion in a two-layer porous media burner by numerical simulation, *Fuel*, vol. 202, 2017, p. 56–65.
42. H. Rahnema, M. Barrufet, D. D. Mamora, Combustion assisted gravity drainage—Experimental and simulation results of a promising in-situ combustion technology to recover extra-heavy oil, *Journal of Petroleum Science and Engineering*, vol. 154, 2017, p. 513–520.
43. R. O. Olayiwola, Modeling and simulation of combustion fronts in porous media, *Journal of the Nigerian Mathematical Society*, vol. 34, 2015, p. 1–10.
44. M. J. S. de Lemos, Numerical simulation of turbulent combustion in porous materials, *International Communications in Heat and Mass Transfer*, vol. 36, 2009, p. 996–1001.
45. D. B. Spalding, Concentration fluctuations in a round turbulent free jet, *Chemical Engineering Science*, vol. 26, 1971, p. 95–107.
46. B. F. Magnussen, B. H. Hjertager, The eddy dissipation concept—A bridge between science and technology, *16th International Symposium on Combustion*, 1976, p. 719.
47. T. Lackmann, A. R. Kerstein, M. Oevermann, A representative linear eddy model for simulating spray combustion in engines (RILEM), *Combustion and Flame*, vol. 193, 2018, p. 1–15.
48. Y. Halouane, A. Dehbi, CFD simulations of premixed hydrogen combustion using the Eddy dissipation and the turbulent flame closure models, *International Journal of Hydrogen Energy*, vol. 42, 2017, p. 21990–22004.
49. H. I. Kassem, K. M. Saqr, H. S. Aly, M. M. Siesa. M. A. Wahid, Implementation of the eddy dissipation model of turbulent non-premixed combustion in OpenFOAM, *International Communications in Heat and Mass Transfer*, vol. 38, 2011, p. 363–367.
50. I. R. Gran, B. F. Magnussen, A numerical study of a bluff-body stabilized diffusion flame. Part 2. Influence of combustion modeling and finite-rate chemistry, *Combustion Science and Technology*, vol. 119, 2007, p. 191–217.

51. B. F. Magnussen, *ECCOMAS Thematic Conference on Computational Combustion*, Lisbon, June 21–24, 2005, p. 1–25.
52. J. Chen, W. Yin, S. Wang, C. Meng, J. Li, B. Qin, G. Yu, Effect of reactions in small eddies on biomass gasification with eddy dissipation concept—Sub-grid scale reaction model, *Bioresource Technology*, vol. 211, 2016, p. 93–100.
53. M. Farokhi, M. Birouk, A new EDC approach for modeling turbulence/chemistry interaction of the gas-phase of biomass combustion, *Fuel*, vol. 220, 2018, p. 420–436.
54. M. Farokhi, M. Birouk, Modeling of the gas-phase combustion of a grate-firing biomass furnace using an extended approach of Eddy dissipation concept, *Fuel*, vol. 227, 2018, p. 412–423.

Index

A

Absorption, 167, 169, 180–181
Accumulation system, 200, 202–203, 206
Adsorption, 165–166
Advection, 323
AES (Annual Economic Saving), 253, 258, 260
Agricultural, 286, 299
Air
    combustion air, 56, 57, 61, 80
    primary air, 55, 57, 65
    secondary air, 51, 56, 57
Air quality guidelines (AQG), 129–130
Amaranth, 68
    amaranth–phytomass, 69
    caloric value of Amaranthus cruentus, 70
Annual net saving, 27, 28
Arrhenius equation, 325, 332, 335
Ash
    ash characteristics, 7, 57
    ash content, 43, 44, 47, 69, 71
    duct, 234
    fly ash, 58, 61, 63, 77, 79
    temperature, 236
Ashes, 291–292, 295, 298, 304–305
Aspen Plus©, 255
Atmospheric thermal load, 60, 61
Avoided
    CO_2, 262
    cost, 252, 260, 262, 265
    disposal, 252
    electricity, 258, 260
    energy, 252, 253
    power, 259
    sludge, 258

B

Batch boiler, 217, 218, 219, 220
Battery, 186, 193, 197
    bank, 173–174
    discharge, 176
    storage, 159, 175
Benzene, 59, 64, 68
Benzo(a)pyrene (B(a)P), 127–131, 133, 135–136, 140
Biocarbon, 3
Biochemical processes, 10
Biofuels, 216

Biomass, 3–16, 101–107, 109–122, 283–291, 294, 296–299, 304, 313–314
    biochemical conversion, 10
    boiler, 22, 24, 25, 31, 36, 190, 193, 194, 201, 211, 216, 231, 235, 236, 242, 245, 284, 313
    briquettes, 101, 104–106, 118–119
    calorific value, *see* Fuel characteristics
    characteristics, *see* Fuel characteristics
    combustion, 46, 51, 55, 186, 193, 195, 211, 215, 231, 237
    conversion, 216, 220
    conversion technologies, 9–13
    device, 186, 193, 210–211
    drying, 232, 237
    energy, 186, 211
    gasification, 216
    gasifier, *see* Gasifiers
    heat source, 198
    installation, 201
    non–woody, 101, 105–106, 118
    physicochemical conversion, 10
    stove, 200
    straw, 101–102, 104–106, 113, 118
    technology, 186, 211
    thermochemical conversion, 1–13
    types, 276, 279; *see also* Fuel characteristics
    unit, 191, 201–202
Bismuth telluride, 189
Boundary conditions, 321, 326, 332
Brayton, 288, 306, 309
Briquettes, 5

C

Calculations, 145, 146
Calorific, 286, 294–298, 302–303, 305–307, 313
Calorific value, 3, 4, 7, 8, 273–275, 277, 280
Capital expenditures, 22, 25, 28
Capital recovery factor, 33, 34
Capstone, 311–313
Carbon, 290, 292, 295, 304–305
Carbon dioxide, 284, 288, 313
Carbon monoxide (CO), 55, 59, 63, 65, 81, 145, 286, 288, 291, 294, 308
Carnot cycle, 189
Case study, 31, 32, 34, 36, 68
Cells, 321, 323, 325
Char combustion, 330, 332
Chemical energy, 7–8
Chemkin-Pro, 137

Classification, 54, 71
Clausius–Rankine cycle, 216, 220
CO_2 emission, 253
CoDeSys software, 221, 224
Cogeneration, 10–11, 283, 287, 304–309, 312–314
    efficiency, 160–161
    system, 169, 176, 179
Cold side
    cooling, 195
    heat stream, 187–188, 191
    temperature, 198, 206, 208
Combined heat and power
    system, 163
    unit, 176, 179
Combustion, 236, 244, 284, 288, 290, 302, 310
    appliances
        automatic fueled, 101, 103, 108, 113, 117
    biomass, 231, 237
    chamber, 146, 190, 193–197, 200, 208, 210, 237, 243, 283, 287, 291, 298, 309
    conditions, 205
    control, 198
    devices, 194
    efficiency, 210
    fixed bed, 102
    fuel, 232
    gas temperatures, 150
    installations, small (SICs), *see* appliances 101, 104, 106–107, 118
    manually fueled, 101, 102, 113–115
    modeling, 326–329
    process, 101–102, 105–110, 118, 120–121, 202, 205, 231, 232, 242, 245
    straw, 231
    technique, 102
Composition biomass, 6–9
    combustion air, 56, 57, 61, 80
    combustion chamber, 46, 47, 55, 65, 91
    full combustion, 65
    incomplete combustion
        causes, 57
        of biomass, 59
        of carbon, 63
        of fossil fuels, 66
        of fuel, 68
        of hydrocarbons, 64
Computational fluid dynamics, 319–323
Conservation principles, 319, 326
Control volumes, 321–322
Convection, 323
    forced, 195
    natural, 195
Convergence, 324, 329
Conversion technologies, 10
    combustion, 10–11
    gasification, 11–13
    pyrolysis, 13

Cooking stove, 186, 199
Cooling
    air, 195, 210
    convective, 194–195
    efficiency, 197, 210
    fan forced, 194–195, 206
    liquid, 201
    medium, 193, 210
    method, 194
    system, 188, 190, 199, 210
    water, 194, 197, 207–208, 210
    water tank, 198
Cost
    of investment, 252, 256, 257, 258, 265
    of maintenance, 249, 250, 252, 253, 257
    of operation, 249, 252, 253, 257, 265
Counter-current combustion, 237
Counter flow combustion system, 219
Cuboidal bale, 232, 235
Cyclone, 237
Cylindrical bale, 232, 235, 237
    hay, 234
    straw, 237, 241, 244

D

Damköhler number, 328, 333
Decommissioning, 22, 28, 29
Degassing, 285, 288, 290, 298, 314
Demonstration plant, 271, 280–281
Density
    bulk density, 43, 53
    energy density, 48, 53
    flue gas density, 63
Density-based approaches, 321
Desalination, 180–182
Desiccant flow, 252, 254, 255, 260
Dew point condition, 151
Diffusion flames, 327
Discrete phase transport, 329–331
Discretization, 321, 323
Disposal, 252, 254, 258, 260, 262
Domains, 321, 331, 335
DRM22, 137
Dry matter, 266
Dryer, 249, 250, 255, 256, 258
    batch, 232, 233
    belt, 232
    dedicated straw, 236, 238, 245
    drum, 232
    duct-grate, 234
    fluid bed, 232
    hybrid, 235, 236
    pneumatic, 232
    straw, 237, 245
    with the expansion chamber, 235

Index

Drying, 248–251, 258, 260
    agent, 231–245
    as a combustion stage, 55
    bales, 233
    biomass, 232, 237
    chamber, 233, 237
    curve, 255
    cylindrical
        hay bales, 232, 234, 235, 236, 245
        straw bales, 233, 236, 237, 240, 241, 244, 245
    devices, 232
    efficiency, 245
    flue gas based, 90
    food material, 232
    fuel, 237, 244
    in forced draft conditions, 81
    in grated combustion systems, 56
    material, 236
    mechanism, 232
    medium, 231, 233–235, 238, 240, 244
    modelling, 255
    of lumber, 88
    on the grate, 47
    phase, 241
    process, 232–234, 236, 237, 239–245, 255, 260
    rate, 245
    temperature, 237, 241–243
    time, 232, 233, 236, 244, 245
    transpiration, 61
Dust
    combustible dusts, 81
    dust particles, 51, 78
    sawdust, 44, 49, 50
Dynamic metrics, 25, 31

E

Eco+ firebox, 146
Economic, 248
    analysis, 21, 22, 36, 164, 166–168, 252, 256, 257, 258, 264
    benefit, 249
    convenience, 262
    cost, 248–250, 264
    evaluation, 21, 28, 31
    feasibility, 249–251, 258, 265
    indexes, 253
    model, 252
    saving, 253
    sustainability, 264
    viability, 250
Eddy dissipation
    concept, 335–336
    model, 334–335
EKOD, *see* Gasifiers

Electrical
    conductivity, 187
    efficiency, 167, 228
    energy, 250, 252, 254
    optimization, 249
    power, 250, 259
    process, 187
    surplus, 258
Electricity, 258, 260, 262–263, 265
    in wastewater treatment, 248, 249
    production, 252–254, 256
Elementary analysis, 6, 7
Emissions, 24, 31, 37, 102–103, 104–107, 110–115, 118, 120, 284, 313
    abatement measures, 118–122
    directives, 112, 114–116
    dust, *see* Particulate Matter
    factors (EFs), 107, 112–113, 114–117, 136
    limit values (ELVs), 106, 114–117
    particulate matter, 101, 107–109, 112, 119
    pollutants, 101–102, 106–115, 121
    regulatory measures, *see* directives
    requirements, 106, 112–118, 146
    residential, 100–101, 104–106, 109, 117
    short stack, *see* Residential Emissions
    standards, 105, 112–119
    storage, 157, 173–175
    testing methods, 118
    testing standards, 116–118
Endothermic reactions, 285–286
Energy, 248, 250, 252, 254
    demand, 172–173, 249–251
    efficiency, 4, 102–103, 113–114, 116, 146, 151, 232, 237, 248, 250
    exploitation, 254
    potential, 251, 254
    production, 248, 249, 253, 255
    purposes, 231, 232, 236
    recovery, 248
    requirements, *see* Energy Efficiency, 114
    saving, 250
    source, 248, 250–253
    willow, 232
Energy resources, 4–9
    primary energy, 4–5
    processed energy, 5
    secondary energy, 5
Enthalpy, 252, 255
Environmental, 265
    analysis, 252, 257
    impact, 262
    restriction, 254
    sustainability, 264
Equilibrium constant, 289
Eulerian reference frames, 331
Evaluation, 22, 24, 36
Evaporator, 223, 224

Excess air, 148
Exhaust
 channel, 200, 203, 208, 211
 gas temperature, 245
 gases, 193, 200, 202–203, 206, 208, 232, 237, 244
Exothermic reactions, 285–286, 288–289, 308
Experience curves, 30, 37
Extinctions, 328

F

Fan, 186, 194–197, 200, 206–207, 210
Fast chemistry, 328, 334
Feasibility, 249–252, 258, 265
Feedstock resources, 68
Finite
 differences, 321
 elements, 321
 volumes, 321, 331
Fireplace, 202, 206–208, 210
Fischer–Tropsch, 288
Fixed carbon, 9
Flowforge, 233
Flue gas, 44, 46, 51, 57, 60
 channel, 146, 189, 200, 203, 205, 210
 heat exchange, 195, 208
 heat transfer, 190, 211
 temperature, 202, 206, 210
Forced
 air, 232, 235
 drying agent, 236
Fuel feeding system, 22, 23, 224
Fuel function control, 151
Fuel price, 23, 32, 35, 36
Fuels, 101–107, 284–300, 302–313, 331
 biogenic, see Biomass Fuels, 104, 105
 briquettes, see Biomass Fuels101–102, 104–106, 118–119
 burner, 236
 coal, 101–107, 110–111, 113, 115
 combustion, 201
 composition, see fuel properties103
 feed rate, 197
 fossil, see Mineral Fuels Fuel Characteristics52, 273, 277
 fuel standards, 54
 gasses, 236, 237, 244
 handling, 198
 humidity, 245
 ignition, 245
 load, 199
 mineral, 104
 pellets, 103, 105, 113, 116–117, 121
 properties, 102, 103, 106–107
 solid, 101–102, 104–108, 112
 wood, 101–103, 105–106, 113, 117–118
 wood fuel composition, 51
Furnace, 202–203

G

Gas, 284–291, 294–297, 299, 300, 302–313
 combustible gases, 56, 57, 80
 exhaust gases, 77, 88
 flammable gases, 57
 flue gas, 51, 60, 68
 odorous gases, 80
 volatile gases, 47
 wood gases, 57, 59
 woody generator gas, 88
Gas burner, 233, 235, 236
Gas characteristics, 275
Gas cleaning system, 271, 279–280
Gas composition, 273–275, 277–279
Gas turbines, 283–284, 304–305, 307–309, 311–314
Gasification, 11–13, 55, 57, 64, 75, 77, 80, 237, 284–291, 294–300, 302–306, 308, 313
Gasifiers, 76–79, 85, 89, 94
 EKOD, 271–272, 274
 fluid, 254–256, 258, 260
 GazEla, 271–272, 275–277, 279–280
 reservoir, 263
 source, 254, 263
Gasifying agents, 284–286, 290–291, 295, 299, 304
Generator gas, 284, 286, 291, 297, 302–313
Geographic information system, 22
Geothermal energy, 158, 177, 179–180, 250, 251, 254, 265
Governing equations, 321–322, 324, 332
Grates, 44, 47, 51
GRI, 137
Grid
 connection, 173–174
 parity, 160

H

Heat, 237, 245
 accumulation system, 200, 202–203, 206
 balance, 252
 carrier, 194
 conductive flow, 188
 conversion, 194
 delivered, 190
 dissipation, 188, 190
 engine, 187
 exchanger, 194–196, 201–202, 204, 211, 236, 237, 249, 256
 flux, 202, 210
 generated–combustion chamber, 190, 198

Index

generated–internal resistance, 188
generation, 245
hot gas, 200, 203, 211
losses, 190
Peltier flow, 188
pipe, 196
recovery, 193, 210
sink, 195, 200
source, 193–194, 198, 200, 210, 236
stream, 186, 188
transfer
    biomass combustor, 194
    calculations, 198
    efficiency, 210, 211
    exchanger, 205; *see also* Beck
    voltage, 188
Heat transfer fluid, 161, 163, 168
Heating
    capacity, 208
    medium, 208
Heating installation, 21, 31, 36
Heating rate, 285
HHV, 297, 304
Hot side
    heat stream, 187–188, 191
    heating, 197
    temperature, 195, 198, 206, 208, 210
HRSG (Heat Recovery Steam Generator), 83
Humidity, 231–245
Hybridization of biomass, 158, 171, 174
Hydrocarbons, 288, 291, 305, 313
Hydrogen, 284, 286, 290–292, 294–297, 302
Hydrogenation, 296–297, 302, 313
Hydrothermal combustion process, 13–14

I

IGCC (Integrated Gasification Combined Cycle), 11, 76, 82, 84, 86–88, 94
Ignition, 237, 242, 243, 245, 328
Injection nozzle, 238, 240
Institute for Chemical Processing of Coal (IChPW), 272, 274–275, 280
Internal rate of return, 22, 27, 28
Investment project, 21, 22, 26, 28
Investment project's profitability, 24, 36

K

Kinetic flames, 327
Kinetic models, 288, 306
Kinetics, 324, 326, 328, 332–335

L

Laboratory tests, 287
Lagrangian reference frame, 330, 331

Lambda sensors, 57
Laminar flow, 319, 327, 334, 335
Learning curves, 31
Learning rate, 30, 31
Length of scale, 336
Levelized cost of heat, 27, 28
Levoglucosan, 138
LHV, 292, 294–298, 303–305, 308
Life cycle cost, 22, 27, 28, 29
Limit value, 130
Low-stack emission, 129, 131

M

Mach number, 321
Maintenance costs, 23, 24
Mass
    conservation, 319, 326
    less transport, 329
    transfer, 329, 330
Mass balance, 252
Maximum fuel load, 145
Maximum power point, 193–194, 206
Meshing, 321
Methanation, 285–286, 288–290, 296–297, 302, 308
Methane, 286–289, 291–294, 299, 304–309, 313
Methane formation, 283, 286, 303
Microturbines, 283, 309, 311–312
Minimum flue length, 146, 147
Mixer, 237, 238
Mixing process, 237
Model, 250–252, 256
Modeling, 251, 256
Moisture, 47–49, 51, 62, 65, 92–94
Moisture, 232, 244, 252, 258, 285–286, 291, 292, 295
    content, 231, 232, 235
    field, 235
Molten salt, 161, 164, 168
Momentum conservation, 319, 321, 326

N

Nernst equation, 289
Net present value, 22, 27, 253, 258
Nitrogen dioxide (NO_2), 66, 129–130
Nitrogen oxide (NO_x), 86, 131, 134–136, 138–139, 145
    nitric oxide, 66
Nodes, 321
Noise, 66, 68, 77, 80, 95
Nominal discount rate, 27
Nominal heat output, 145
Nonmethane volatile organic compounds (NMVOC), 133–136
Non-premixed combustion, 327

O

Odor, 77, 80
Oil circuit, 220, 222, 224
Operation costs, 23, 24, 29
Organic
  matter, 248, 256
  waste, 249
Organic Rankine cycle, 189, 216, 254–255, 263–264
  modelling, 255–256
Other costs, 23, 24
Oxidation zones, 57, 65
Oxygen, 284, 290–292, 295, 299, 302
Ozone (O_3), 127, 129, 139

P

Parabolic trough
  collector, 158, 161
  solar field, 165, 170
Part load, 146
Partial oxidation, 12
Particles, 329–331
Particulate emissions, 145
Particulate matter (PM), 47, 59, 61, 77
Particulates, 87
Peltier cell, 238
Photovoltaic panel, 158–159
Physical lifetime, 24
Physical model, 321
Physicochemical process, 10
Pilot plant, 271, 275, 280
Pollutants, 106–112, 113–114
Pollution emissions, 59, 64, 67, 84
Polygeneration, 180–182
Porous media, 332
Post-processing, 320, 323
Power
  electric, 190, 194, 197, 200
  extracted, 191
  generation, 186, 190–194, 197–198, 200–201, 209, 225, 227, 228
  maximal, 206
  net, 206
  nominal, 198, 208
  output, 197, 200, 208
  surplus, 186, 193, 211
  thermal, 199, 203
Power plant, 159, 165, 177–178
Power-voltage characteristics, 206, 209
Premixed combustion, 327, 333
Pre-processing, 320, 321
Pressure-based solver, 321
Pressure condition, 151
Preventive measures, 76, 78
Primary
  energy, 250, 253
  sludge, 250
Profitability calculations, 25
Programmable logic controller, 221, 223
Pump
  auxiliary elements, 186, 194
  power, 197, 198, 201
  water, 197, 198, 207, 210
Pyrolysis, 13, 216, 285

R

R245fa, 255, 256
RDFs, 283, 304, 306
Radiation, 326, 334
Radiator, 195, 203, 205
Rapeseed meal, 283, 297–304, 314
Reaction chambers, 287, 291, 295, 300–302, 305
Real discount rate, 27
Recirculations, 58, 60, 65
Reduced reaction mechanisms, 324
Reference levels (RL), 129–130
Regulation, 24, 32, 37
Renewable energy, 247, 248, 250–252
Renewable energy sources, 157–158, 180
Retene, 138
Reynolds
  averaged Navier–Stokes, 322
  number, 57, 328

S

Sabatier–Senderens, 288
Salvage value, 23, 24, 28, 29, 32
Sawdust, 290–294
Self-powering, 186, 211
Seebeck
  cell, 187
  coefficient, 187
  voltage, 188
Sensitivity analysis, 36
Simple payback time, 25, 253, 258, 262
Simultaneous thermal analyses, 335
Slags, 44, 57, 65
Slow chemistry, 328, 333, 335
Sludge, 247–260, 262, 264, 265
  disposal, 248, 251, 258
  drying, 248–250, 256, 257, 260
  LHV, 257
  properties, 257
  treatment, 249, 262, 265
Solar
  collectors, 250
  dryers, 250
  drying, 250

Index

energy, 250
heater, 250
source, 251
thermal collectors, 158, 171
Solid pollution matter, 66
Solvers, 321, 329–332
Source terms, 332
Species transport, 325, 333
Stack emissions, 87
Standardizations, 54
Steady state processes, 322
Steam
　circuit, 226
　engine, 225
Stirling engines, 190, 216
Storage, 22, 23, 24
Stove
　biomass, 200
　camp, 196–197
　cast iron, 196
　convective cooling, 195
　cooking, 199
　forced air cooling, 195
　heat accumulation, 202, 206
　open fire, 186
　operation, 210
　pellet, 198–199
　plate–steel, 208
　traditional, 186
　water cooling, 197
　wood, 186, 211
Stove–fireplace, 206
Straw, 217, 218, 224, 232, 233, 236, 241, 242, 243, 244, 245
　bales, 237, 239, 240
　boiler, 201, 211, 237
　combustion, 231
　dryer, 245
　drying, 236, 239, 240, 244, 245
　pellets, 276–278, 280
Superheater, 223, 224
Syngas, 12, 284, 294, 296, 308, 309
Syngaz S.A., 275, 278

T

Tank, 22, 24
Target value, 129–130
Tars, 55, 61–62, 65, 78–79, 93, 285–288, 290, 291, 295, 306
Temperature gradient, 187
Tetrahedrons, 321
Thermal
　demand, 248, 249, 259, 264
　drying, 248–249, 251–252, 260
　efficiency, 168, 181–182, 218

energy, 7, 254, 260
oil boiler, 163
oil Circuit, 172
treatments, 248
Thermal conductivity, 187, 205
Thermal cracking, 286, 288, 290
Thermal oil, 194, 201
Thermochemical conversion, 55, 93
Thermochemical processes, 10–13
Thermodynamic, 265, 283, 286, 311
　analysis, 162, 164, 167, 180–181, 258
　cycle, 161, 256
　model, 160, 173
　properties, 255
Thermoelectric, 186–187
　generator, 186, 190, 197, 205
　modules, 187, 201–202, 211
Thermosyphon, 201
Tiled stoves, 146, 151
Time steps, 323, 332
Tires, 273–275
Too dry, 65
Too wet, 64
Torrefaction, 5
Total cost, 22, 23, 24, 29
Transient
　processes, 322
　systems, 322, 323
Transport equations, 323, 325
Trigeneration, 165–168, 171
Turbines, 284, 287, 304–305, 307–312
Turbulent
　flow, 322, 323, 327, 334, 335
　transport, 322

V

Validation, 323, 326, 336
Venturi flowmeter, 238
Volatile matter, 9, 257, 266
Volatile organic carbon, 145

W

Waste heat, 237, 245
Wastes, 284–286, 290, 291, 292, 311
Wastewater, 255, 262, 265
　sludge, 249
　treatment, 248, 250, 251, 255, 265
Water, 249, 250, 252, 255
　air exchanger, 236
　blocks, 201
　circulation, 197–198
　container, 197–198, 200
　content, 248, 252, 255, 264
　cooling, 194, 210

Water (*Continued*)
  evaporation, 232
  heating, 194
  jacket, 190, 201
  pump, 197–198, 207
  reservoir, 200
  storage tank, 197, 200
  system, 197, 210
  vapor, 52, 55, 59–61, 65, 87, 231
Weighted average cost of capital, 26
Well, 256
Willows, 42, 48, 232
Wind turbine, 172–176
Wood
  boards, 273–275
  devolatization, 203
  pellets, 276–278
  stove, 186, 211
  waste, 272–276
Wood biomass, 283, 290–291, 294, 297–300, 302–303
Woodchips, 273–279
Working fluid, 255
World Health Organization, 63, 128, 130

Y

Years of life lost (YOLL), 129

Z

Z figure-of-merit, 187
Zones, 288, 289, 290